THE LOEB CLASSICAL LIBRARY
FOUNDED BY JAMES LOEB, LL.D.

EDITED BY
E. H. WARMINGTON, M.A., F.R.HIST.SOC.

FORMER EDITORS
† T. E. PAGE, C.H., LITT.D.   † E. CAPPS, PH.D., LL.D.
† W. H. D. ROUSE, LITT.D.   L. A. POST, L.H.D.

CELSUS

I

CELSUS

DE MEDICINA

WITH AN ENGLISH TRANSLATION BY
W. G. SPENCER
MS. LOND., F.R.C.S. ENG.

IN THREE VOLUMES

I

CAMBRIDGE, MASSACHUSETTS
HARVARD UNIVERSITY PRESS
LONDON
WILLIAM HEINEMANN LTD
MCMLXXI

American ISBN 0-674-99322-5
British ISBN 0 434 99292 5

First printed 1935
Reprinted 1940, 1948, 1960, 1971

PA
6156
.C4
1935a
v.1   /45,955

Printed in Great Britain

CONTENTS

	PAGE
INTRODUCTION	vii
SUMMARY OF THE ARGUMENTS FOR REGARDING CELSUS AS A MEDICAL PRACTITIONER	xi
BIBLIOGRAPHY	xiii
BOOK I	1
BOOK II	83
BOOK III	217
BOOK IV	353
APPENDIX	463
LIST OF CHAPTER HEADINGS	467
LIST OF *ALIMENTA*	483

INTRODUCTION

(By Dr. W. H. S. Jones, Fellow of St. Catharine's College, Cambridge)

Practically nothing is known of the life of the man whose name appears as the author of the medical treatise *De Medicina*. Recent research has made it likely that he was acquainted with the later poems of Ovid, and that he probably lived in Narbonensis. Quintilian, who describes Celsus as *vir mediocri ingenio*, informs us that he wrote on many other subjects besides medicine.[a] From this passage in Quintilian Marx has inferred that Celsus was a mere general editor of an encyclopaedia, and that he did not himself write the *De Medicina*, which can scarcely be regarded as the work of a man of mediocre intellect. Be this as it may, it appears likely that this medical book was but the second part of a large treatise containing six parts, the other five being:

(1) Agriculture;
(3) Military Arts;
(4) Rhetoric;
(5) Philosophy;
(6) Jurisprudence.

Not only Quintilian but also Pliny the elder refers to Celsus, who therefore lived in all probability in the reign of the Emperor Tiberius. Some would place his birth in the year 25 b.c.

[a] See Quintilian XII. 11. 24: non solum de his omnibus conscripserit artibus, sed amplius rei militaris et rusticae et medicinae.

INTRODUCTION

The complete name of the author was probably Aulus Cornelius Celsus. The tradition that A stands for Aurelius must be wrong, as Aurelius is not a *praenomen*.

It is a disputed question whether the author of the work was a practising physician or not. It may be remarked in passing that in ancient times there was not such a sharp distinction between the professional and the amateur as there is to-day. The amount of medical knowledge was not so great as to be out of the range of an ordinary, educated man of average intelligence. On the one hand, it may be said that a work so complete and so accurate as the *De Medicina* must have come from the pen of a man with professional experience.[a] On the other hand, several reasons may be urged making the other view more probable.

(1) The elder Pliny puts Celsus among the *auctores*, not among the *medici*.

(2) Unseemly details are less frankly expressed than is usual among practising physicians.

(3) A professional reader finds the handling of the subject in parts superficial and formal.

(4) Certain passages dealing with surgical operations seem unlike the account of a man familiar with them from practical experience.

(5) It was not unusual for a Roman gentleman to have a fairly intimate knowledge of medicine, which was useful to him as the head of a large household of slaves and freedmen.

It is a possible, but an unproved hypothesis, that the *De Medicina* is a translation or rather adaptation of a Greek medical work, and that the seeming

---

[a] For a further discussion of this point cf. p. xi.

INTRODUCTION

lapses from the professional standard are due to misunderstandings or to ignorant attempts at comment or explanation.

The most obvious sources, however, are the Hippocratic *Corpus*, which still survives, and the lost works of Asclepiades, Heracleides, Erasistratus, and Meges of Sidon, a surgeon who lived a little earlier than Celsus. Wellmann supposes that Celsus translated a Greek treatise written by his friend Cassius; Marx believes that the Greek original was by Titus Aufidius Siculus, a pupil of Asclepiades. Of course the supposition that Celsus translated either of these two works is not inconsistent with references to Hippocrates and to the other physicians mentioned above. These may have been embodied in the text that Celsus is supposed to have had before him.

The *prooemium* to the *De Medicina* is a most fair and judicious summary of the history of medicine, and deals at some length with the Dogmatic, the Methodic and the Empiric Schools. The writer himself tries to follow a *via media* between the Dogmatics and the Empirics.

The work recognizes the importance of anatomy as a basis of medicine, and the anatomical knowledge displayed is sound. Stress is laid on diagnosis and prognosis, which it is said must precede treatment—a true Hippocratic touch. Drugs are recommended more than they are by Greek writers on medicine.[a] On the other hand, all due importance is attached to general hygiene and to physical exercises. Scholars have noticed that sport is preferred to gymnastics, wherein the writer agrees with both

[a] A list of drugs (*medicamenta*) is given at the beginning of Volume II, pp. xv–lxvii.

INTRODUCTION

Roman feeling and Roman practice. In the treatment of fevers the *De Medicina* is more empirical than usual. It "regards exclusively the clinical picture and the empirical remedy."

The style of the work has won the praise of Latinists in all ages. The whole book is remarkable for its symmetry and completeness, and the language is strong, lucid and elegant. It has been said with justice that the writer did for science what Cicero did for philosophy.

SUMMARY OF THE ARGUMENTS FOR REGARDING CELSUS AS A MEDICAL PRACTITIONER

THE view that the author of the *De Medicina* was a learned and experienced medical practitioner is based upon a number of passages in his work which, even allowing for the less sharp distinction drawn between the professional and amateur in his day, seem to indicate that he himself regularly attended patients, and wrote with the authority of a practising physician.

(1) In many passages he expresses his opinion as to a treatment or symptom by using the first person singular or plural of the verb. See Prooem. 73, 74, 75; Bk. I, Ch. 3, sect. 3, 17 and 21, Ch. 8, sect. 2; Bk. II, Ch. 6, sect. 13 and 17, Ch. 10, sect. 13 and 14, Ch. 12, sect. 2; Bk. III, Ch. 3, sect. 3, Ch. 21, sect. 15; Bk. IV, Ch. 7, sect. 5, Ch. 26, sect. 8; Bk. V, Ch. 17, sect. 1 B, Ch. 18, sect. 13, Ch. 19, sect. 12 and Ch. 28, sect. 12 N.

In other passages the personal note is made more marked by the use of the emphatic "ego." See Prooem. 50, Bk. III, Ch. 1, sect. 3, Ch. 4, sect. 3, Ch. 11, sect. 2, Ch. 24, sect. 3; Bk. IV, Ch. 26, sect. 4; Bk. VI, Ch. 6, sect. 2; Bk. VII, Prooem. 5, Ch. 6, sect. 1, Ch. 7, sect. 6 c, Ch. 12, sect. 4.

(2) He writes of patients whom he knew personally and attended sedulously, even by night, Bk. II, Ch. 4, sect. 1 and 6; Bk. III, Ch. 4, sect. 9 and 10, Ch. 5,

CELSUS AS A MEDICAL PRACTITIONER

sect. 6, Ch. 6, sect. 6–8. His description of the treatment of the insane seems to imply personal attendance on such cases, Bk. III, Ch. 18, sect. 10, 11.

(3) He was evidently well acquainted with the leading medical writers of his own age as well as with the older Greek authorities (Prooem. 49, 54, 69; Bk. II, Ch. 1, sect. 1, Ch. 12, sect. 2, Ch. 14, sect. 1; Bk. III, Ch. 4, sect. 1–9, Ch. 9, sect. 2 and 3; Bk. IV, Ch. 7, sect. 1, Ch. 9, sect. 2, Ch. 11, sect. 3; Bk. V, Ch. 27, sect. 5 B, Ch. 28, sect. 10; Bk. VI, Ch. 6, sect. 8, Ch. 18, sect. 2 and 7). Some knowledge of them he must, of course, have had to be able to write his book at all, but his references to their opinions and practice seem to indicate personal knowledge and experience as well, and it is significant that his numerous citations of Hippocrates are almost always accurate and appropriate.[a]

[a] A list of these parallel passages is given at the end of Volume III, p. 624, ff.

W. G. S.

BIBLIOGRAPHY

The chief MSS are these:

(1) F, Codex Florent., Laurentian Library, 73, 1. IX cent. and in parts defective.
(2) V, Codex Romanus, Vatican Library, 5951. IX cent. and in parts defective.
(3) P, Codex Parisinus, National Library, 7928. X cent.; copied from V when this was less defective.
(4) J, Codex Florent., Laurentian 73, 7, copied by Nicolao Niccoli from a very old codex now no longer extant. XV cent.

Down to 1841 there were 49 printed editions, the *editio princeps* appearing at Florence in 1478. Recent works on Celsus include:—

1756; J. Greive: *Aulus Cornelius Celsus*, of medicine in eight books.
1859; C. Daremberg: *A. Cornelii Celsi de medicina libri octo*.
1907; J. Ilberg: *A. Cornelius Celsus und die Medizin in Rom*.
1913; M. Wellmann: *A. Cornelius Celsus, eine Quellenuntersuchung*.
1915: F. Marx: *A. Cornelii Celsi quae supersunt*. This is a very valuable work, especially the Prolegomena.

BIBLIOGRAPHY

1926; W. G. Spencer in *Proceedings of the Royal Society of Medicine*, XIX. 129.

There are also some valuable comments on the *De Medicina* in Sir Clifford Allbutt's *Greek Medicine in Rome*, and in Dr. J. F. Payne's article on Roman medicine in *Cambridge Companion to Latin Studies*.

For Hippocrates references are given for the treatises included in the translation published in the Loeb Classical Library (Jones and Withington) to the volume and page of that edition; for other treatises reference is to the numeration adopted by Littré in his edition. For Galen the numeration referred to is that of Kühn's edition.

The Editors and Translator have to acknowledge with very many thanks the generous permission of Messrs. G. B. Teubner of Leipzig to use the text edited by F. Marx, published by them in 1915. The numeration given in this text has been followed throughout and it will be found that in some chapters Marx, following Codex J, has substituted letters (B C D, etc.) for numbers in marking the sections (cf. p. 168 and elsewhere).

CELSUS

DE MEDICINA

A. CORNELII CELSI

DE MEDICINA

PROOEMIUM

Ut alimenta sanis corporibus agricultura, sic sanitatem aegris Medicina promittit. Haec nusquam quidem non est, siquidem etiam inperitissimae gentes herbas aliaque promta in auxilium vulnerum 2 morborumque noverunt. Verum tamen apud Graecos aliquanto magis quam in ceteris nationibus exculta est, ac ne apud hos quidem a prima origine, sed paucis ante nos saeculis. Ut pote cum vetustissimus auctor Aesculapius celebretur, qui quoniam adhuc rudem et vulgarem hanc scientiam paulo subtilius 3 excoluit, in deorum numerum receptus est. Huius deinde duo filii Podalirius et Machaon bello Troiano ducem Agamemnonem secuti non mediocrem opem commilitonibus suis attulerunt; quos tamen Homerus non in pestilentia neque in variis generibus morborum aliquid adtulisse auxilii, sed vulneribus tantummodo ferro et medicamentis mederi solitos esse proposuit. 4 Ex quo apparet has partes medicinae solas ab iis esse tentatas, easque esse vetustissimas. Eodem

---

a The reference is to Celsus' preceding treatise on Agriculture, of which only fragments remain. Cf. Introduction.
b Cf. *Il.* XI. 833.   *c* For the pestilence cf. *Il.* I. 43 ff.

CELSUS

ON MEDICINE

PROOEMIUM

Just as agriculture[a] promises nourishment to healthy bodies, so does the Art of Medicine promise health to the sick. Nowhere is this Art wanting, for the most uncivilized nations have had knowledge of herbs, and other things to hand for the aiding of wounds and diseases. This Art, however, has been cultivated among the Greeks much more than in other nations—not, however, even among them from their first beginnings, but only for a few generations before ours. Hence Aesculapius is celebrated as the most ancient authority, and because he cultivated this science, as yet rude and vulgar, with a little more than common refinement, he was numbered among the gods. After him his two sons, Podalirius and Machaon, who followed Agamemnon as leader to the Trojan War, gave no inconsiderable help to their comrades.[b] Homer stated, however, not that they gave any aid in the pestilence[c] or in the various sorts of diseases, but only that they relieved wounds by the knife and by medicaments. Hence it appears that by them those parts only of the Art were attempted, and that they were the oldest.

vero auctore disci potest morbos tum ad iram deorum inmortalium relatos esse, et ab iisdem opem posci solitam verique simile est inter[1] nulla auxilia adversae valetudinis, plerumque tamen eam bonam contigisse ob bonos mores, quos neque desidia neque luxuria vitiarant; siquidem haec duo corpora prius in Graecia, deinde apud nos adflixerunt ideoque multiplex ista medicina, neque olim neque apud alias gentes necessaria, vix aliquos ex nobis ad senectutis principia perducit.

Ergo etiam post eos, de quibus rettuli, nulli clari viri medicinam exercuerunt, donec maiore studio litterarum disciplina agitari coepit; quae ut animo praecipue omnium necessaria, sic corpori inimica est. Primoque medendi scientia sapientiae pars habebatur, ut et morborum curatio et rerum naturae contemplatio sub iisdem auctoribus nata sit: scilicet iis hanc maxime requirentibus, qui corporum suorum robora inquieta[2] cogitatione nocturnaque vigilia minuerant. Ideoque multos ex sapientiae professoribus peritos eius fuisse accipimus, clarissimos vero ex iis Pythagoran et Enpedoclen et Democritum. Huius autem, ut quidam crediderunt, discipulus Hippocrates Cous, primus ex omnibus memoria dignus, a studio sapientiae disciplinam hanc separavit, vir et arte et facundia insignis. Post quem Diocles

---

[1] *Marx notes that the text is corrupt and suggests* interisse quidem tum morbis plurimos, cum fuerint nulla auxilia etc. . . . "*it is probable that then most perished from their diseases, since there were no aids against bad health.*"

[2] inquieta *one MS.* quieta *Marx, following the majority.*

---

[a] For particulars about these (and other persons referred to in the course of the work), see index of proper names, Vol. III, p. 628.

PROOEMIUM 4–8

From the same authority, indeed, it can be learnt that diseases were then ascribed to the anger of the immortal gods, and from them help used to be sought; and it is probable that with no aids against bad health, none the less health was generally good because of good habits, which neither indolence nor luxury had vitiated: since it is these two which have afflicted the bodies of men, first in Greece, and later amongst us; and hence this complex Art of Medicine, not needed in former times, nor among other nations even now, scarcely protracts the lives of a few of us to the verge of old age.

Therefore even after these I have mentioned, no distinguished men practised the Art of Medicine until literary studies began to be pursued with more attention, which more than anything else are a necessity for the spirit, but at the same time are bad for the body. At first the science of healing was held to be part of philosophy, so that treatment of disease and contemplation of the nature of things began through the same authorities; clearly because healing was needed especially by those whose bodily strength had been weakened by restless thinking and night-watching. Hence we find that many who professed philosophy became expert in medicine, the most celebrated being Pythagoras, Empedocles and Democritus.[a] But it was, as some believe, a pupil of the last, Hippocrates of Cos, a man first and foremost worthy to be remembered, notable both for professional skill and for eloquence, who separated this branch of learning from the study of philosophy.[b] After him

---

[b] *Disciplina rationalis*, λογισμός (λόγος), Book I. Prooem. 10. *Disciplina chirurgica*, Book VII. Prooem. 1–3. Philosophy is used in the wider sense, to include natural science.

Carystius, deinde Praxagoras et Chrysippus, tum Herophilus et Erasistratus sic artem hanc exercuerunt, ut etiam in diversas curandi vias processerint.

9 Iisdemque temporibus in tres partes medicina diducta est, ut una esset quae victu, altera quae medicamentis, tertia quae manu mederetur. Primam Διαιτητικήν secundam Φαρμακευτικήν tertiam Χειρουργίαν Graeci nominarunt. Eius autem, quae victu morbos curat, longe clarissimi auctores etiam altius quaedam agitare conati, rerum quoque naturae sibi cognitionem vindicarunt, tamquam sine ea
10 trunca et debilis medicina esset. Post quos Serapion, primus omnium nihil hanc rationalem disciplinam pertinere ad medicinam professus, in usu tantum et experimentis eam posuit. Quem Apollonius et Glaucias et aliquanto post Heraclides Tarentinus et aliqui non mediocres viri secuti ex ipsa
11 professione se empiricos appellaverunt. Sic in duas partes ea quoque, quae victu curat, medicina divisa est, aliis rationalem artem, aliis usum tantum sibi vindicantibus, nullo vero quicquam post eos, qui supra comprehensi sunt, agitante, nisi quod acceperat, donec Asclepiades medendi rationem ex magna parte mutavit. Ex cuius successoribus Themison nuper ipse quoque quaedam in senectute deflexit. Et per hos quidem maxime viros salutaris ista nobis professio increvit.

12 Quoniam autem ex [tribus] medicinae partibus ut

Diocles of Carystus, next Praxagoras and Chrysippus, then Herophilus and Erasistratus, so practised this art that they made advances even towards various methods of treatment.

During the same times the Art of Medicine was divided into three parts: one being that which cures through diet, another through medicaments, and the third by the hand. The Greeks termed the first Διαιτητική, the second Φαρμακευτική, the third Χειρουργία. But of that part which cures diseases by diet those who were by far the most famous authorities, endeavouring to go more deeply into things, claimed for themselves also a knowledge of nature, without which it seemed that the Art of Medicine would be stunted and weak. After them first of all Serapion, declaring that this kind of reasoning method was in no way pertinent to Medicine, based it only upon practice and upon experience. To him followed Apollonius and Glaucias, and somewhat later Heraclides of Tarentum, and other men of no small note, who in accordance with what they professed called themselves Empirici (or Experimentalists). Thus this Art of Medicine which treats by diet was also divided into two parts, some claiming an Art based upon speculation, others on practice alone. But after those mentioned above no one troubled about anything except what tradition had handed down to him until Asclepiades changed in large measure the way of curing. Of his successors, Themison, late in life, diverged from Asclepiades in some respects. And it is through these men in particular that this healthgiving profession of ours has grown up.

Since of the divisions of the Art of Medicine, the

CELSUS

difficillima, sic etiam clarissima est ea, quae morbis medetur, ante omnia de hac dicendum est. Et quia prima in eo dissensio est, quod alii sibi experimentorum tantummodo notitiam necessariam esse contendunt, alii nisi corporum rerumque ratione comperta non satis potentem usum esse proponunt, indicandum est, quae maxime ex utraque parte dicantur, quo facilius nostra quoque opinio interponi possit.

13 Igitur ii, qui rationalem medicinam profitentur, haec necessaria esse proponunt: abditarum et morbos continentium causarum notitiam, deinde evidentium; post haec etiam naturalium actionum, novissime partium interiorum.

14 Abditas causas vocant, in quibus requiritur, ex quibus principiis nostra corpora sint, quid secundam, quid adversam valetudinem faciat. Neque enim credunt posse eum scire, quomodo morbos curare conveniat, qui unde sint ignoret; neque esse dubium quin alia curatione opus sit, si ex quattuor principiis vel superans aliquod vel deficiens adversam valetudinem creat, ut quidam ex sapientiae professoribus 15 dixerunt: alia, si in umidis omne vitium est, ut Herophilo visum est; alia, si in spiritu, ut Hippocrati; alia, si sanguis in eas venas, quae spiritui accommodatae sunt, transfunditur et inflammationem, quam

---

[a] The four principles or elements of Empedocles with their associations. See Plato, *Timaeus*, 82; Hippocrates, IV. p. xvii.

(a) Fire associated with: hot, rough, red blood; spring; choleric temperament.
(b) Water associated with: cold, smooth, white phlegm; winter; phlegmatic temperament.
(c) Air associated with: dry, salt, sweet, black bile; autumn; melancholic temperament.
(d) Earth associated with: moist, acid, yellow bile; summer; sanguineous temperament.

one which heals diseases, as it is the most difficult, is also the most famous, we must speak about it first. And because there is a primary difference of opinion, some holding that the sole knowledge necessary is derived from experience, others propounding that practice is not efficient enough except after acquiring a reasoned knowledge of human bodies and of nature, I must indicate which are the principal statements on either side, so that I may the more easily interpose my own opinion also.

They, then, who profess a reasoned theory of medicine propound as requisites, first, a knowledge of hidden causes involving diseases, next, of evident causes, after these of natural actions also, and lastly of the internal parts.

They term hidden, the causes concerning which inquiry is made into the principles composing our bodies, what makes for and what against health. For they believe it impossible for one who is ignorant of the origin of diseases to learn how to treat them suitably. They say that it does not admit of doubt that there is need for differences in treatment, if, as certain of the professors of philosophy have stated, some excess, or some deficiency, among the four elements,[a] creates adverse health; or, if all the fault is in the humours,[b] as was the view of Herophilus; or in the breath, according to Hippocrates; or if blood is transfused into those blood-vessels which are fitted for pneuma,[c] and excites inflammation

[b] The four humours: health was supposed to depend upon the proportions of blood, phlegm, choler, and black bile. See Hippocr. IV. pp. xvii, 63 and Index.

[c] Πνεῦμα, in the Greek writers whom Celsus followed, meant breath with the addition of some vital spirit—almost the Greek equivalent to oxygen.

CELSUS

Graeci φλεγμόνην nominant, excitat, eaque inflammatio talem motum efficit, qualis in febre est, ut 16 Erasistrato placuit; alia, si manantia corpuscula per invisibilia foramina subsistendo iter claudunt, ut Asclepiades contendit: eum vero recte curaturum, quem prima origo causae non fefellerit. Neque vero infitiantur experimenta quoque esse necessaria, sed ne ad haec quidem aditum fieri potuisse nisi ab 17 aliqua ratione contendunt: non enim quidlibet antiquiores viros aegris inculcasse, sed cogitasse quid maxime conveniret, et id usu explorasse, ad quod ante coniectura aliqua duxisset. Neque interesse, an nunc iam pleraque explorata sint . . .[1] si a consilio tamen coeperunt. Et id quidem in multis ita se habere. Saepe vero etiam nova incidere genera morborum, in quibus nihil adhuc usus ostenderit et ideo necessarium sit animadvertere, unde ea coeperint; sine quo nemo reperir emortalium possit, cur hoc quam illo potius utatur. Et ob haec quidem in obscuro positas causas persecuntur.

18 Evidentes vero has appellant, in quibus quaerunt, initium morbi calor attulerit an frigus, fames an satietas, et quae similia sunt: occursurum enim vitio dicunt eum, qui originem non ignorarit.

19 Naturales vero corporis actiones appellant, per quas spiritum trahimus et emittimus, cibum

---

[1] *Marx suggests* si quotannis tamen nova remedia inveniuntur, neque dicendum esse antiquiores experimentis esse usos . . ., "*if new remedies nevertheless are found every year, nor must we say that the ancients went by experience. . . .*"

PROOEMIUM 15-19

which the Greeks term φλεγμόνη, and that inflammation effects such a disturbance as there is in fever, which was taught by Erasistratus; or if little bodies by being brought to a standstill in passing through invisible pores block the passage, as Asclepiades contended—his will be the right way of treatment, who has not failed to see the primary origin of the cause. They do not deny that experience is also necessary; but they say it is impossible to arrive at what should be done unless through some course of reasoning. For the older men, they say, did not cram the sick anyhow, but reasoned out what might be especially suitable, and then put to the test of experience what conjecture of a sort had previously led up to. Again they say that it makes no matter whether by now most remedies have been well explored already . . . if, nevertheless, they started from a reasoned theory; and that in fact this has also been done in many instances. Frequently, too, novel classes of disease occur about which hitherto practice has disclosed nothing, and so it is necessary to consider how such have commenced, without which no one among mortals can possibly find out whether this rather than that remedy should be used; this is the reason why they investigate the occult causes.

But they call evident those causes, concerning which they inquire, as to whether heat or cold, hunger or surfeit, or such like, has brought about the commencement of the disease; for they say that he will be the one to counter the malady who is not ignorant of its origin.

Further, they term natural actions of the body, those by which we draw in and emit breath, take in

CELSUS

potionemque et adsumimus et concoquimus, itemque per quas eadem haec in omnes membrorum partes digeruntur. Tum requirunt etiam, quare venae nostrae modo summittant se, modo attollant; quae ratio somni, quae vigiliae sit; sine quorum notitia neminem putant vel occurrere vel mederi morbis
20 inter haec nascentibus posse. Ex quibus quia maxime pertinere ad rem concoctio videtur, huic potissimum insistunt; et duce alii Erasistrato teri cibum in ventre contendunt, alii Plistonico Praxagorae discipulo putrescere; alii credunt Hippocrati per calorem cibos concoqui; acceduntque Asclepiadis aemuli, qui omnia ista vana et supervacua esse proponunt: nihil enim concoqui, sed crudam materiam, sicut adsumpta est, in corpus omne diduci.
21 Et haec quidem inter eos parum constant: illud vero convenit, alium dandum cibum laborantibus, si hoc, alium, si illud verum est: nam si teritur intus, eum quaerendum esse, qui facillime teri possit; si putrescit, eum, in quo hoc expeditissimum est; si calor
22 concoquit, eum, qui maxime calorem movet: at nihil ex his esse quaerendum, si nihil concoquitur, ea vero sumenda, quae maxime manent, qualia adsumpta sunt. Eademque ratione, cum spiritus gravis est, cum somnus aut vigilia urguet, eum mederi posse arbitrantur, qui prius illa ipsa qualiter eveniant perceperit.
23 Praeter haec, cum in interioribus partibus et

---

[a] Caelius Aurelianus, *Morb. Acut.* i. 14, § 113, mentions a division or solution, dividing up the food into minute particles: *neque ullam digestionem in nobis esse sed solutionem ciborum in ventre fieri et per singulas particulas corporis*, etc.

and digest food and drink, as also those actions through which food and drink are distributed into every part of the members. Moreover, they also inquire why our blood-vessels now subside, now swell up; what is the explanation of sleep and wakefulness: for without knowledge of these they hold that no one can encounter or remedy the diseases which spring up in connexion with them. Among these natural actions digestion seems of most importance, so they give it their chief attention. Some following Erasistratus hold that in the belly the food is ground up; others, following Plistonicus, a pupil of Praxagoras, that it putrefies; others believe with Hippocrates, that the food is cooked up by heat. In addition there are the followers of Asclepiades, who propound that all such notions are vain and superfluous, that there is no concoction at all, but that the material is transmitted throughout the body, crude as swallowed.[a] And on these points there is little agreement indeed among them; but what does follow is that a different food is to be given to patients according as this or that view is true. For if it is ground up inside, that food should be selected which can be ground up the most readily; if it putrefies, that which does so most expeditiously; if heat concocts it, that which most excites heat. But none of these points need be inquired into if there be no concoction but such things be taken which persist most in the state in which they were when swallowed. In the same way, when breathing is laboured, when sleep or wakefulness disturbs, they deem him able to remedy it who has understood beforehand how these same natural actions happen.

Moreover, as pains, and also various kinds of

CELSUS

dolores et morborum varia genera nascantur, neminem putant his adhibere posse remedia, qui ipsas ignoret. Ergo necessarium esse incidere corpora mortuorum, eorumque viscera atque intestina scrutari; longeque optime fecisse Herophilum et Erasistratum, qui nocentes homines a regibus ex carcere
24 acceptos vivos inciderint, considerarintque etiamnum spiritu remanente ea, quae natura ante clausisset, eorumque positum, colorem, figuram, magnitudinem, ordinem, duritiem, mollitiem, levorem, contactum, processus deinde singulorum et recessus, et sive quid inseritur alteri, sive quid partem alterius in se
25 recipit: neque enim, cum dolor intus incidit, scire quid doleat eum, qui, qua parte quodque viscus intestinumve sit, non cognoverit neque curari id, quod aegrum est, posse ab eo, qui quid sit ignoret; et cum per volnus alicuius viscera patefacta sunt, eum, qui sanae cuiusque colorem partis ignoret,
26 nescire quid integrum, quid corruptum sit; ita ne succurrere quidem posse corruptis. Aptiusque extrinsecus inponi remedia conpertis interiorum et sedibus et figuris cognitaque eorum magnitudine; similesque omnia, quae posita supra[1] sunt, rationes habere. Neque esse crudele, sicut plerique proponunt, hominum nocentium et horum quoque paucorum suppliciis remedia populis innocentibus saeculorum omnium quaeri.
27 Contra ii, qui se Empiricos ab experientia nominant, evidentes quidem causas ut necessarias

[1] supra *added by Marx.*

diseases, arise in the more internal parts, they hold that no one can apply remedies for these who is ignorant about the parts themselves; hence it becomes necessary to lay open the bodies of the dead and to scrutinize their viscera and intestines. They hold that Herophilus and Erasistratus did this in the best way by far, when they laid open men whilst alive—criminals received out of prison from the kings—and whilst these were still breathing, observed parts which beforehand nature had concealed, their position, colour, shape, size, arrangement, hardness, softness, smoothness, relation, processes and depressions of each, and whether any part is inserted into or is received into another. For when pain occurs internally, neither is it possible for one to learn what hurts the patient, unless he has acquainted himself with the position of each organ or intestine; nor can a diseased portion of the body be treated by one who does not know what that portion is. When a man's viscera are exposed in a wound, he who is ignorant of the colour of a part in health may be unable to recognize which part is intact, and which part damaged; thus he cannot even relieve the damaged part. External remedies too can be applied more aptly by one acquainted with the position, shape and size of the internal organs, and like reasonings hold good in all the instances mentioned above. Nor is it, as most people say, cruel that in the execution of criminals, and but a few of them, we should seek remedies for innocent people of all future ages.

On the other hand, those who are called "Empirici" because they have experience, do indeed accept evident causes as necessary; but they contend

amplectuntur: obscurarum vero causarum et naturalium actionum quaestionem ideo supervacuam esse contendunt, quoniam non conprehensibilis natura sit. 28 Non posse vero conprehendi patere ex eorum, qui de his disputarunt, discordia, cum de ista re neque inter sapientiae professores, neque inter ipsos medicos conveniat. Cur enim potius aliquis Hippocrati credat quam Herophilo? cur huic potius quam 29 Asclepiadi? Si rationes sequi velit, omnium posse videri non inprobabiles; si curationes, ab omnibus his aegros perductos esse ad sanitatem. Ita neque disputationi neque auctoritati cuiusquam fidem derogari oportuisse. Etiam sapientiae studiosos maximos medicos esse, si ratiocinatio hoc faceret: nunc illis 30 verba superesse, deesse medendi scientiam. Differre quoque pro natura locorum genera medicinae, et aliud opus esse Romae, aliud in Aegypto, aliud in Gallia. Quod si morbos haec [causae] facerent, quae ubique eadem essent, eadem remedia quoque ubique esse debuisse. Saepe etiam causas apparere, ut puta lippitudinis, vulneris, neque ex his patere 31 medicinam. Quod si scientiam hanc non subiciat evidens causa, multo minus eam posse subicere, quae in dubio est. Cum igitur illa incerta, inconprehensibilis sit, a certis potius et exploratis petendum esse praesidium, id est is, quae experientia in ipsis curationibus docuerit, sicut in ceteris omnibus

that inquiry about obscure causes and natural actions is superfluous, because nature is not to be comprehended. That nature cannot be comprehended is in fact patent, they say, from the disagreement among those who discuss such matters; for on this question there is no agreement, either among professors of philosophy or among actual medical practitioners. Why, then, should anyone believe rather in Hippocrates than in Herophilus, why in him rather than in Asclepiades? If one wants to be guided by reasoning, they go on, the reasoning of all of them can appear not improbable; if by method of treatment, all of them have restored sick folk to health: therefore one ought not to derogate from anyone's credit, either in argument or in authority. Even philosophers would have become the greatest of medical practitioners, if reasoning from theory could have made them so; as it is, they have words in plenty, and no knowledge of healing at all. They also say that the methods of practice differ according to the nature of localities, and that one method is required in Rome, another in Egypt, another in Gaul; but that if the causes which produce diseases were everywhere the same, the same remedies should be used everywhere; that often, too, the causes are apparent, as, for example, of ophthalmia, or of wounds, yet such causes do not disclose the treatment: that if the evident cause does not supply the knowledge, much less can a cause which is in doubt yield it. Since, therefore, the cause is as uncertain as it is incomprehensible, protection is to be sought rather from the ascertained and explored, as in all the rest of the Arts, that is, from what experience has taught in the actual course of treatment: for

32 artibus. Nam ne agricolam quidem aut gubernatorem disputatione sed usu fieri. Ac nihil istas cogitationes ad medicinam pertinere eo quoque disci, quod qui diversa de his senserint, ad eandem tamen sanitatem homines perduxerint: id enim fecisse, quia non ab obscuris causis neque a naturalibus actionibus, quae apud eos diversae erant, sed ab experimentis, prout cuique responderant, medendi
33 vias traxerint. Ne inter initia quidem ab istis quaestionibus deductam esse medicinam, sed ab experimentis: aegrorum enim, qui sine medicis erant, alios propter aviditatem primis diebus protinus cibum adsumpsisse, alios propter fastidium abstinuisse; levatumque magis eorum morbum esse, qui
34 abstinuerant. Itemque alios in ipsa febre aliquid edisse, alios paulo ante eam, alios post remissionem eius; optime deinde iis cessisse, qui post finem febris id fecerant; eademque ratione alios inter principia protinus usos esse cibo pleniore, alios exiguo;
35 gravioresque eos factos, qui se implerant. Haec similiaque cum cottidie inciderent, diligentes homines notasse quae plerumque melius responderent; deinde aegrotantibus ea praecipere coepisse. Sic medicinam ortam, subinde aliorum salute, aliorum interitu perniciosa discernentem a salutaribus.

36 Repertis deinde iam remediis, homines de rationibus eorum disserere coepisse; nec post rationem medicinam esse inventam, sed post inventam medicinam rationem esse quaesitam. Requirere etiam

even a farmer, or a pilot, is made not by disputation but by practice. That such speculations are not pertinent to the Art of Medicine may be learned from the fact that men may hold different opinions on these matters, yet conduct their patients to recovery all the same. This has happened, not because they deduced lines of healing from obscure causes, nor from the natural actions, concerning which different opinions were held, but from experiences of what had previously succeeded. Even in its beginnings, they add, the Art of Medicine was not deduced from such questionings, but from experience; for of the sick who were without doctors, some in the first days of illness, longing for food, took it forthwith; others, owing to distaste, abstained; and the illness was more alleviated in those who abstained. Again, some partook of food whilst actually under the fever, some a little before, others after its remission, and it went best with those who did so after the fever had ended; and similarly some at the beginning adopted at once a rather full diet, others a scanty one, and those were made worse who had eaten plentifully. When this and the like happened day after day, careful men noted what generally answered the better, and then began to prescribe the same for their patients. Thus sprang up the Art of Medicine, which, from the frequent recovery of some and the death of others, distinguished between the pernicious and the salutary.

It was afterwards, they proceed, when the remedies had already been discovered, that men began to discuss the reasons for them: the Art of Medicine was not a discovery following upon reasoning, but after the discovery of the remedy, the reason for it was

se, ratio idem doceat quod experientia an aliud: si idem, supervacuam esse; si aliud, etiam contrariam. Primo tamen remedia exploranda summa cura fuisse; nunc vero iam explorata esse; neque aut nova genera morborum reperiri, aut novam desiderari
37 medicinam. Quod si iam incidat mali genus aliquod ignotum, non ideo tamen fore medico de rebus cogitandum obscuris, sed eum protinus visurum cui morbo id proximum sit, temptaturumque remedia similia illis, quae vicino malo saepe succurrerint, et
38 per eius similitudines opem reperturum. Neque enim se dicere medicum consilio non egere et inrationale animal hanc artem posse praestare; sed has latentium rerum coniecturas ad rem non pertinere, quia non intersit, quid morbum faciat, sed quid tollat; neque ad rem pertineat, quomodo, sed quid optime digeratur, sive hac de causa concoctio incidat sive illa, et sive concoctio sit illa sive tantum
39 digestio. Neque quaerendum esse quomodo spiremus, sed quid gravem et tardum spiritum expediat; neque quid venas moveat, sed quid quaeque motus genera significent. Haec autem cognosci experimentis. Et in omnibus eiusmodi cogitationibus in utramque partem disseri posse; itaque ingenium et facundiam vincere, morbos autem non eloquentia sed remediis curari. Quae si quis elinguis usu discreta bene norit, hunc aliquanto maiorem medicum futurum, quam si sine usu linguam suam excoluerit.

sought out. They ask, too, does reasoning teach the same as experience? If the same, it was needless; if something else, then it was even opposed to it: nevertheless, at first remedies had to be explored with the greatest care; now, however, they have been explored already; there were neither new sorts of diseases to be found out, nor was a novel remedy wanted. For even if there happened nowadays some unknown form of malady, nevertheless the practitioner had not to theorize over obscure matters, but straightway would see to which disease it came nearest, then would make trial of remedies similar to those which have succeeded often in a kindred affection, and so through its similarities find help; that was not to say that a practitioner had no need to take counsel, and that an irrational animal was capable of exhibiting this art, but that these conjectures about concealed matters are of no concern because it does not matter what produces the disease but what relieves it; nor does it matter how digestion takes place, but what is best digested, whether concoction comes about from this cause or that, and whether the process is concoction or merely distribution. We had no need to inquire in what way we breathe, but what relieves laboured breathing; not what may move the blood-vessels, but what the various kinds of movements signify. All this was to be learnt through experiences; and in all theorizing over a subject it is possible to argue on either side, and so cleverness and fluency may get the best of it; it is not, however, by eloquence but by remedies that diseases are treated. A man of few words who learns by practice to discern well, would make an altogether better practitioner than he who, unpractised, over-cultivates his tongue.

40 Atque ea quidem, de quibus est dictum, supervacua esse tantummodo: id vero, quod restat, etiam crudele, vivorum hominum alvum atque praecordia incidi, et salutis humanae praesidem artem non solum pestem alicui, sed hanc etiam atrocissimam inferre; cum praesertim ex his, quae tanta violentia quaerantur, alia non possint omnino cognosci, alia
41 possint etiam sine scelere. Nam colorem, levorem, mollitiem, duritiem, similiaque omnia non esse talia inciso corpore, qualia integro fuerint, quia, cum corpora inviolata sint, haec tamen metu, dolore, inedia, cruditate, lassitudine, mille aliis mediocribus adfectibus saepe mutentur; multo magis veri simile esse interiora, quibus maior mollities, lux ipsa nova sit, sub gravissimis vulneribus et ipsa trucidatione
42 mutari. Neque quicquam esse stultius, quam quale quidque vivo homine est, tale existimare esse moriente, immo iam mortuo. Nam uterum quidem, qui minus ad rem pertineat, spirante homine posse diduci: simul atque vero ferrum ad praecordia accessit et discissum transversum saeptum est, quod membrana quaedam est quae[1] superiores partes ab inferioribus diducit (διάφραγμα Graeci vocant), hominem animam protinus amittere: ita mortui demum praecordia et viscus omne in conspectum latrocinantis medici dari utique necesse est
43 tale, quale mortui sit, non quale vivi fuit. Itaque consequi medicum, ut hominem crudeliter iugulet,

[1] est quae *added by Marx.*

[a] The strong terms of condemnation in §§ 40–43 represent the opinions of these practitioners (the Empirics), but Celsus endorses them himself in § 74.

[b] For *praecordia* see note [a] on p. 100, and index, Vol. III, p. 644.

PROOEMIUM 40–43

Now the matters just referred to they [a] deem to be superfluous; but what remains, cruel as well, to cut into the belly and chest [b] of men whilst still alive, and to impose upon the Art which presides over human safety someone's death, and that too in the most atrocious way. Especially is this true when, of things which are sought for with so much violence, some can be learnt not at all, others can be learnt even without a crime. For when the body had been laid open, colour, smoothness, softness, hardness and all similars would not be such as they were when the body was untouched; because bodies, even when uninjured, yet often change in appearance, they note, from fear, pain, want of food, indigestion, weariness and a thousand other mediocre affections; it is much more likely that the more internal parts, which are far softer, and to which the very light is something novel, should under the most severe of woundings, in fact mangling, undergo changes. Nor is anything more foolish, they say, than to suppose that whatever the condition of the part of a man's body in life, it will also be the same when he is dying, nay, when he is already dead; for the belly indeed, which is of less importance, can be laid open with the man still breathing; but as soon as the knife really penetrates to the chest, by cutting through the transverse septum, a sort of membrane which divides the upper from the lower parts (the Greeks call it διάφραγμα), the man loses his life at once: so it is only when the man is dead that the chest and any of the viscera come into the view of the medical murderer, and they are necessarily those of a dead, not of a living man. It follows, therefore, that the medical man just plays the cut-throat, not that he learns

CELSUS

non ut sciat, qualia vivi viscera habeamus. Si quid tamen sit, quod adhuc spirante homine conspectu subiciatur, id saepe casum offerre curantibus. Interdum enim gladiatorem in harena vel militem in acie vel viatorem a latronibus exceptum sic vulnerari, ut eius interior aliqua pars aperiatur, et in alio alia; ita sedem, positum, ordinem, figuram, similiaque alia cognoscere prudentem medicum, non caedem sed sanitatem molientem, idque per misericordiam
44 discere, quod alii dira crudelitate cognorint. Ob haec ne mortuorum quidem lacerationem necessariam esse (quae etsi non crudelis, tamen foeda sit), cum aliter pleraque in mortuis se habeant; quantum vero in vivis cognosci potest, ipsa curatio ostendat.

45 Cum haec per multa volumina perque magnas contentionis [disputationes] a medicis saepe tractata sint atque tractentur, subiciendum est, quae proxima vero videri possint. Ea neque addicta alterutri opinioni sunt, neque ab utraque nimium abhorrentia, sed[1] media quodammodo inter diversas sententias; quod in plurimis contentionibus deprehendere licet sine ambitione verum scrutantibus: ut in hac ipsa re.

46 Nam quae demum causae vel secundam valetudinem praestent, vel morbos excitent, quo modo spiritus aut cibus vel trahatur vel digeratur, ne sapientiae quidem professores scientia conprehendunt, sed coniectura persecuntur. Cuius autem rei non est certa notitia, eius opinio certum reperire remedium

[1] sed *added by v. d. Linden.*

what our viscera are like when we are alive. If, however, there be anything to be observed whilst a man is still breathing, chance often presents it to the view of those treating him. For sometimes a gladiator in the arena, or a soldier in battle, or a traveller who has been set upon by robbers, is so wounded that some or other interior part is exposed in one man or another. Thus, they say, an observant practitioner learns to recognize site, position, arrangement, shape and such like, not when slaughtering, but whilst striving for health; and he learns in the course of a work of mercy, what others would come to know by means of dire cruelty. That for these reasons, since most things are altered in the dead, some hold that even the dissection of the dead is unnecessary; although not cruel, it is none the less nasty; but all that is possible to come to know in the living, the actual treatment exhibits.

Since all these questions have been discussed often by practitioners, in many volumes and in large and contentious disputations, and the discussion continues, it remains to add such views as may seem nearest the truth. These are neither wholly in accord with one opinion or another, nor exceedingly at variance with both, but hold a sort of intermediate place between divers sentiments, a thing which may be observed in most controversies when men seek impartially for truth, as in the present case. For as regards the causes which either favour health or excite disease, how breath is drawn in or food distributed, not even philosophers attain to full knowledge, but seek it out by conjecture. But where there is no certain knowledge about a thing, mere opinion about it cannot find a certain remedy.

CELSUS

47 non potest. Verumque est ad ipsam curandi rationem nihil plus conferre quam experientiam. Quamquam igitur multa sint ad ipsas artes proprie non pertinentia, tamen eas adiuvant excitando artificis ingenium: itaque ista quoque naturae rerum contemplatio, quamvis non faciat medicum, aptiorem tamen medicinae reddit perfectumque. Verique simile est et Hippocraten et Erasistratum, et quicumque alii non contenti [sint] febres et ulcera agitare rerum quoque naturam aliqua parte scrutati sunt, non ideo quidem medicos fuisse, verum ideo 48 quoque maiores medicos extitisse. Ratione vero opus est ipsi medicinae, etsi non inter obscuras causas neque inter naturales actiones, tamen saepe . . . :[1] est enim haec ars coniecturalis. Neque respondet ei plerumque non solum coniectura sed etiam experientia et interdum non febris, non cibus,[2] non somnus subsequitur, sicut adsuevit. 49 Rarius sed aliquando morbus quoque ipse novus est: quem non incidere manifeste falsum est, cum aetate nostra . . .[3] quae ex naturalibus partibus carne prolapsa et arente intra paucas horas exspiraverit, sic ut nobilissimi medici neque genus mali neque 50 remedium invenerint. Quos ego nihil temptasse iudico, quia nemo in splendida persona periclitari coniectura sua voluerit, ne occidisse, nisi servasset, videretur: veri tamen simile est potuisse aliquid

[1] *Marx suggests inserting* in eorum qui evidentibus causis nati sunt morborum curatione, "*in treating those diseases which come from evident causes.*"

[2] *M. suggests* pus, *suppuration: see* IV. 11, 3; VIII. 10, 1. c.

[3] *M. suggests e.g.* fuerit matrona equiti Romano nupta.

PROOEMIUM 46–50

And it is true that nothing adds more to a really rational treatment than experience. Although, therefore, many things, which are not strictly pertinent to the Arts as such, are yet helpful by stimulating the minds of those who practise them, so also this contemplation of the nature of things, although it does not make a practitioner, yet renders him more apt and perfected in the Art of Medicine. And it is probable that Hippocrates, Erasistratus and certain others, who were not content to busy themselves over fevers and ulcerations, but also to some extent searched into the nature of things, did not by this become practitioners, but by this became better practitioners. But reasoning is necessary to the Art of Medicine, not only when dealing with obscure causes, or natural actions, but often . . . for it is an art based on conjecture. However, in many cases not only does conjecture fail, but experience as well; and at times, neither fever, nor appetite, nor sleep follow their customary course. More rarely, yet now and again, a disease itself is new. That this does not happen is manifestly untrue, for in our time a lady, from whose genitals flesh had prolapsed and become gangrenous,[a] died in the course of a few hours, whilst practitioners of the highest standing found out neither the class of malady nor a remedy. I conclude that they attempted nothing because no one was willing to risk a conjecture of his own in the case of a distinguished personage, for fear that he might seem to have killed, if he did not save her;

---

[a] For this meaning of *arente*, see V. **26**, 31 C. *caro arida*; VII. **12**, 1, propter gingivarum arescentium vitium. Morgagni, Ep. IV., thought an inversion of the uterus to be indicated. See Hippocrates, *Prolapse of the uterus and its reduction* (Littré, VIII. 143–147).

cogitare, detracta tali verecundia, et fortasse responsurum fuisse id, quod aliquis esset expertus. Ad quod medicinae genus neque semper similitudo aliquid confert, et si quando confert, tamen id ipsum rationale est, inter similia genera et morborum et remediorum cogitare, quo potissimum medicamento sit utendum. Cum igitur talis res incidit, medicus aliquid oportet inveniat, quod non utique fortasse sed saepius tamen etiam respondeat. Petet autem novum quodque consilium non ab rebus latentibus (istae enim dubiae et incertae sunt), sed ab iis, quae explorari possunt, id est evidentibus causis. Interest enim fatigatio morbum an sitis, an frigus an calor, an vigilia an fames fecerit, an cibi vinique abundantia, an intemperantia libidinis. Neque ignorare hunc oportet, quae sit aegri natura, umidum magis an magis siccum corpus eius sit, validi nervi an infirmi, frequens adversa valetudo an rara, eaque, cum est, vehemens esse soleat an levis, brevis an longa; quod is vitae genus sit secutus, laboriosum an quietum, cum luxu an cum frugalitate : ex his enim similibusque saepe curandi nova ratio ducenda est.

Quamvis ne haec quidem sic praeteriri debent, quasi nullam controversiam recipiant. Nam et Erasistratus non ex illis causis[1] fieri morbos dixit, quoniam et alii et idem alias post istas non febricitarent;

[1] causis *added by Marx.*

[a] For allusions by Celsus to his own practice, see Introduction, p. xi., and IV. 26, 4, etc.
[b] Considered in Books I. and II. 1-7.

yet it is probable that something might possibly have been thought of, had no such timidity prevented, and perchance this might have been successful had one but tried it.[a] In this sort of practice similarity is not always of service, and when it does prove serviceable, nevertheless there has been a process of reasoning, in the theorizing over similar classes of diseases and of remedies, as to which is the best remedy to use. When, therefore, such an incident occurs, the practitioner ought to arrive at something which may answer, even if perhaps not always, yet nevertheless more often than not. He will seek, however, every novel plan, not from hidden things, for these are dubious and unascertainable, but from those which can be explored, that is, from evident causes.[b] For what matters is this: whether fatigue or thirst, whether heat or cold, whether wakefulness or hunger, whether abundance in food or wine, whether intemperance in venery, has produced the disease. Nor should there be ignorance of the sick man's temperament; whether his body is rather humid or rather dry, whether his sinews are strong or weak, whether he is frequently or rarely ill; and when ill whether so severely or slightly, for a short or long while; the kind of life he has lived, laborious or quiet, accompanied by luxury or frugality. From such and similar data, one may often deduce a novel mode of treatment.

None the less the foregoing statements ought not to be passed by as if they did not admit of controversy. For Erasistratus himself has affirmed that diseases were not produced by such causes, since other persons, and even the same person at different times, were not rendered feverish by them. Further,

CELSUS

et quidam medici saeculi nostri sub auctore, ut ipsi videri volunt, Themisone contendunt nullius causae notitiam quicquam ad curationes pertinere; satisque esse quaedam communia morborum intueri.
55 Siquidem horum tria genera esse, unum adstrictum, alterum fluens, tertium mixtum. Nam modo parum excernere aegros, modo nimium, modo alia parte parum, alia nimium: haec autem genera morborum modo acuta esse, modo longa, et modo increscere,
56 modo consistere, modo minui. Cognito igitur eo, quod ex his est, si corpus adstrictum est, digerendum esse; si profluvio laborat, continendum; si mixtum vitium habet, occurrendum subinde vehementiori malo. Et aliter acutis morbis medendum, aliter vetustis, aliter increscentibus, aliter subsistentibus,
57 aliter iam ad sanitatem inclinatis. Horum observationem medicinam esse; quam ita finiunt, ut quasi viam quandam quam μέθοδον nominant, eorumque, quae in morbis communia sunt, contemplatricem esse contendant. Ac neque rationalibus se neque experimenta tantum spectantibus adnumerari volunt, cum ab illis eo nomine dissentiant, quod in coniectura rerum latentium nolunt esse medicinam; ab his eo, quod parum artis esse in observatione experimentorum credunt.
58 Quod ad Erasistratum pertinet, primum ipsa

---

[a] Hippocrates, I. 180. (Epid. I. xxiii.).

[b] Commonest examples: *strictum*, constipation advancing to intestinal obstruction; *laxum*, loose stools advancing to diarrhoea and dysentery; *mixtum*, μεμιγμένον, a makeshift description for conditions not fitting into the two other categories.

[c] Here *digerere* = διαφορεῖν, disperse the *materies morbi*: I. 9, 6, by a sweat; II. 14, 3, by rubbing; II. 33, 6, by topical applications; V. 28, 11 B., C., VII. 2, 3, by abscession, or gathering of matter, maturing to abscess.

certain practitioners of our time, following, as they would have it appear, the authority of Themison, contend that there is no cause whatever, the knowledge of which has any bearing on treatment: they hold that it is sufficient to observe certain general characteristics of diseases;[a] that of these there are three classes, one a constriction, another a flux, the third a mixture.[b] For the sick at one time excrete too little, at another time too much; again, from one part too little, from another too much; and these classes of diseases are sometimes acute, sometimes chronic, at times on the increase, at times constant, at times diminishing. Once it has been recognized, then, which it is of these, if the body is constricted, it has to be relaxed;[c] if suffering from a flux, that has to be controlled; if a mixed lesion, the more severe malady must be countered first. Moreover, there must be treatment of one kind for acute diseases, another kind for chronic ones, another for increasing, stationary, or for those already tending to recovery. They hold that the Art of Medicine consists of such observations; which they define as a sort of way, which they name μέθοδος, and maintain that medicine should examine those characteristics which diseases have in common. They do not want to be classed with reasoners from theory, nor with those who look to experience only; for in so naming themselves Methodici, they dissent from the former because they are unwilling that the Art should consist in conjecture about hidden things, and from the latter because they think that in the observation of experience there is little of an Art of Medicine.

As relates to Erasistratus, in the first place the

CELSUS

evidentia eius opinioni repugnat[1] quia raro nisi post horum aliquid morbus venit; deinde non sequitur, ut, quod alium non adficit aut eundem alias, id ne alteri quidem aut eidem tempore alio noceat. Possunt enim quaedam subesse corpori vel ex infirmitate eius vel ex aliquo adfectu, quae vel in alio non sunt, vel in hoc alias non fuerunt eaque per se non tanta, ut[2] concitent morbum, tamen obnoxium
59 magis aliis iniuriis corpus efficiant. Quod si contemplationem rerum naturae, quam temere medici isti[3] sibi vindicant, satis conprehendisset, etiam illud scisset, nihil omnino ob unam causam fieri, sed id pro causa adprehendi, quod contulisse plurimum videtur. Potest autem id, dum solum est, non
60 movere, quod iunctum aliis maxime moveat. Accedit ad haec, quod ne ipse quidem Erasistratus, qui transfuso in arterias sanguine febrem fieri dicit idque nimis repleto corpore incidere, repperit, cur ex duobus aeque repletis alter in morbum incideret, alter omni periculo vacaret; quod cotidie fieri apparet.
61 Ex quo disci potest, ut vera sit illa transfusio, tamen illam non per se, cum plenum corpus est, fieri, sed cum horum aliquid accesserit.
62 Themisonis vero aemuli, si perpetua quae

[1] *Marx thinks there is a gap after* repugnat *and suggests* esse enim quaedam quae morbos efficiant apparet, "*for it is clear that there are certain causes which can produce diseases*," *etc.*
[2] ut *Constantine*, vi. *MSS.*
[3] isti *added by Marx*; *v. d. Linden (followed by Targa and Daremberg) reads* quae non temere medici sibi vindicant, "*which the practitioners with very good reason claim.*"

actual evidence is against his opinion, because seldom does a disease occur unless following upon one of these; secondly, it does not follow that what has done no harm to one patient, or to that same patient upon one occasion, may not harm another patient, or the same one at another time. For it is possible that there are certain underlying conditions in the body, whether related to infirmity, or to an actual affection of some kind, which either are not present in another person, or were not existent in that patient on another occasion, and which of themselves are not enough to constitute a disease, yet they may render the body more liable to other injurious affections. But if Erasistratus had been sufficiently versed in the study of the nature of things, as those practitioners rashly claim themselves to be, he would have known also that nothing is due to one cause alone, but that which is taken to be the cause is that which seems to have had the most influence. Indeed it is possible that when one cause acts alone, it may not disturb, yet when acting in conjunction with other causes it may produce a very great disturbance. Moreover, even Erasistratus himself, who says that fever is produced by blood transfused into the arteries, and that this happens in an over-replete body, failed to discover why, of two equally replete persons, one should lapse into disease, and the other remain free from anything dangerous; and that clearly happens every day. Hence, however true this transfusion, one can learn that it does not occur of itself when there is bodily fullness, but when there is added something else.

But disciples of Themison, if they hold their

CELSUS

promittunt habent, magis etiam quam ulli rationales sunt. Neque enim, si quis non omnia tenet, quae rationalis alius probat, protinus alio [novo] nomine artis indiget, si modo (quod primum est) non 63 memoriae soli sed rationi quoque insistit. Si, vero quod propius est, vix ulla perpetua praecepta medicinalis ars recipit, idem sunt quod ii, quos experimenta sola sustinent; eo magis quoniam, conpresserit aliquem morbus an fuderit, quilibet etiam inperitissimus videt: quid autem conpressum corpus resolvat, quid solutum teneat, si a ratione tractum est, rationalis est medicus; si, ut ei, qui se rationalem negat, confiteri necesse est, ab experi- 64 entia, empiricus. Ita apud eum morbi cognitio extra artem, medicina intra usum est; neque adiectum quicquam empiricorum professioni, sed demptum est, quoniam illi multa circumspiciunt, hi 65 tantum facillima, et non plus quam vulgaria. Nam et ii, qui pecoribus ac iumentis medentur, cum propria cuiusque ex mutis animalibus nosse non possint, communibus tantummodo insistunt; et exterae gentes, cum subtilem medicinae rationem non noverint, communia tantum vident; et qui ampla valetudinaria nutriunt, quia singulis summa cura consulere non sustinent, ad communia ista

precepts to be of constant validity, are reasoners even more than anybody else; for if a man does not hold all the tenets that another reasoner approves, he does not forthwith have to assume a different name for his art, if (and this is the essential point) he does rely not only on written authority, but also upon reasoning from theory. But if, which is nearer to the truth, the Art of Medicine admits of scarcely any universal precepts, reasoners are in the same position as those who depend upon experience alone, all the more because whether the disease has braced or relaxed is what the most uninstructed can see. But if a remedy which loosens a body braced up, or tightens a loosened body, has been deduced by a reasoning from theory, the practitioner is a reasoner; if (as the man who denies himself to be a reasoner must admit) he acts from experience, he is an Empiric. Thus according to Themison, knowledge of a disease is outside the Art, and medicine is confined to practice; nor has there been added anything to what Empirics profess, but something taken away; for reasoners from theory gaze about over a multiplicity of matters, Empirics look to circumstances the most simple, and nothing more than commonplaces. For in like manner those who treat cattle and horses, since it is impossible to learn from dumb animals particulars of their complaints, depend only upon common characteristics; so also do foreigners as they are ignorant of reasoning subtleties look rather to common characteristics of disease. Again, those who take charge of large hospitals, because they cannot pay full attention to individuals, resort to these common characteristics.

CELSUS

66 confugiunt. Neque Hercules istud antiqui medici nescierunt, sed his contenti non fuerunt. Ergo etiam vetustissimus auctor Hippocrates dixit mederi oportere et communia et propria intuentem. Ac ne isti quidem ipsi intra suam professionem consistere ullo modo possunt: siquidem et conpressorum et fluentium morborum genera diversa sunt; faciliusque
67 id in iis, quae fluunt, inspici potest. Aliud est enim sanguinem, aliud bilem, aliud cibum vomere; aliud deiectionibus, aliud torminibus laborare; aliud sudore digeri, aliud tabe consumi. Atque in partes quoque umor erumpit, ut oculos aurisque; quo periculo nullum humanum membrum vacat. Nihil autem horum sic ut aliud curatur.

68 Ita protinus in his a communi fluentis morbi contemplatione ad propriam medicina descendit. Atque in hac quoque rursus alia proprietatis notitia saepe necessaria est; quia non eadem omnibus etiam in similibus casibus opitulantur: siquidem certae quaedam res sunt, quae in pluribus ventrem aut adstringunt aut resolvunt. Inveniuntur tamen, in quibus aliter atque in ceteris idem eveniat: in his ergo communium inspectio contraria est, propriorum tantum salutaris.
69 Et causae quoque aestimatio saepe morbum solvit. Ergo etiam ingeniosissimus saeculi nostri medicus, quem nuper vidimus, Cassius febricitanti cuidam et magna siti adfecto, cum post ebrietatem eum premi

---

[a] ὁ ἐπιλογισμός—a reckoning up of factors to arrive at a *cognitio* or διάγνωσις, II. 14, 3 *et al.*

I vow, the ancients knew all this, but were not content therewith; therefore even the oldest authority, Hippocrates, said that in healing it was necessary to take note both of common and of particular characteristics. Indeed these very Methodici, even within their professed limitations, cannot be consistent; for there are divers kinds of constricting and of relaxing diseases, those in which there is a flux being the more easy to observe. For it is one thing to vomit blood, another bile, another food; it is one thing to suffer from diarrhoea, another from dysentery; one thing to be relaxed through sweating, another to be wasted by consumption. Humour may break out into particular parts, such as the eyes or the ears; from a risk of this kind there is no human member free. No one of these occurrences is treated in the same way as another.

Hence the Art descends straight down from a consideration of the common characteristics of a flux to the particular case. Moreover, because the same remedies do not meet with success in all, even of similar cases, additional knowledge of peculiarities in such a case is often necessary. Although certain things act upon the bowels in most cases, whether as astringents or as laxatives, yet there are to be found some in whom the same thing acts differently than it does in others. In such instances, therefore, investigation of particular characteristics is salutary, that of common characteristics the reverse. Moreover, a reckoning up of the cause often solves the malady.[a] Thus Cassius, the most ingenious practitioner of our generation, recently dead, in a case suffering from fever and great thirst, when he learnt that the man had begun to feel oppressed after intoxication,

coepisse cognosset, aquam frigidam ingessit; qua ille epota cum vini vim miscendo fregisset, protinus 70 febrem somno et sudore discussit. Quod auxilium medicus opportune providit non ex eo, quod aut adstrictum corpus erat aut fluebat, sed ex ea causa, quae ante praecesserat. Estque etiam proprium aliquid et loci et temporis istis quoque auctoribus: qui, cum disputant, quemadmodum sanis hominibus agendum sit, praecipiunt, ut gravibus aut locis aut temporibus magis vitetur frigus, aestus, satietas, labor, libido; magisque ut conquiescat isdem locis aut temporibus, si quis gravitatem corporis sensit, ac neque vomitu stomachum neque purgatione alvum 71 sollicitet. Quae vera quidem sunt; a communibus tamen ad quaedam propria descendunt, nisi persuadere nobis volunt sanis quidem considerandum esse, quod caelum, quod tempus anni sit, aegris vero non esse; quibus tanto magis omnis observatio necessaria est, quanto magis obnoxia offensis infirmitas est. Quin etiam morborum in isdem hominibus aliae atque aliae proprietates sunt; et qui secundis aliquando frustra curatus est, contrariis 72 saepe restituitur. Plurimaque in dando cibo discrimina reperiuntur, ex quibus contentus uno ero. Nam famem facilius adulescens quam puer, facilius in denso caelo quam in tenui, facilius hieme quam aestate, facilius uno cibo quam prandio quoque

administered cold water, by which draught, when by the admixture he had broken the force of the wine, he forthwith dispersed the fever by means of a sleep and a sweat. He, as a practitioner, provided an opportune remedy, not out of consideration whether the man's body was constricted or relaxed, but from what had happened beforehand to cause it. Besides, according to these very authorities there are particulars relating to locality and to season. When they are discussing what should be done by men in health, they prescribe the avoidance of cold, heat, surfeit, fatigue, venery, especially in sickly localities and seasons; in such places and seasons rest is to be taken, particularly when one feels a sense of oppression, and neither the stomach is to be disturbed by an emetic, nor the bowels by a purge. Such generalities are indeed true: none the less they descend from them to certain particular characteristics, unless they would persuade us that climate and season are to be taken into consideration by those in health but not by the sick, the very persons in whom all such observance is by so much the more necessary, the more that their weakness is liable to all attacks. Nay, even in the same patient, the particular characteristics of a disease are very variable, and those who have been treated for a time in vain by the ordinary remedies have been often restored by contrary ones. And in the giving of food too there are many distinctions to be noted; I will content myself with one instance. For hunger is more easily borne by an adult than by a boy, more easily in a dense than in a thin atmosphere, more easily in winter than in summer, more easily by one accustomed to a single meal than by one used in addition to one at midday,

adsuetus, facilius inexercitatus quam exercitatus
73 homo sustinet: saepe autem in eo magis necessaria
cibi festinatio est, qui minus inediam tolerat. Ob
quae conicio eum, qui propria non novit, communia
tantum debere intueri; eumque, qui nosse proprietates potest, non illas quidem oportere neglegere,
sed his quoque insistere; ideoque, cum par scientia
sit, utiliorem tamen medicum esse amicum quam
extraneum.

74 Igitur, ut ad propositum meum redeam, rationalem
quidem puto medicinam esse debere, instrui vero ab
evidentibus causis, obscuris omnibus non ab cogitatione artificis sed ab ipsa arte reiectis. Incidere
autem vivorum corpora et crudele et supervacuum
est, mortuorum discentibus necessarium: nam positum et ordinem nosse debent, quae cadaver melius
75 quam vivus et vulneratus homo repraesentat. Sed
et cetera, quae modo in vivis cognosci possunt, in
ipsis curationibus vulneratorum paulo tardius sed
aliquanto mitius usus ipse monstrabit.

His propositis, primum (*lib. I*) dicam, quemadmodum sanos agere conveniat, tum ad ea transibo
(*lib. II*, **1-8**), quae ad morbos curationesque eorum
(*lib. II*, **9-33**) pertinebunt.

more easily when sedentary than when in active exercise; and often it is necessary to hurry on the meal in the case of one who is intolerant of hunger. Hence I conjecture that he who is not acquainted with the peculiar characteristics has merely to consider the general ones; and he who can become acquainted with peculiarities, whilst insistent upon them, ought not to neglect generalities as well; and consequently, presuming their state to be equal, it is more useful to have in the practitioner a friend rather than a stranger.

Therefore, to return to what I myself propound, I am of opinion that the Art of Medicine ought to be rational, but to draw instruction from evident causes, all obscure ones being rejected from the practice of the Art, although not from the practitioner's study. But to lay open the bodies of men whilst still alive is as cruel as it is needless; that of the dead is a necessity for learners, who should know positions and relations, which the dead body exhibits better than does a living and wounded man. As for the remainder, which can only be learnt from the living, actual practice will demonstrate it in the course of treating the wounded in a somewhat slower yet much milder way.

With these premises I will first speak of how those in health should act (Book I), then I will pass on to what pertains to diseases (Book II, **1–8**), and to their treatments (Book II, **9–33**).

CELSUS

LIBER I

1. Sanus homo, qui et bene valet et suae spontis est, nullis obligare se legibus debet, ac neque medico neque iatroalipta egere. Hunc oportet varium habere vitae genus: modo ruri esse, modo in urbe, saepiusque in agro; navigare, venari, quiescere interdum, sed frequentius se exercere; siquidem ignavia corpus hebetat, labor firmat, illa maturam senectutem, hic longam adulescentiam reddit.

2 Prodest etiam interdum balineo, interdum aquis frigidis uti; modo ungui, modo id ipsum neglegere; nullum genus cibi fugere, quo populus utatur; interdum in convictu esse, interdum ab eo se retrahere; modo plus iusto, modo non amplius adsumere; bis die potius quam semel cibum capere, et semper quam plurimum, dummodo hunc concoquat. Sed ut

3 huius generis exercitationes cibique necessariae sunt, sic athletici supervacui: nam et intermissus propter civiles aliquas necessitates ordo exercitationis corpus adfligit, et ea corpora, quae more eorum repleta sunt, celerrime et senescunt et aegrotant.

4 Concubitus vero neque nimis concupiscendus, neque nimis pertimescendus est. Rarus corpus

---

[a] Book II. **14.**
[b] Hipp., IV. 346 (Regimen II. 60).

BOOK I

1. A MAN in health, who is both vigorous and his own master, should be under no obligatory rules, and have no need, either for a medical attendant, or for a rubber and anointer.[a] His kind of life should afford him variety; he should be now in the country, now in town, and more often about the farm; he should sail, hunt, rest sometimes, but more often take exercise; for whilst inaction[b] weakens the body, work strengthens it; the former brings on premature old age, the latter prolongs youth.

It is well also at times to go to the bath, at times to make use of cold waters;[c] to undergo sometimes inunction, sometimes to neglect that same; to avoid no kind of food in common use; to attend at times a banquet, at times to hold aloof; to eat more than sufficient at one time, at another no more; to take food twice rather than once a day, and always as much as one wants provided one digests it. But whilst exercise and food of this sort are necessaries, those of the athletes are redundant; for in the one class any break in the routine of exercise, owing to necessities of civil life, affects the body injuriously, and in the other, bodies thus fed up in their fashion age very quickly and become infirm.[d]

Concubitus indeed is neither to be desired overmuch, nor overmuch to be feared; seldom used it

[c] Book II. **17,** 1; III. **21,** 6.
[d] Both training and diet were carried to excess by athletes.

excitat, frequens solvit. Cum autem frequens non numero sit sed natura . . .,[1] ratione aetatis et corporis, scire licet eum non inutilem esse, quem corporis neque languor neque dolor sequitur. Idem interdiu peior est, noctu tutior, ita tamen, si neque illum cibus, neque hunc cum vigilia labor statim sequitur. Haec firmis servanda sunt, cavendumque ne in secunda valetudine adversae praesidia consumantur.

2. At imbecillis, quo in numero magna pars urbanorum omnesque paene cupidi litterarum sunt, observatio maior necessaria est, ut, quod vel corporis 2 vel loci vel studii ratio detrahit, cura restituat. Ex his igitur qui bene concoxit, mane tuto surget; qui parum, quiescere debet, et si mane surgendi necessitas fuit, redormire; qui non concoxit, ex toto conquiescere ac neque labori se neque exercitationi neque negotiis credere. Qui crudum sine praecordiorum dolore ructat, is ex intervallo aquam frigidam bibere, et se nihilo minus continere.

3 Habitare vero aedificio lucido, perflatum aestivum, hibernum solem habente; cavere meridianum solem, matutinum et vespertinum frigus, itemque auras fluminum atque stagnorum; minimeque nubilo caelo soli aperienti se . . .[2] committere, ne modo frigus, modo calor moveat; quae res maxime gravidines

---

[1] *Some MSS. omit* natura; *Marx retains it and suggests that* aestimandus, habita *has dropped out and this reading is translated.*

[2] *Marx proposes to insert* subinde se *before* committere; *the meaning would then be :* " to expose himself as little as possible to a cloudy sky, to a sun that occasionally breaks through, lest," *etc.*

braces the body, used frequently it relaxes. Since, however, nature and not number should be the standard of frequency, regard being had to age and constitution, concubitus can be recognized as harmless when followed neither by languor nor by pain. The use is worse in the day-time, and safer by night; but care should be taken that by day it be not immediately followed by a meal, and at night not immediately followed by work and watching. Such are the precautions to be observed by the strong, and they should take care that whilst in health their defences against ill-health are not used up.

2. The weak, however, among whom are a large portion of townspeople, and almost all those fond of letters, need greater precaution, so that care may re-establish what the character of their constitution or of their residence or of their study detracts. Anyone therefore of these who has digested well may with safety rise early; if too little, he must stay in bed, or if he has been obliged to get up early, must go to sleep again; he who has not digested, should lie up altogether, and neither work nor take exercise nor attend to business. He who without heartburn eructates undigested food should drink cold water at intervals and none the less exercise self-control.

He should also reside in a house that is light, airy in summer, sunny in winter; avoid the midday sun, the morning and evening chill, also exhalations from rivers and marshes; and he should not often expose himself when the sky is cloudy to a sun that breaks through . . ., lest he should be affected alternately by cold and heat—a thing which excites

destillationesque concitat. Magis vero gravibus locis ista servanda sunt, in quibus etiam pestilentiam faciunt.

4 Scire autem licet integrum corpus esse, quo die mane urina alba, dein rufa est: illud concoquere, hoc concoxisse significat. Ubi experrectus est aliquis, paulum intermittere; deinde, nisi hiemps est, fovere
5 os multa aqua frigida debet; longis diebus meridiari potius ante cibum; si minus, post eum. Per hiemem potissimum totis noctibus conquiescere; sin lucubrandum est, non post cibum id facere, sed post concoctionem. Quem interdiu vel domestica vel civilia officia tenuerunt, huic tempus aliquod servandum curationi corporis sui est. Prima autem eius curatio exercitatio est, quae semper antecedere cibum debet, in eo, qui minus laboravit et bene concoxit, amplior; in eo, qui fatigatus est et minus concoxit, remissior.

6 Commode vero exercent clara lectio, arma, pila, cursus, ambulatio, atque haec non utique plana commodior est, siquidem melius ascensus quoque et descensus cum quadam varietate corpus moveat, nisi tamen id perquam inbecillum est: melior autem est sub divo quam in porticu; melior, si caput patitur, in sole quam in umbra, melior in umbra[1] quam paries aut viridia efficiunt, quam quae tecto
7 subest; melior recta quam flexuosa. Exercitationis autem plerumque finis esse debet sudor aut certe

---

[1] *Some MSS. omit* melior in umbra.

---

a IV. 5.   *b* See I. 10.
c See II. 3, 3.

BOOK I. 2. 3–7

particularly choked nostrils and running colds.[a] Much more indeed are these things to be watched in unhealthy localities, where they even produce pestilence.[b]

He can tell that his body is sound,[c] if his morning urine is whitish, later reddish; the former indicates that digestion is going on, the latter that digestion is complete. On waking one should lie still for a while, then, except in winter time, bathe the face freely with cold water; when the days are long the siesta should be taken before the midday meal, when short, after it. In winter, it is best to rest in bed the whole night long; if there must be study by lamp-light, it should not be immediately after taking food, but after digestion. He who has been engaged in the day, whether in domestic or on public affairs, ought to keep some portion of the day for the care of the body. The primary care in this respect is exercise, which should always precede the taking of food; the exercise should be ampler in the case of one who has laboured less and digested well; it should be lighter in the case of one who is fatigued and has digested less well.

Useful exercises are: reading aloud, drill, handball, running, walking; but this is not by any means most useful on the level, since walking up and down hill varies the movement of the body, unless indeed the body is thoroughly weak; but it is better to walk in the open air than under cover; better, when the head allows of it, in the sun than in the shade; better under the shade of a wall or of trees than under a roof; better a straight than a winding walk. But the exercise ought to come to an end with sweating, or at any rate lassitude, which should be

CELSUS

lassitudo, quae citra fatigationem sit, idque ipsum modo minus, modo magis faciendum est. Ac ne his quidem athletarum exemplo vel certa esse lex vel inmodicus labor debet. Exercitationem recte sequitur modo unctio, vel in sole vel ad ignem; modo balineum, sed conclavi quam maxime et alto et lucido et spatioso. Ex his vero neutrum semper fieri oportet, sed saepius alterutrum pro corporis natura. Post haec paulum conquiescere opus est.

8 Ubi ad cibum ventum est, numquam utilis est nimia satietas, saepe inutilis nimia abstinentia: si qua intemperantia subest, tutior est in potione quam in esca. Cibus a salsamentis, holeribus similibusque rebus melius incipit; tum caro adsumenda est, quae 9 assa optima aut elixa est. Condita omnia duabus causis inutilia sunt, quoniam et plus propter dulcedinem adsumitur, et quod modo par est, tamen aegrius concoquitur. Secunda mensa bono stomacho nihil nocet, in inbecillo coacescit. Si quis itaque hoc parum valet, palmulas pomaque et similia melius primo cibo adsumit. Post multas potiones, quae aliquantum sitim excesserunt, nihil edendum est, 10 post satietatem nihil agendum. Ubi expletus est aliquis, facilius concoquit, si, quicquid adsumpsit, potione aquae frigidae includit, tum paulisper invigilat, deinde bene dormit. Si quis interdiu se

---

[a] See II. 21, 1.
[b] *Condita:* the context suggests sweets preceding dessert. The word is not found elsewhere in Celsus, but such sweets are referred to by Columella, de R. R. II. 22, *uvas itemque*

BOOK I. 2. 7–10

well this side of fatigue; and sometimes less, sometimes more, is to be done. But in these matters, as before, the example of athletes should not be followed, with their fixed rules and immoderate labour. The proper sequel to exercise is: at times an anointing, whether in the sun or before a brazier; at times a bath, which should be in a chamber as lofty, well lighted and spacious as possible. However, neither should be made use of invariably, but one of the two the oftener, in accordance with the constitution. There is need of a short rest afterwards.

Coming to food, a surfeit is never of service, excessive abstinence is often unserviceable; if any intemperance is committed, it is safer in drinking than in eating. It is better to begin a meal with savouries,[a] salads and such-like; and after that meat is to be eaten, best either when roasted or boiled. All preserved fruits[b] are unserviceable for two reasons, because more is taken owing to their sweetness, and even what is moderate is still digested with some difficulty. Dessert does no harm to a good stomach, in a weak one it turns sour. Whoever then in this respect has too little strength, had better eat dates, apples and such-like at the beginning of the meal. After many drinkings which have somewhat exceeded the demands of thirst, nothing should be eaten; after a surfeit of food there should be no exertion. Anyone who has had his fill digests the more readily if he concludes the meal with a drink of cold water, then after keeping awake for a time has a sound sleep. When a full meal is taken at

olivas conditui legere licet: raisins and olives can be gathered for candying (in honey).

CELSUS

inplevit, post cibum neque frigori neque aestui neque labori se debet committere: neque enim tam facile haec inani corpore quam repleto nocent. Si quibus de causis futura inedia est, labor omnis vitandus est.

3. Atque haec quidem paene perpetua sunt. Quasdam autem observationes desiderant et novae res et corporum genera et sexus et aetates et tempora anni. Nam neque ex salubri loco in gravem, neque ex gravi in salubrem transitus satis tutus est. Ex salubri in gravem prima hieme, ex gravi in eum, qui salubris est, prima aestate transire 2 melius est. Neque ex multa vero fame nimia satietas neque ex nimia satietate fames idonea est. Periclitaturque et qui semel et qui bis die cibum incontinenter contra consuetudinem adsumit. Item neque ex nimio labore subitum otium neque ex nimio otio subitus labor sine gravi noxa est. Ergo cum quis mutare aliquid volet, paulatim debebit adsuescere. Omnem etiam laborem facilius vel puer vel senex 3 quam insuetus homo sustinet. Atque ideo quoque nimis otiosa vita utilis non est, quia potest incidere laboris necessitas. Si quando tamen insuetus aliquis laboravit, aut si multo plus quam solet etiam si qui adsuevit, huic ieiuno dormiendum est, multo magis etiam si os amarum est vel oculi caligant, aut venter perturbatur: tum enim non dormiendum tantummodo ieiuno est, sed etiam . . .[1] in posterum diem permanendum, nisi cito id quies sustulit. Quod si

[1] *Marx supplies* quieto *after* etiam *and this reading is translated.*

midday, after it there should be no exposure to cold, heat or fatigue, which do not harm the body so easily when it is empty as when it is full. When from whatever causes there is prospective want of food, everything laborious should be avoided.

3. Now the foregoing precepts indeed almost always hold good; but some particular notice requires to be taken of changes of surroundings and varieties of constitution and sex and age and seasons. For it is not safe to remove either from a salubrious to an oppressive locality, or from an oppressive to a salubrious one. It is better to make the move from a salubrious into an oppressive place at the beginning of winter, from an oppressive into a salubrious one in early summer. It is not good indeed to overeat after a long fast, nor to fast after overeating. And he runs a risk who goes contrary to his habit and eats immoderately whether once or twice in the day. Again, neither sudden idleness after excessive labour, nor sudden labour after excessive idleness, is without serious harm. Therefore when a man wishes to make a change, he ought to habituate himself little by little; indeed any work is easier even for a boy or an old man than for an unaccustomed adult. Hence also too idle a life is inexpedient, because there may come up some necessity for labour. But if at any time a man has had to undergo unaccustomed labour, or at any rate much more than he is used to, he should go to bed on an empty stomach, more especially if he has a bitter taste in his mouth, or his eyes are dimmed, or his bowels disturbed; for then he must not only sleep with his stomach empty, but even remain at rest over the next day, unless rest has quickly removed the trouble; in this case

CELSUS

factum est, surgere oportet et lente paulum ambulare. At si somni necessitas non fuit, quia modice magis aliquis laboravit, tamen ingredi aliquid eodem modo
4 debet.   Communia deinde omnibus sunt post fatigationem cibum sumpturis : ubi paulum ambulaverunt, si balneum non est, calido loco vel in sole vel ad ignem ungui atque sudare; si est, ante omnia in tepidario sedere, deinde ubi paululum conquierunt, intrare et descendere in solium; tum multo oleo ungui leniterque perfricari, iterum in solium descendere, post haec os aqua calida, deinde frigida
5 fovere.   Balneum his fervens idoneum non est. Ergo si nimium alicui fatigato paene febris est, huic abunde est loco tepido demittere se inguinibus tenus in aquam calidam, cui paulum olei sit adiectum, deinde totum quidem corpus, maxime tamen eas partes, quae in aqua fuerunt, leviter perfricare ex oleo, cui
6 vinum et paulum contriti salis sit adiectum.   Post haec omnibus fatigatis aptum est cibum sumere, eoque umido uti, aqua vel certe diluta potione esse contentos, maximeque ea, quae moveat urinam. Illud quoque nosse oportet, quod ex labore sudanti frigida potio perniciosissima est atque etiam, cum
7 sudor se remisit, itinere fatigatis inutilis.   A balineo quoque venientibus Asclepiades inutilem eam iudicavit; quod in iis verum est, quibus alvus facile nec tuto resolvitur quique facile inhorrescunt; perpetuum

---

[a] Order of the rooms at the Bath : *elaeothesium*, anointing room; *frigidarium*, cool room; *apodyterium*, undressing room; *tepidarium*, warm room; *calidarium*, hot room; *laconicum*, for dry sweating.—See further II. **17**.

BOOK I. 3. 3-7

he should get up and take slowly a short walk. But even when there has been no necessity for a sleep, because a man has only done more moderate work, still he ought, all the same, to take a little walk. This then should be the rule for everyone after incurring fatigue before taking food: first to walk about a little, then, if no bath is at hand, to undergo anointing and sweating in a warm place whether in the sun or before a fire; where there is a bath,[a] he should first sit in the warm room, then, after resting there a while, go down into the tubs; next, after being anointed freely with oil and gently rubbed down, again descend into the tub; finally he should foment the face, first with warm, then with cold water. A very hot bath does not suit such cases. Therefore if one's excessive fatigue almost amounts to a fever, it is quite sufficient for him to sit in warm water, to which a little oil may be added, up to the groins, in a tepid room; next his whole body, and especially the parts which have been under water, should be rubbed gently with oil to which a little wine and pounded salt have been added. This done, anybody who has undergone fatigue is ready for food, in particular food of a fluid consistency; he should be content with water to drink, or if wine, certainly diluted, of the sort to promote diuresis. Further it should be recognized that after labour accompanied by sweating a cold drink is most pernicious, and even although sweating after a fatiguing journey has passed off, it is unserviceable. After coming out of the bath, too, Asclepiades held it unserviceable; and this is true in the case of those whose bowels are loose at uncertain moments, and who readily shiver; but it is not the universal rule

in omnibus non est, cum potius naturale sit potione
aestuantem stomachum refrigerari, frigentem cale-
fieri. Quod ita praecipio, ut tamen fatear, ne ex
hac quidem causa sudanti adhuc frigidum bibendum
8 esse. Solet etiam prodesse post varium cibum
frequentesque dilutas potiones vomitus, et postero
die longa quies, deinde modica exercitatio. Si
adsidua fatigatio urguet, in vicem modo aquam,
modo vinum bibendum est, raro balineo utendum.
Levatque lassitudinem etiam laboris mutatio; eumque, quem novum genus eiusdem laboris pressit, id
9 quod in consuetudine est, reficit. Fatigato coti-
dianum cubile tutissimum est: lassat enim quod
contra consuetudinem, seu molle seu durum est.
Proprie quaedam ad eum pertinent, qui ambulando
fatigatur. Hunc reficit in ipso quoque itinere
frequens frictio, post iter primum sedile, deinde
unctio; tum calida aqua in balineo magis superiores
10 partes quam inferiores foveat. Si quis vero exustus
in sole est, huic in balneum protinus eundum per-
fundendumque oleo corpus et caput; deinde in
solium bene calidum descendendum est; tum multa
aqua per caput infundenda, prius calida, deinde
frigida. At ei, qui perfrixit, opus est in balineo
primum involuto sedere, donec insudet; tum ungui,
deinde lavari; cibum modicum, deinde potiones
11 meracas adsumere. Is vero, qui navigavit et nausea
pressus est, si multam bilem evomuit, vel abstinere
a cibo debet vel paulum aliquid adsumere. Si

BOOK I. 3. 7–11

in all cases, since it is more natural that a heated stomach should be cooled, and a cold one warmed by a drink. I grant so much, but I hesitate to give this as a rule, for as a matter of fact a cold drink is bad while sweating. It also happens that after a dinner of many courses and many drinks of diluted wine a vomit is even advantageous; the next day there should be a prolonged rest followed by exercise in moderation. If there is oppression due to a persistence of fatigue, water and wine should be drunk alternately, but the bath seldom used. A change of work, too, relieves lassitude; and when a novel form of customary work has tired a man, that form to which he is accustomed restores him. To one who is fatigued that couch is best which he uses every day; for whether soft or hard, one to which he is unaccustomed wearies him. Certain things are specially applicable to one who is fatigued whilst travelling on foot. To be rubbed often while actually on the way restores him; after the journey he should sit awhile, then undergo anointing; next at the bath foment with hot water his upper rather than his lower parts. But anyone who has become overheated in the sun should go at once to the bath, and there have oil poured over the head and body; next go down to a thoroughly hot tub; then have water poured over his head freely, first hot, next cold. On the other hand, he who has become much chilled should first sit in the calidarium, well wrapped up, until he sweats; next be anointed, afterwards laved, then take food in moderation and after that drinks of undiluted wine. He too who on a voyage is troubled by seasickness, if he has vomited out a quantity of bile, should fast or take very little food. If he has spewed

pituitam acidam effudit, utique sumere cibum, sed
adsueto leviorem: si sine vomitu nausea fuerit, vel
12 abstinere vel post cibum vomere. Qui vero toto
die vel in vehiculo vel in spectaculis sedit, huic nihil
currendum sed lente ambulandum est. Lenta
quoque in balineo mora, dein cena exigua prodesse
consuerunt. Si quis in balineo aestuat, reficit hunc
ore exceptum et in eo retentum acetum; si id non
est, eodem modo frigida aqua sumpta.

13 Ante omnia autem norit quisque naturam sui
corporis, quoniam alii graciles, alii obessi sunt, alii
calidi, alii frigidiores, alii umidi, alii sicci; alios
adstricta, alios resoluta alvus exercet. Raro quisquam non aliquam partem corporis inbecillam habet.
14 Tenuis vero homo inplere se debet, plenus extenuare;
calidus refrigerare, frigidus calefacere; madens
siccare, siccus madefacere; itemque alvum firmare
is, cui fusa, solvere is, cui adstricta est: succur-
rendumque semper parti maxime laboranti est.

15 Implet autem corpus modica exercitatio, frequentior quies, unctio et, si post prandium est, balineum;
contracta alvus, modicum frigus hieme, somnus et
plenus et non nimis longus, molle cubile, animi securitas, adsumpta per cibos et potiones maxime dulcia et
pinguia; cibus et frequentior et quantus plenissimus
16 potest concoqui. Extenuat corpus aqua calida, si
quis in ea descendit, magisque si salsa est; ieiuno
balineum, inurens sol ut omnis calor, cura, vigilia;

BOOK I. 3. 11–16

out sour phlegm, he may take food notwithstanding, but lighter than usual; if he has nausea without vomiting, he should either fast, or after food excite a vomit. But he who has spent all day sitting in a carriage or at the games should not after that hurry but walk slowly; also it is of service to linger somewhat in the bath, and then take a small dinner afterwards. When overheated in the bath, taking vinegar and holding it in the mouth restores him; if that is not at hand, cold water may be taken in the same way.

But above all things everyone should be acquainted with the nature of his own body, for some are spare, others obese; some hot, others more frigid; some moist, others dry; some are costive, in others the bowels are loose. It is seldom but that a man has some part of his body weak. So then a thin man ought to fatten himself up, a stout one to thin himself down; a hot man to cool himself, a cold man to make himself warmer; the moist to dry himself up, the dry to moisten himself; he should render firmer his motions if loose, relax them if costive; treatment is to be always directed to the part which is mostly in trouble.

Now the body is fattened: by moderate exercise, by oftener resting, by anointing, and by the bath if after a meal at midday; by the bowels being confined, by winter cold in moderation, by sleep adequate but not over long, by a soft couch, by a tranquil spirit, by food whether solid or fluid which is sweet and fatty; by meals rather frequent and as large as it is possible to digest. The body is thinned: by hot water if one bathes in it and especially if salt; by the bath on an empty stomach, by a scorching sun, by heat of all kinds, by worry, by late nights;

somnus nimium vel brevis vel longus, per aestatem durum cubile; cursus, multa ambulatio, omnisque vehemens exercitatio; vomitus, deiectio, acidae res et austerae; et semel die adsumptae epulae; et vini non praefrigidi ieiuno potio in consuetudinem adducta.

17 Cum vero inter extenuantia posuerim vomitum et deiectionem, de his quoque proprie quaedam dicenda sunt. Reiectum[1] esse ab Asclepiade vomitum in eo volumine, quod *De tuenda sanitate* composuit, video; neque reprehendo, si offensus eorum est consuetudine, qui cotidie eiciendo vorandi facultatem moliuntur. Paulo etiam longius processit; idem purgationes quoque eodem volumine expulit: et sunt eae perniciosae, si nimis valentibus medica-
18 mentis fiunt. Sed haec tamen summovenda esse non est perpetuum, quia corporum temporumque ratio potest ea facere necessaria, dum et modo et non nisi cum opus est adhibeantur. Ergo ille quoque ipse, si quid iam corruptum esset, expelli debere confessus est: ita non ex toto res condemnanda est. Sed esse eius etiam plures causae possunt; estque in ea quaedam paulo subtilior observatio adhibenda.
19 Vomitus utilior est hieme quam aestate: nam tunc et pituitae plus et capitis gravitas maior subest. Inutilis est gracilibus et inbecillum stomachum habentibus: utilis plenis, biliosis omnibus, si vel nimium se replerunt, vel parum concoxerunt. Nam sive plus est quam quod concoqui possit, periclitari ne

---

[1] Eiectum *MSS.*; reiectum *Constantine*.

---

[a] See II. 12, 13.

by sleep unduly short or overlong, by a hard bed throughout the summer; by running or much walking or any violent exercise; by a vomit, by purgation, by sour and harsh things consumed; by a single meal a day; by the custom of drinking wine not too cold upon an empty stomach.

But as I have mentioned a vomit and a purge among thinning measures, there are some things to be said in particular concerning them.[a] I note that a vomit was rejected by Asclepiades in the book written by him, entitled *De tuenda sanitate*; I do not blame him for being disquieted with the custom of those, who by ejecting every day achieve a capacity for gormandizing. He has even gone somewhat further; for from the same volume he has expelled likewise purgings; which indeed are pernicious when procured by too powerful medicaments. Such measures, however, are not to be dispensed with entirely, because regard for different constitutions and times can make them necessary, provided that they are employed in moderation and only when needed. Hence Asclepiades has himself allowed that what is already corrupted ought to be expelled: so this kind of treatment is not wholly to be condemned. But there may be more than one reason for this too; and so a somewhat closer consideration may be given to the matter.

A vomit is more advantageous in winter than in summer, for then more phlegm and severer stuffiness in the head occur. It is unsuitable for the thin and for those with a weak stomach, but suitable for the plethoric, and all who have become bilious, whether after overeating or imperfect digestion. For if the meal has been larger than can be digested, it is not

conrumpatur non oportet: si ve[ro][1] corruptum est, nihil commodius est quam id, qua via primum expelli 20 potest, eicere. Itaque ubi amari ructus cum dolore et gravitate praecordiorum sunt, ad hunc protinus confugiendum est. Item prodest ei, cui pectus aestuat et frequens saliva vel nausea est, aut sonant aures, aut madent oculi, aut os amarum est; similiterque ei, qui vel caelum vel locum mutat; isque, quibus, si per plures dies non vomuerunt, dolor 21 praecordia infestat. Neque ignoro inter haec praecipi quietem, quae non semper contingere potest agendi necessitatem habentibus, nec in omnibus idem facit. Itaque istud luxuriae causa fieri non oportere confiteor: interdum valetudinis causa recte fieri experimentis credo cum eo tamen, ne quis, qui valere et senescere volet, hoc cottidianum habeat. 22 Qui vomere post cibum volt, si ex facili facit, aquam tantum tepidam ante debet adsumere; si difficilius, aquae vel salis vel mellis paulum adicere. At qui mane vomiturus est, ante bibere mulsum vel hysopum, aut esse radiculam debet, deinde aquam tepidam, ut supra scriptum est, bibere. Cetera, quae antiqui medici praeceperunt, stomachum 23 omnia infestant. Post vomitum, si stomachus infirmus est, paulum cibi, sed huius idonei, gustandum, et aquae frigidae cyathi tres bibendi sunt, nisi tamen fauces vomitus exasperarint. Qui vomuit, si mane

[1] sive *v. d. Linden for the* si vero *of MSS.*

BOOK I. 3. 19-23

well to risk its corruption; and if it has already become corrupted, nothing is more to the purpose than to eject it by whatever way its expulsion is first possible. When, therefore, there are bitter eructations, with pain and weight over the heart, recourse should be had at once to a vomit, which is likewise of service to anyone who has heartburn and copious salivation or nausea, or ringing in the ears, or watering of the eyes, or a bitter taste in the mouth; similarly in the case of one who is making a change of climate or locality; as well as in the case of those who become troubled by pain over the heart when they have not vomited for several days. Nor am I unaware that in such cases there is prescribed rest, but that is not always within the reach of those who are obliged to be busy; nor does rest act in the same way with everybody. Accordingly I allow that vomiting should not be practised for the sake of luxury; on account of health I believe from experiment that it is sometimes rightly practised, nevertheless with this reservation, that no one who wants to keep well, and live to old age, should make it a daily habit. He who after a meal wants to vomit, if he does so easily should first take tepid water by itself; when there is more difficulty, a little salt or honey should be added. To cause a vomit on getting up in the morning, he should first drink some honey or hyssop in wine, or eat a radish, and after that drink tepid water as described above. The other emetics prescribed by the ancient practitioners all disturb the stomach. After a vomit, when the stomach is weak, a little suitable food should be taken, and for drink, unless the vomiting has made the throat raw, three cupfuls of cold water. He who has provoked

CELSUS

id fecit, ambulare debet, tum ungi, dein cenare; si post cenam, postero die lavari et in balneo sudare.

24 Inde proximus cibus mediocris utilior est isque esse debet cum pane hesterno, vino austero meraco et carne assa cibisque omnibus quam siccissimis. Qui vomere bis in mense vult, melius consulet, si biduo continuarit, quam si post quintum decimum diem vomuerit, nisi haec mora gravitatem pectori faciet.

25 Deiectio autem medicamento quoque petenda est, ubi venter suppressus parum reddit, ex eoque inflationes, caligines, capitis dolores, aliaque superioris partis mala increscunt. Quid enim inter haec adiuvare possunt quies et inedia . . .[1] quae per illas maxime eveniunt? Qui deicere volet, primum cibis vinisque utetur iis, qui hoc praestant; dein, si parum

26 illa proficient, aloen sumat. Sed purgationes quoque, ut interdum necessariae sunt, sic, ubi frequentes sunt, periculum adferunt: adsuescit enim non ali corpus, cum omnibus morbis obnoxia maxime infirmitas sit.

27 Calefacit autem unctio, aqua salsa, magisque si calida est, omnia salsa, amara, carnosa; si post cibum est, balneum, vinum austerum. Refrigerant in ieiunio et balneum et somnus, nisi nimis longus est, omnia acida, aqua quam frigidissima, oleum, si aqua miscetur.

28 Umidum autem corpus efficit labor maior quam ex consuetudine, frequens balineum, cibus plenior,

[1] *Marx suggests that some words have been omitted such as* quae nonnulli commendant horum malorem remedia, *"which some commend as remedies in such ills." Van d. Linden, followed by Daremberg, reads* per quae illa, *"through which (rest and fasting) these conditions (i.e. constipation) arise."*

• See further II. 12.

a vomit, if it be early in the day, should after that take a walk, next undergo anointing, then dine; if after dining, he should the next day bathe, or sweat in the baths. After that the following meal had better be a light one, consisting of bread a day old, harsh undiluted wine, roasted meat, all food being of the dryest. Whoever aims to provoke a vomit twice a month, had better arrange to do so on two consecutive days, rather than once a fortnight, unless this longer interval causes heaviness in the chest.

Now defaecation is to be procured also by a medicament, when, the bowels being costive, too little is passed, with the result that there is increase of flatulence, dizziness of vision, headaches, and other disturbances in the upper parts. For what can rest and fasting help in such circumstances which come about so much through them? He who wants to defaecate should in the first place make use of such food and wine as will promote it; then if these have little effect, he should take aloes. But purgatives also, whilst necessary at times, when frequently used entail danger; for the body becomes subject to malnutrition, since a weakened state leaves it exposed to maladies of all sorts.[a]

The body is heated: by anointing, by salt-water affusion and the more so when hot; by all food which is salt, bitter and fleshy; and after meals by the bath and harsh wine. On the contrary it is cooled: by the bath and sleep on an empty stomach, if not too prolonged; by all sour food; by the coldest water to drink, by oil affusion when mixed with water.

The body is rendered humid: by more than customary exertion, by a frequent bath, by food in

multa potio, post hanc ambulatio et vigilia; per se quoque ambulatio multa et matutina et vehemens, exercitationi non protinus cibus adiectus; ea genera escae, quae veniunt ex locis frigidis et pluviis et
29 inriguis. Contra siccat modica exercitatio, fames, unctio sine aqua, calor, sol modicus, frigida aqua, cibus exercitationi statim subiectus, et is ipse ex siccis et aestuosis locis veniens.

30 Alvum adstringit labor, sedile, creta figularis corpori inlita, cibus inminutus, et is ipse semel die adsumptus ab eo, qui bis solet; exigua potio neque adhibita, nisi cum cibi quis, quantum adsumpturus
31 est, cepit, post cibum quies. Contra solvit aucta ambulatio atque esca potusque, motus, qui post cibum est, subinde potiones cibo inmixtae. Illud quoque scire oportet, quod ventrem vomitus solutum conprimit, compressum solvit; itemque conprimit is vomitus, qui statim post cibum est, solvit is, qui tarde supervenit.

32 Quod ad aetates vero pertinet, inediam facillime sustinent mediae aetates, minus iuvenes, minime pueri et senectute confecti. Quo minus fert facile quisque, eo saepius debet cibum adsumere, maxime-que eo eget, qui increscit. Calida lavatio et pueris et senibus apta est. Vinum dilutius pueris, senibus meracius: neutri aetati, quae inflationes movent.
33 Iuvenum minus quae adsumant et quomodo curentur,

---

[a] See further, II. 1, 5, 17–22.

increased amount, by copious drinking, followed by walking and late hours; much walking, early and forced, has by itself the same effect, food being taken not immediately after exercise; so also those classes of edibles which come from cold and rainy and irrigated localities. On the contrary the body is dried: by moderate exercise, hunger, anointing without the addition of water, summer heat with moderate exposure to the sun, cold water to drink, food immediately after exercise, and all edibles coming from hot and dry districts.

The bowels are confined by exertion, by sitting still, by besmearing the body with potter's clay, by a scanty diet, and that taken once a day in the case of one accustomed to two meals, by drinking little and that only after the consumption of whatever food is to be taken, also by rest after food. On the contrary they are rendered loose: by increasing the length of the walk, more food and drink; by moving about after the meal; by frequently drinking during the meal. This too should be recognized, that a vomit confines the bowels when relaxed, and relaxes them when costive: again, a vomit immediately after the meal confines the bowels, later it relaxes them.

As to what pertains to age:[a] the middle-aged sustain hunger more easily, less so young people, and least of all children and old people. The less readily one supports it, the more often should food be taken; one who is growing needs it most. Children and the old should bathe in warm water. Wine should be diluted for children; for the old men it should be rather undilute: but at neither age be of a kind to cause flatulence. It matters less for the young what they take and the way they are treated. Those who

CELSUS

interest. Quibus iuvenibus fluxit alvus, plerumque in senectute contrahitur: quibus in adulescentia fuit adstricta, saepe in senectute solvitur. Melior est autem in iuvene fusior, in sene adstrictior.

34 Tempus quoque anni considerare oportet. Hieme plus esse convenit, minus sed meracius bibere; multo pane uti, carne potius elixa, modice holeribus; semel die cibum capere, nisi si nimis venter adstrictus est. Si prandet aliquis, utilius est exiguum aliquid, et ipsum siccum sine carne, sine potione sumere. Eo tempore anni calidis omnibus potius utendum est vel calorem moventibus. Venus tum
35 non aeque perniciosa est. At vere paulum cibo demendum, adiciendum potioni, sed dilutius tamen bibendum est; magis carne utendum, magis holeribus; transeundum paulatim ad assa ab elixis.
36 Venus eo tempore anni tutissima est. Aestate vero et potione et cibo saepius corpus eget; ideo prandere quoque commodum est. Ei tempori aptissima sunt et caro et holus, potio quam dilutissima, ut et sitim tollat nec corpus incendat; frigida lavatio, caro assa,
37 frigidi cibi vel qui refrigerent. Ut saepius autem cibo utendum, sic exiguo est. Per autumnum propter caeli varietatem periculum maximum est. Itaque neque sine veste neque sine calceamentis prodire oportet, praecipueque diebus frigidioribus, neque sub divo nocte dormire, aut certe bene operiri. Cibo vero iam paulo pleniore uti licet, minus sed
38 meracius bibere. Poma nocere quidam putant,

---

[a] See further, II. 1, 1–4.

when young are relaxed, when old are generally costive; those constipated in youth are often relaxed when old. It is better to be rather relaxed when young, rather costive when old.

The season *a* of the year also merits consideration. In winter it is fitting to eat more, and to drink less but of a stronger wine, to use much bread, meat preferably boiled, vegetables sparingly; to take a single meal unless the bowels are too costive. If a meal is taken at midday, it is better that it should be somewhat scanty, and that dry, without meat, and without drinking. At that season everything taken should be hot or heat-promoting. Venery then is not so pernicious. But in spring food should be reduced a little, the drink added to, but, however, of wine more diluted; more meat along with vegetables should be taken, passing gradually from boiled to roast. Venery is safest at this season of the year. But in summer the body requires both food and drink oftener, and so it is proper in addition to take a meal at midday. At that season both meat and vegetables are most appropriate; wine that is much diluted in order that thirst may be relieved without heating the body; laving with cold water, roasted meat, cold food or food which is cooling. But just as food is taken more frequently, so there should be less of it. In autumn owing to changes in the weather there is most danger. Hence it is not good to go out of doors unless well covered, and with thick shoes, especially on the colder days; nor at night to sleep in the open air, or at any rate to be well covered. A little more food may now be taken, the wine less in quantity but stronger. Some think orchard fruit injurious, which is generally the

CELSUS

quae inmodice toto die plerumque sic adsumuntur, ne quid ex densiore cibo remittatur. Ita non haec sed consummatio omnium nocet; ex quibus in nullo tamen minus quam in his noxae est. Sed his uti non saepius quam alio cibo convenit. Denique aliquid densiori cibo, cum hic accedit, necessarium est demi. Neque aestate vero neque autumno utilis venus est, tolerabilior tamen per autumnum: aestate in totum, si fieri potest, abstinendum est.

4. Proximum est, ut de iis dicam, qui partes aliquas corporis inbecillas habent. Cui caput infirmum est, is si bene concoxit, leniter perfricare id mane manibus suis debet; numquam id, si fieri potest, veste velare; ad cutem tonderi. Utileque lunam vitare, maximeque ante ipsum lunae solisque concursum; sed nusquam post cibum . . .[1] Si cui capilli sunt, cotidie pectere, multum ambulare, sed, si licet, neque sub tecto neque in sole; ubique autem vitare solis ardorem, maximeque post cibum et vinum; potius ungui quam lavari, numquam ad flammam ungui, interdum ad prunam. Si in balineum venit, sub veste primum paulum in tepidario insudare, ibi ungui; tum transire in caldarium; ubi sudavit, in solium non descendere, sed multa calida aqua per caput se totum perfundere, tum tepida, deinde frigida, diutiusque ea caput quam ceteras partes perfundere; deinde id aliquamdiu perfricare, novissime detergere et unguere. Capiti nihil aeque

[1] *Marx notes that some MSS. insert* progredi *and this reading is translated. Targa noted the lacuna and proposed to omit the sentence* utile . . . concursum. *As it stands the passage has no meaning.*

case when eaten immoderately all day, without reducing more substantial food. Hence it is not the fruit but the heaping of all things together which does harm, but in none of them all is there less harm than in the fruit. But it is not fitting to eat of it oftener than other kinds of food, and when eaten, it is necessary to subtract some of the more substantial food. But venery is useful neither in summer nor in autumn; it is more tolerable nevertheless in autumn, in summer it is to be abstained from entirely, if that possibly can be done.

4. I have next to speak of those who have some parts of the body weak. He whose head is infirm ought, after he has digested well, to rub it gently in the morning with his own hands; never if possible cover it with a wrap; have it shaved to the skin. It is well to avoid moonlight, and especially before the actual conjunction of the moon and sun, and to walk nowhere after dinner. If he has retained his hair, he should comb it every day, walk much, but, if possible, not under cover nor in the sun; everywhere, however, he should avoid the sun's blaze, especially after taking food and wine; undergo anointing rather than affusion, but that never before a flaming fire, on occasion before a brazier. If he goes to the bath he should first sweat for a while, in the tepidarium, wrapped up, and then undergo anointing there; next pass into the calidarium; after a further sweat he should not go down into the hot bath, but have himself sluiced freely from the head downwards, first with hot, next with tepid, then with cold water, which should be poured for longer on the head than upon other parts, after which it should be rubbed for a while, lastly wiped dry and anointed. Nothing is so

prodest atque aqua frigida: itaque is, cui hoc infirmum est, per aestatem id bene largo canali cotidie debet aliquamdiu subicere. Semper autem, etiamsi sine balineo unctus est neque totum corpus refrigerare sustinet, caput tamen aqua frigida perfundere: sed cum ceteras partes adtingi nolit, demittere id, ne ad cervices aqua descendat; eamque, ne quid oculis aliisve partibus noceat, defluentem subinde manibus [ad os] regerere. Huic modicus cibus necessarius est, quem facile concoquat; isque, si ieiuno caput laeditur, adsumendus etiam medio die est; si non laeditur, semel potius. Bibere huic adsidue vinum dilutum leve quam aquam magis expedit, ut, cum caput gravius esse coeperit, sit quo confugiat. Eique ex toto neque vinum neque aqua semper utilia sunt: medicamentum utrumque est, cum in vicem adsumitur. Scribere, legere, voce contendere huic opus non est, utique post cenam; post quam ne cogitatio quidem ei satis tuta est; maxime tamen vomitus alienus est.

5. Neque vero iis solis, quos capitis inbecillitas torquet, usus aquae frigidae prodest, sed iis etiam, quos adsiduae lippitudines, gravidines, destillationes tonsillaeque male habent. His autem non caput tantum cottidie perfundendum est, sed os quoque multa frigida aqua fovendum est; praecipueque omnibus, quibus hoc utile auxilium est, eo utendum est, ubi gravius caelum austri reddiderunt. Cumque

---

a Lippitudo is used to translate the ὀφθαλμία of Hippocrates (cf. esp. *Prorrhet.* II. 18. Littré, IX. 44). See further, VI. 6.

beneficial to the head as cold water, and so he who has a weak head should every day throughout the summer hold it for a while under the stream from a large conduit. But even if he undergoes anointing without going into the bath, and cannot bear cooling of the whole body, he should always nevertheless douche his head with cold water; but since he does not want the rest of his body wetted, he bends forward for the water not to run down his neck, and with his hands directs the flow to his face, that his eyes or other parts may not be irritated. He must take food in moderation and such as he can easily digest; and if fasting affects his head, he should take a meal at midday; if it does not so suffer, the single meal is preferable. It is more expedient for him to drink a light wine, well diluted, rather than water, in order that he may have something in reserve when his head begins to become heavier; and to him, on the whole, neither wine nor water is proper always; each constitutes a remedy when taken in its turn. To write, to read, to argue, is not beneficial to him, particularly after dinner; after which, indeed, even cogitation is not sufficiently safe; worst of all, however, is a vomit.

5. Nor indeed is it only those who are troubled by a weakness of the head that find the use of cold water beneficial, but also those who suffer with persistent running from the eyes,[a] choked nostrils and running from the nose, and tonsillar maladies. In these cases, not only is the head to be douched every day, but also the face bathed with abundance of cold water; especially should this be carried out by all those benefited by it, whenever the south wind renders the weather more oppressive. And whereas

omnibus inutilis sit post cibum aut contentio aut agitatio animi, tum iis praecipue, qui vel capitis vel arteriae dolores habere consuerunt, vel quoslibet alios oris adfectus. Vitari etiam gravidines destillationesque possunt, si quam minime qui his opportunus est loca aquasque mutet; si caput in sole protegit, ne incendatur, neu subito ex repentino nubilo frigus id moveat; si post concoctionem ieiunus caput radit; si post cibum neque legit neque scribit.

6. Quem vero frequenter cita alvus exercet, huic opus est pila similibusque superiores partes exercere; dum ieiunus est, ambulare; vitare solem, continua balinea; ungi citra sudorem; non uti cibis variis, minimeque iurulentis, aut leguminibus holeribusque, iisque, quae celeriter descendunt; omnia denique
2 fugere,[1] quae tarde concocuntur. Venatio durique pisces et ex domesticis animalibus assa caro maxime iuvant. Numquam vinum salsum bibere expedit, ne tenue quidem aut dulce, sed austerum et plenius, neque id ipsum pervetus. Si mulso uti volet, id ex decocto melle faciendum est. Si frigidae potiones ventrem eius non turbant, his utendum potissimum est. Si quid offensae in cena sensit, vomere debet, idque postero quoque die facere; tertio modici ponderis panem ex vino esse, adiecta uva ex olla vel ex defruto similibusque aliis; deinde ad consuetudinem redire. Semper autem post cibum

[1] fugere, *so Marx following the MSS. ; the advice is to avoid food which passing through quickly is ill digested. Cf.* I. 8, 1. *v. d. Linden (followed by Daremberg) reads* sumere, *" to take all such things as are digested slowly."*

after dinner either wrangling or mental worry is injurious to everybody, it is especially so to those who are disposed to pains in the head or windpipe, or to other forms of oral affections. Choked nostrils and running from the nose can be also avoided, or minimized, if one who is liable to these makes as little change as possible in respect to residence and water; if he protects the head, that it may not be scorched by the sun, or be chilled by a passing cloud; if the head be shaved after digestion and the stomach empty; if there be no reading or writing after a meal.

6. But one who is often troubled by an urgent motion should exercise his upper parts at handball and the like; walk on an empty stomach, avoid the sun and continual bathing; undergo anointing even without sweating, not make use of multifarious foods, least of all stews or pulse or greens, and of those things which pass through quickly; in a word, to avoid all things which are digested slowly. Especially advantageous are: venison and hard fish and meat of domestic animals roasted. It is never expedient to drink wine treated with sea-water, nor indeed thin or sweet wine, but that which is dry and fuller-bodied, and not too old. If one desires to use honeyed wine, it should be made from boiled honey. Cold drinks are to be used whenever possible, so long as they do not disturb the bowels. When anything in the dinner is felt to disagree, he should provoke a vomit, repeating it next day; on the third day should be eaten a small quantity of bread soaked in wine with the addition of grapes preserved in a jar or in must which has been boiled down and such like; then he should return to his accustomed habit; but he must always rest after the meal,

conquiescere, ac neque intendere animum, neque ambulatione quamvis levi dimoveri.

7. At si laxius intestinum dolere consuevit, quod colum nominant, cum id nihil nisi genus inflationis sit, id agendum est, ut concoquat aliquis; ut lectione aliisque generibus exerceatur; utatur balineo calido, cibis quoque et potionibus calidis, denique omni modo frigus vitet, item dulcia omnia leguminaque et quicquid inflare consuevit.

8. Si quis vero stomacho laborat, legere clare debet et post lectionem ambulare; tum pila et armis aliove quo genere, quo superior pars movetur, exerceri; non aquam sed vinum calidum bibere ieiunus; cibum bis die adsumere, sic tamen ut facile concoquat; uti vino tenui et austero, et si post
2 cibum, frigidis potius potionibus. Stomachum autem infirmum indicant pallor, macies, praecordiorum dolor, nausea, et nolentium vomitus, ieiuno dolor capitis; quae in quo non sunt, is firmi stomachi est. Neque credendum utique nostris est, qui cum in adversa valetudine vinum aut frigidam aquam concupiverunt, deliciarum patrocinium in accusatione[m]
3 non merentis stomachi habent. At qui tarde concocunt, et quorum ideo praecordia inflantur, quive propter ardorem aliquem noctu sitire consuerunt, ante quam conquiescant duos tresve cyathos per tenuem fistulam bibant. Prodest etiam adversus tardam concoctionem clare legere, deinde ambulare, tum vel ungui vel lavari; adsidue vinum frigidum

there must be no tension of mind, no moving about in a walk however short.

7. When the more lax intestine, which they name colon, tends to be painful, and when it is nothing more than an inflation of a sort, the aim should be to promote digestion; to practise reading aloud and other exercises; to use a hot bath, also hot food and drink, and in short, to avoid all manner of cold, all sweets and pulse, and whatever else tends to flatulence.

8. But if anyone suffers from his stomach, he should read out loud, and after the reading take a walk, then exercise himself at handball and at drill or at anything else which brings the upper part of the body into play; on an empty stomach he should not drink water but hot wine; if he digests readily he should take two meals a day; drink light and dry wine, and after a meal drinks should preferably be cold. Weakness of the stomach is indicated by pallor, wasting, pain over the heart, nausea, and involuntary vomiting, headache when the stomach is empty; where these symptoms are absent, the stomach is sound. Nor must one absolutely trust those of our patients who when very unwell have conceived a longing for wine or cold water, and in backing up their desires, lay the blame on their perfectly innocent stomach. But those who digest slowly, and whose parts below the ribs on that account become inflated, or who on account of heat of some kind become thirsty at night, may drink before going to bed three or four cupfuls of wine through a fine reed. Also, to counter slow digestion, it is well to read aloud, next to take a walk, then to be either anointed or laved, taking care to

bibere, et post cibum magnam potionem, sed, ut supra dixi, per siphonem; deinde omnes potiones 4 aqua frigida includere. Cui vero cibus acescit, is ante eum bibere aquam egelidam debet et vomere: at si cui ex hoc frequens deiectio incidit, quotiens alvus ei constiterit, frigida potione potissimum utatur.

9. Si cui vero dolere nervi solent, quod in podagra cheiragrave esse consuevit, huic, quantum fieri potest, exercendum id est, quod adfectum est, obiciendumque labori et frigori, nisi cum dolor increvit. Sub divo 2 quies optima est. Venus semper inimica est; concoctio, sicut in omnibus corporis adfectibus, necessaria: cruditas enim id maxime laedit, et quotiens offensum corpus est, vitiosa pars maxime sentit.

3 Ut concoctio autem omnibus vitiis occurrit, sic rursus aliis frigus, aliis calor; quae sequi quisque pro habitu corporis sui debet. Frigus inimicum est seni, tenui, vulneri, praecordiis, intestinis, vesicae, auribus, coxis, scapulis, naturalibus, ossibus, dentibus, nervis, 4 vulvae, cerebro. Idem summam cutem facit pallidam, aridam, duram, nigram; ex hoc horrores tremoresque nascuntur. At prodest iuvenibus et omnibus plenis; erectiorque mens est, et melius concoquitur, ubi frigus quidem est sed cavetur. 5 Aqua vero frigida infusa, praeterquam capiti, etiam stomacho prodest, etiam articulis doloribusque, qui sunt sine ulceribus, item rubicundis nimis hominibus, si dolore vacant. Calor autem adiuvat omnia, quae frigus infestat, item lippientis, si nec dolor nec lacrimae sunt, nervos quoque, qui contrahuntur,

---

See IV. 31, 32.

drink wine cold, a large drink after dinner, but as I have said through a tube, ending all by drinking cold water. He whose food tends to turn sour should beforehand take a draught of tepid water and vomit; but if as a consequence he has frequent motions, he should, whenever possible after each stool, take a draught of cold water.

9. When sinews tend to become painful, as is common in foot or hand ache,[a] the affected part should be exercised as far as possible, even exposing it to work or to cold, unless when pain is increasing. Rest in the open air is best. Venery is always inimical; as in all other bodily affections, digestion is a necessity, for indigestion is most harmful, and whenever the body is attacked, the faulty part feels it most.

But just as digestion has to do with all sorts of troubles, so has cold with some, heat with others; each person should be guided by his own bodily habit. Cold is inimical to the aged, and to the thin; to wounds, to the parts below the ribs, intestines, bladder, ears, hips, bladebones, genitals, bones, teeth, sinews, womb, brain. It renders the skin pale, dry, hard, black; from it are developed shiverings and tremors. But cold is beneficial to the young and to stout people; in cold weather, with due precautions, the mind is more vigorous and the digestion better. Cold water affusion is of service, not only to the head, but also to the stomach, and to painful joints not accompanied by ulcerations, also for those who are too rubicund, when pain is absent. But heat benefits all that cold harms, such as dimness of eyesight when there is neither pain nor lacrimation, also contracted sinews, and particularly

praecipueque ea ulcera, quae ex frigore sunt. Idem
6 corporis colorem bonum facit, urinam movet. Si
nimius est, corpus effeminat, nervos emollit, stomachum solvit. Minime vero frigus et calor tuta
sunt, ubi subita insuetis sunt: nam frigus lateris
dolores aliaque vitia, frigida aqua strumas excitat.
Calor concoctionem prohibet, somnum aufert, sudorem digerit, obnoxium morbis pestilentibus corpus
efficit.

10. Est etiam observatio necessaria, qua quis
in pestilentia utatur adhuc integer, cum tamen
securus esse non possit. Tum igitur oportet
peregrinari, navigare, ubi id non licet, gestari,
ambulare sub diu ante aestum leniter eodemque
modo ungui; et, ut supra (**9**, 1—3, 6) conprensum
est, vitare fatigationem, cruditatem, frigus, calorem,
libidinem, multoque magis se continere, si qua
2 gravitas in corpore est. Tum neque mane surgendum neque pedibus nudis ambulandum est,
minime post cibum aut balineum; neque ieiuno
neque cenato vomendum est, neque movenda alvus;
atque etiam, si per se mota est, conprimenda est.
3 Abstinendum potius, si plenius corpus est, itemque
vitandum balneum, sudor, somnus meridianus,
utique si cibus quoque antecessit; qui tamen semel
die tum commodius adsumitur, insuper etiam
modicus, ne cruditatem moveat. Alternis diebus in
vicem modo aqua modo vinum bibendum est. Quibus servatis ex reliqua victus consuetudine quam
4 minimum mutari debet. Cum vero haec in omni

ᵃ See further, II. **1**, 19; V. **18**, 5, 13; **28**, 7.

those ulcerations which are due to cold. It gives the surface of the body a good colour; it promotes diuresis. If excessive it weakens the body, mollifies sinews, relaxes the stomach. Yet cold and heat are both least safe when applied suddenly to persons unaccustomed to them; for cold gives rise to pain in the side and other diseases, cold water excites swelling in the neck.[a] Heat hinders concoction, prevents sleep, exhausts by sweating, renders liable the body to pestilential illnesses.

10. There are also observances necessary for a healthy man to employ during a pestilence, although in spite of them he cannot be secure. At such a time, then, he will do well to go abroad, take a voyage; when this cannot be, to be carried in a litter, walk in the open before the heat of the day, gently, and to be anointed in like manner; further as stated above he should avoid fatigue, indigestion, cold, heat, venery, and keep all the more to rule, should he feel any bodily oppression. At such a time he should not get up early in the morning nor walk about barefoot, and least so after a meal or bath. Neither on an empty stomach nor after a meal should he provoke a vomit, or set up a motion; indeed if the bowels tend to be loose, they are to be restrained. The fuller his habit of body, the more abstinence; he should avoid the bath, sweating, a midday siesta, and in any case if food has been taken previously; at such times, however, it is better then to take only one meal a day, and that a moderate one, lest indigestion be provoked. He should drink, one day water, the next day wine; if he observes these rules, there should be the least possible alteration as to the rest of his accustomed dietary. Such then are the things

pestilentia facienda sint, tum in ea maxime, quam austri excitarint. Atque etiam peregrinantibus eadem necessaria sunt, ubi gravi tempore anni discesserunt ex suis sedibus, vel ubi in graves regiones venerunt. Ac si cetera res aliqua prohibebit, utique retineri debebit a vino ad aquam, ab hac ad vinum qui supra (§ 3) positus est transitus.

to be done in pestilence of all sorts, and particularly in one brought by south winds. And the same precautions are needed by those who travel, when they have left home during an unhealthy season, or when entering an unhealthy district. Even when something prevents observance of other rules, yet he ought to keep up the alternation, mentioned above, from wine to water, and from water to wine.

BOOK II

LIBER II

PROOEMIUM

Instantis autem adversae valetudinis signa conplura sunt. In quibus explicandis non dubitabo auctoritate antiquorum virorum uti, maximeque Hippocratis, cum recentiores medici, quamvis quaedam in curationibus mutarint, tamen haec illos 2 optime praesagisse fateantur. Sed antequam dico (*capp. 2 seqq.*), quibus praecedentibus morborum timor subsit, non alienum videtur exponere, quae tempora anni (**1**, 1–2), quae tempestatum genera (**1**, 3–4), quae partes aetatis, qualia corpora maxime tuta vel periculis oportuna sint (**1**, 5), quod genus adversae valetudinis in quo timeri maxime possit (**1**, 6–23), non quo non omni tempore, in omni tempestatum genere omnis aetatis, omnis habitus homines per omnia genera morborum et aegrotent et moriantur, sed quo minus . . .[1] frequentius tamen quaedam eveniant, ideoque utile sit scire unumquemque, quid et quando maxime caveat.

[1] *Some words have fallen out here. Marx suggests* frequenter alia genera alio tempore in aliis, in aliis, "*but as some kinds occur less frequently at some times in some people, but in others more frequently.*" *One MS. gives* quod *for* quo, *omitting* minus.

BOOK II

PROOEMIUM

OF impending disorders there are many signs, in explaining which I shall not hesitate to make use of the authority of ancient men, and especially of Hippocrates; for although more recent practitioners have made some changes in methods of treatment, they allow none the less that the ancients prognosticated best. Before I note,[a] however, those preceding symptoms which suggest fear of disease, it does not seem unfitting to set out: the seasons[b] of the year (**1**, 1–2), the sorts of weather (**1**, 3–4), periods of life and temperaments which may be in particular safe or open to risks (**1**, 5), and what kind of disorders is most to be apprehended in each (**1**, 6–23). Not that men may not sicken and die at any season, in any sort of weather, at any age, whatever their temperament, from any kind of disease, but since certain kinds occur less [c] . . . but some kinds occur more often, so it is of use that everyone should recognize against what, and when, he should be most on his guard.

[a] Cf. chap. **2** ff.
[b] See below: and I. **3**, 1, 34–39. III. **4**, 7; **7**. 1. B. VII. 7. 4. D.; **26**. 2.
[c] Some words have fallen out here. See critical note.

CELSUS

II

1. Igitur saluberrimum ver est, proxime deinde ab hoc hiemps; periculosior quam salubrior aestas, autumnus longe periculosissimus. Ex tempestatibus vero optimae aequales sunt, sive frigidae sive calidae; pessimae, quae maxime variant; quo fit, ut autumnus plurimos opprimat. Nam fere meridianis temporibus calor, nocturnis atque matutinis simulque etiam vespertinis frigus est. Corpus ergo, et aestate et subinde meridianis caloribus relaxatum, subito frigore excipitur. Sed ut eo tempore id maxime fit, sic, quandocumque evenit, noxium est.

Ubi aequalitas autem est, tamen saluberrimi sunt sereni dies; meliores pluvii quam tantum nebulosi nubilive, optimique hieme qui omni vento vacant, aestate quibus favonii perflant. Si genus aliud ventorum est, salubriores septentrionales quam subsolani vel austri sunt, sic tamen haec, ut interdum regionum sorte mutentur. Nam fere ventus ubique a mediterraneis regionibus veniens salubris, a mari gravis est. Neque solum in bono tempestatium habitu certior valetudo est, sed priores[1] morbi quoque, si qui inciderunt, leviores sunt et promptius finiuntur. Pessimum aegro caelum est, quod aegrum

---

[1] *The MSS. here vary between* prior *and* priores. *Marx conjectures* praeterea.

II

1. So then spring is the most salubrious, next after it comes winter; summer is rather more dangerous than salubrious, autumn is by far the most dangerous. But as regards weather the best is that which is settled, whether cold or hot, the worst that which is the most changeable, and that is why autumn brings down the greatest number. For generally about midday there is heat, but at night and in the early morning, cold, as also in the evening. Thus the body, relaxed by the preceding summer, and now by the midday heat, is caught by the sudden cold. But while this chiefly occurs at this season, so whenever the like happens harm is done.

In settled weather fine days are the most salubrious, rainy better than foggy or cloudy days; and in winter the best days are those in which there is an entire absence of wind, in summer those in which westerly winds blow. As for the other winds, the northerly are more salubrious than those from the sunrising or south, nevertheless, these vary somewhat according to the character of the district. For almost everywhere wind when coming from inland is salubrious, and injurious when from the sea. And not only is health more assured in settled weather, but pre-existing diseases too, if there have been any, are milder and more quickly terminated. But the worst weather for the sick man is that which has

CELSUS

fecit, adeo ut in id quoque genus, quod natura peius est, in hoc statu salubris mutatio sit.

5 At aetas media tutissima est, quae neque iuventae calore, neque senectutis frigore infestatur. Longis morbis senectus, acutis adulescentia magis patet. Corpus autem habilissimum quadratum est, neque gracile neque obesum. Nam longa statura, ut in iuventa decora est, sic matura senectute conficitur, gracile corpus infirmum, obesum hebes est.

6 Vere tamen maxime, quae cum umoris motu novantur, in metu esse consuerunt. Ergo tum lippitudines, pustulae, profusio sanguinis, abscessus corporis, quae ἀποστήματα Graeci nominant, bilis atra, quam μελαγχολίαν appellant, insania, morbus comitialis, angina, gravidines, destillationes oriri solent. I quoque morbi, qui in articulis nervisque modo urguent modo quiescunt, tum maxime et inchoantur et repetunt.

7 At aestas non quidem vacat plerisque his morbis sed adicit febres vel continuas vel ardentis vel tertianas, vomitus, alvi deiectiones, auricularum dolores, ulcera oris, cancros et in ceteris quidem partibus, sed maxime obscenis, et quicquid sudore hominem resolvit.

8 Vix quicquam ex his in autumnum non incidit: sed oriuntur quoque eo tempore febres incertae, lienis dolor, aqua inter cutem, tabes, quam Graeci φθίσιν nominant, urinae difficultas, quam στραγγουρίαν

---

[a] All the diseases mentioned in this and the following sections are described in detail in Book IV.

[b] *Abscessus*, ἀπόστημα, is not our 'abscess,' but the inflammatory condition, 'congestion,' which precedes it; (cf. II. 7. 26, and Hippocrates, I. liii); this may resolve, or go on to suppuration and abscess.

[c] Celsus included under *cancri* foul and gangrenous ulcerations, now distinguished from cancer.

caused his sickness, so much so that a change to weather of a naturally worse sort may be, in his condition, salutary.

The middle period of life is the safest, for it is not disturbed by the heat of youth, nor by the chill of age. Old age is more exposed to chronic diseases, youth to acute ones. The square-built frame, neither thin nor fat, is the fittest; for tallness, as it is graceful in youth, shrinks in the fulness of age; a thin frame is weak, a fat one sluggish.

In spring those diseases[a] are usually to be apprehended which are stirred up anew by movement of humor. Consequently there tend to arise running from the eyes, pustules, haemorrhages, congestions[b] in the body, which the Greeks call apostemata, black bile which they call μελανχολία, madness, fits, angina, choked nostrils, runnings from the nose. Also those diseases which affect joints and sinews, being at one time troublesome, at another quiescent, then especially both begin and recur.

But summer, while not wholly exempt from most of the foregoing maladies, adds to them fevers whether continued or ardent or tertian, vomitings, diarrhoeas, earaches, oral ulcerations, cankers[c] which occur on other parts but especially upon the pudenda, and whatever exhausts the patient by sweating.

In autumn there is scarcely one of the foregoing which does not happen; but at this season in addition there arise irregular fevers, splenic pain, subcutaneous dropsy, consumption, called by the Greeks phthisis, urinary difficulty, which they call strangury,[d] the

[d] στραγγουρία, from στράγξ, a drop, and οὖρον, urine. See further, VII. 26.

appellant, tenuioris intestini morbus quem ileon nominant, levitas intestinorum, qui lienteria voca-
9 tur, coxae dolores, morbi comitiales. Idemque tempus et diutinis malis fatigatos, et ab aestate tantum proxima pressos interemit, et alios novis morbis conficit; et quosdam longissimis inplicat, maximeque quartanis, quae per hiemem quoque exerceant. Nec aliud magis tempus pestilentiae patet, cuiuscumque ea generis est; quamvis variis rationibus nocet.

Hiemps autem capitis dolores, tussim et quicquid in faucibus in lateribus in visceribus mali contrahitur, inritat.

10 Ex tempestatibus aquilo tussim movet, fauces exasperat, ventrem adstringit, urinam sup⟨p⟩rimit, horrores excitat, item dolores lateris et pectoris. Sanum tamen corpus spissat et mobilius atque
11 expeditius reddit. Auster aures hebetat, sensum tardat, capitis dolores movet, alvum solvit, totum corpus efficit hebes, umidum, languidum. Ceteri venti, quo vel huic vel illi propiores sunt, eo magis vicinos his illisve affectus faciunt. Denique omnis calor iecur et lienem inflat, mentem hebetat; ut
12 anima deficiat, ut sanguis prorumpat, efficit. Frigus modo nervorum distentionem, modo rigorem infert; illud spasmos, hoc tetanos Graece nominatur; nigritiem in ulceribus, horrores in febribus excitat. In siccitatibus acutae febres, lippitudines, tormina, urinae difficultas, articulorum dolores oriuntur; per imbres longae febres, alvi deiectiones, angina, cancri,

---

[a] II. 7, 6; 8. 17, 34. IV. 20. Cf. Hipp. IV. 136 (*Aph.* IV. 11).
[b] λειεντερία, λεῖος, smooth, and ἔντερον, intestine.

small intestine malady which they term ileos,[a] the intestinal lubricity which they call leienteria,[b] hip-pains, fits. Autumn too is a season fatal to those exhausted by chronic diseases and overwhelmed by the heat just past, others it weakens by fresh maladies; and it involves some in very chronic ones, especially quartan fevers, which may last even through the winter. Nor is any other period of the year more exposed to pestilence of whatever sort; although it is harmful in a variety of ways.

Winter provokes headache, coughs, and all the affections which attack the throat, and the sides of the chest and lungs.

Of the various sorts of weather, the north wind excites cough, irritates the throat, constipates the bowels, suppresses the urine, excites shiverings, as also pain of the lungs and chest. Nevertheless it is bracing to a healthy body, rendering it more mobile and brisk. The south wind dulls hearing, blunts the senses, produces headache, loosens the bowels; the body as a whole is rendered sluggish, humid, languid. The other winds, as they approximate to the north or south wind, produce affections corresponding to the one or other. Moreover, any hot weather inflates the liver and spleen, and dulls the mind; the result is that there are faintings, that there is an outburst of blood. Cold on the other hand brings about: at times tenseness of sinews which the Greeks call spasmos, at times the rigor which they call tetanos, the blackening of ulcerations, shiverings in fevers. In times of drought there arise acute fevers, runnings from the eyes, dysenteries, urinary difficulty, articular pains. In wet weather there occur chronic fevers, diarrhoeas, angina, canker,

morbi comitiales, resolutio nervorum (paralysin
13 Graeci nominant). Neque solum interest, quales
dies sint, sed etiam, quales ante praecesserint. Si
hiemps sicca septentrionales ventos habuit, ver autem
austros et pluvias exhibet, fere subeunt lippitudines,
tormina, febres, maximeque in mollioribus corporibus,
14 ideoque praecipue in muliebribus. Si vero austri
pluviaeque hiemem occuparunt, ver autem frigidum
et siccum est, gravidae quidem feminae, quibus tum
adest partus, abortu periclitantur; eae vero, quae
gignunt, inbecillos vixque vitales edunt. Ceteros
lippitudo arida et, si seniores sunt, gravedines atque
15 destillationes male habent. At si a prima hieme
austri ad ultimum ver continuarint, laterum dolores
et insania febricitantium, quam phrenesin appellant,
celerrime rapiunt. Ubi vero calor a primo vere orsus
aestatem quoque similem exhibet, necesse est multum sudorem in febribus subsequi. At si sicca aestas
aquilones habuit, autumno vero imbres austrique
sunt, tota hieme, quae proxima est, tussis, destillatio,
16 raucitas, in quibusdam etiam tabes oritur. Sin
autem autumnus quoque aeque siccus iisdem aquilonibus perflatur, omnibus quidem mollioribus corporibus, inter quae muliebria esse proposui, secunda
valetudo contingit : durioribus vero instare possunt et
aridae lippitudines, et febres partim acutae partim
longae, et ii morbii, qui ex atra bile nascuntur.
17   Quod ad aetates vero pertinet, pueri proximique

a § 13.

fits, and the loosening of sinews which the Greeks call paralysis. Not only does the weather of the day but also of the preceding days matter. If a dry winter has been accompanied by north winds, or again a spring by south winds and rain, generally there ensue runnings from the eye, dysenteries, fevers, and most of all in more delicate bodies, hence especially in women. If on the other hand south winds and rain have prevailed during winter, and the spring is cold and dry, pregnant women near their confinement are in danger of miscarrying; those indeed who reach term, give birth only to weaklings hardly alive. Other people are attacked by dry ophthalmia, and if elderly by choked nostrils and runnings from the nose. But when the south wind prevails from the beginning of winter to the end of spring, side pains, also the insanity of those in fever which is called phrenesis, are very rapidly fatal. And when hot weather begins in the spring, and lasts through the summer, severe sweating must ensue in cases of fever. If a summer has been kept dry by northerly winds, but in the autumn there are showers and south winds, there may then arise cough, runnings from the nose, hoarseness, and indeed in some, consumption. But if the autumn is dry owing to a north wind continuing to blow, all those with more delicate bodies, among whom, as I have mentioned, are women,[a] enjoy good health. The harder constitutions, however, may possibly be attacked by dry ophthalmias, and by fevers, some acute, some chronic, also by those maladies which arise from black bile.

As regards the various times of life, children and

his vere optime valent, et aestate prima tutissimi sunt, senes aestate et autumni prima parte, iuvenes hieme quique inter iuventam senectutemque sunt. Inimicior senibus hiemps, aestas adulescentibus 18 est. Tum si qua inbecillitas oritur, proximum est, ut infantes tenerosque adhuc pueros serpentia ulcera oris, quae ἄφθας Graeci nominant, vomitus, nocturnae vigiliae, aurium umor, circa umbilicum inflammationes exerceant. Propriae etiam dentientium gingivarum exulcerationes, febriculae, interdum nervorum distentiones, alvi deiectiones; maximeque caninis dentibus orientibus male habent; quae pericula plenissimi cuiusque sunt, et cui 19 maxime venter adstrictus est. At ubi aetas paulum processit, glandulae, et vertebrarum, quae in spina sunt, aliquae inclinationes, struma, verrucarum quaedam genera dolentia (ἀκροχορδόνας Graeci appellant) et plura alia tubercula oriuntur. Incipiente vero iam pube, et ex iisdem multa, et 20 longae febres, sanguinis ex naribus cursus. Maximeque omnis pueritia, primum circa quadragesimum diem, deinde septimo mense, tum septimo anno, postea circa pubertatem periclitatur. Si qua etiam genera morborum in infantem inciderunt, ac neque pubertate neque primis coitibus neque in femina primis menstruis finita sunt, fere longa sunt: saepius tamen morbi pueriles, qui diutius manserunt, ter-21 minantur. Adulescentia morbis acutis item comitialibus tabique maxime obiecta est; fereque iuvenes sunt, qui sanguinem expuunt. Post hanc

---

[a] ἄφθα, thrush (from ἅπτω). Cf. Hipp. IV. 130 (*Aph.* III. 29).
[b] ἀκροχορδών, a pedunculated wart (hanging by a cord—χορδή). Cf. Hipp. IV. 132 (*Aph.* III. 26).

adolescents enjoy the best health in spring, and are safest in early summer; old people are at their best during summer and the beginning of autumn; young and middle-aged adults in winter. Winter is worst for the aged, summer for young adults. At these periods should any indisposition arise, it is very probable that infants and children still of tender age should suffer from the creeping ulcerations of the mouth which the Greeks call aphthas,[a] vomiting, insomnia, discharges from the ear, and inflammations about the navel. Especially in those teething there arise ulcerations of the gums, slight fevers, sometimes spasms, diarrhoea; and they suffer as the canine teeth in particular are growing up; the most well-nourished children, and those constipated, are especially in danger. In those somewhat older there occur affections of the tonsils, various spinal curvatures, swelling in the neck, the painful kind of warts which the Greeks call acrochordones,[b] and a number of other swellings. At the commencement of puberty, in addition to many of the above troubles, there occur chronic fevers and also nose-bleedings. Throughout childhood there are special dangers, first about the fortieth day, then in the seventh month, next in the seventh year, and after that about puberty. The sorts of affections which occur in infancy, when not ended by the time of puberty, or of the first coitions, or of the first menstruations in the females, generally become chronic; more often, however, puerile affections, after persisting for a rather long while, come to an end. Adolescence is liable to acute diseases, such as fits, especially to consumption; those who spit blood are generally youths. After that age come on

aetatem laterum et pulmonis dolores, lethargus, cholera, insania, sanguinis per quaedam velut ora venarum (αἱμορροΐδιας Graeci appellant) pro-
22 fusio. In senectute spiritus et urinae difficultas, gravedo, articulorum et renum dolores, nervorum resolutiones, malus corporis habitus (καχεξίαν Graeci appellant), nocturnae vigiliae, vitia longiora aurium, oculorum, etiam narium, praecipueque soluta alvus, et quae secuntur hanc, tormina vel levitas intestinorum ceteraque ventris fusi mala.
23 Praeter haec graciles tabes, deiectiones, destillationes, item viscerum et laterum dolores fatigant. Obesi plerumque acutis morbis et difficultate spirandi strangulantur, subitoque saepe moriuntur; quod in corpore tenuiore vix evenit.

2. Ante adversam autem valetudinem, ut supra (*II prooem*. 1) dixi, quaedam notae oriuntur, quarum omnium commune est aliter se corpus habere atque consuevit, neque in peius tantum sed etiam in melius. Ergo si plenior aliquis et speciosior et coloratior factus est, suspecta habere bona sua debet; quae quia neque in eodem habitu subsistere neque ultra progredi possunt, fere retro quasi ruina
2 quadam revolvuntur. Peius tamen signum est, ubi aliquis contra consuetudinem emacuit et colorem decoremque amisit, quoniam in iis, quae superant, est quod morbus demat; in iis, quae desunt, non est quod ipsum morbum ferat. Praeter haec protinus timeri debet, si graviora membra sunt, si crebra ulcera oriuntur, si corpus supra consuetudinem inca-

---

[a] Book II, Prooem. 1.

BOOK II. 1. 21–2. 2

pain in the side and lung, lethargy, cholera,[a] madness, and outpourings of blood from certain mouths of veins which the Greeks call haemorrhoids. In old age there occur breathing and urinary difficulties, choked nostrils, joint and renal pains, paralysis, the bad habit of body which the Greeks call cachexia, insomnias, the more chronic maladies of the ears, eyes, also of the nostrils, and especially looseness of the bowels with its sequences, dysentery, intestinal lubricity, and the other ills due to bowel looseness. In addition thin people are fatigued by consumption, diarrhoea, running from the nose, pain in the lung and side. The obese, many of them, are throttled by acute diseases and difficult breathing; they die often suddenly, which rarely happens in a thinner person.

2. Now antecedent to illness, as I have stated above,[a] certain signs arise, all of which have this in common, that the body becomes altered from its accustomed state, and that not only for the worse, but it may be even for the better. Hence when a man has become fatter and better looking and with a higher colour, he should regard with suspicion these gains of his; for, because they can neither remain in the same state nor advance further, as a rule they fall back in a sort of collapse. Still it is a worse sign when anyone, contrary to his habit, becomes thinner, and loses his colour and good looks; for when there is a superfluity of flesh there is something for the disease to draw upon; when there is a deficiency, there is nothing to hold out against the disease itself. Further, there should be apprehension at once: if the limbs become heavier, if frequent ulcerations arise, if the body feels hotter

CELSUS

luit; si gravior somnus pressit, si tumultuosa somnia fuerunt, si saepius expergiscitur aliquis quam adsuevit, deinde iterum soporatur; si corpus dormientis circa partes aliquas contra consuetudinem insudat, maximeque si circa pectus, aut cervices, aut 3 crura vel genua vel coxas. Item si marcet animus, si loqui et moveri piget, si corpus torpet; si dolor praecordiorum est aut totius pectoris aut, qui in plurimis evenit, capitis; si salivae plenum os est, si oculi cum dolore vertuntur, si tempora adstricta sunt, si membra inhorrescunt, si spiritus gravior est; si circa frontem intentae venae moventur, si frequentes oscitationes; si genua quasi fatigata sunt, 4 totumve corpus lassitudinem sentit. Ex quibus saepe plura, numquam non aliqua febrem antecedunt. In primis tamen illud considerandum est, num cui saepius horum aliquid eveniat neque ideo corporis ulla difficultas subsequatur. Sunt enim quaedam proprietates hominum, sine quarum notitia non facile quicquam in futurum praesagiri potest. Facile itaque securus est in iis aliquis, quae saepe sine periculo evasit: ille sollicitari debet, cui haec nova sunt, aut qui ista numquam sine custodia sui tuta habuit.

3. Ubi vero febris aliquem occupavit, scire licet non periclitari, si in latus aut dextrum aut sinistrum, ut ipsi visum est, cubat, cruribus paulum reductis, qui fere sani quoque iacentis habitus est; si facile

BOOK II. 2. 2–3. 1

than customary; if heavier sleep oppresses, if there are tumultuous dreams, if anyone wakes up oftener than usual, then falls asleep again; if the body of the sleeper has partial sweats in unaccustomed places, and especially about the chest or neck or legs or knees or hips. Again, if the spirit flags, if he is reluctant to talk or move about, if the body is torpid; if there is pain over the heart or over all the chest, or of the head as happens in most; if the mouth becomes filled with saliva, if there is pain in turning the eyes, if the temples are constricted, when the limbs shiver, if the breathing becomes more laboured; if the blood-vessels of the forehead are distended and throb, if there are frequent yawns; if the knees feel as if fatigued, or the whole body feels weary. Of these signs, many are often, some always, antecedents of fever. The first thing, however, to be considered is, whether any of these signs happen somewhat frequently, yet no bodily trouble has followed it. For there are some peculiarities of persons, without knowledge of which it is not easy for anybody to prognosticate what is going to happen. Consequently anyone may readily be at ease in the case of happenings which he has frequently escaped without harm: the man who ought to be anxious is the one to whom these signs are new, or who has never found them free from danger unless he has taken precautions.

3. But when fever has actually seized upon a man, it may be known that he is not in danger: if he lies upon his side, whether on his right or left, just as suits him, with his legs a little drawn up, as is generally the way with a healthy person when lying

CELSUS

convertitur, si noctu dormit, interdiu vigilat; si ex facili spirat, si non conflictatur; si circa umbilicum et pubem cutis plena est; si praecordia eius sine ullo sensu doloris aequaliter mollia in utraque parte sunt:
2 quod si paulo tumidiora sunt, sed tamen digitis cedunt et non dolent, haec valetudo, ut spatium aliquod habebit, sic tuta erit. Corpus quoque, quod aequaliter molle et calidum est, quodque aequaliter totum insudat, et cuius febricula eo sudore finitur,
3 securitatem pollicetur. Sternumentum etiam inter bona indicia est, et cupiditas cibi vel a primo servata, vel etiam post fastidium orta. Neque terrere debet ea febris, quae eodem die finita est, ac ne ea quidem, quae, quamvis longiore tempore evanuit, tamen ante alteram accessionem ex toto quievit, sic ut corpus integrum, quod εἰλικρινές Graeci vocant,
4 fieret. Si quis autem incidit vomitus, mixtus esse et bile et pituita debet, et in urina subsidere album, leve, aequale, sic ut etiam, si quae quasi nubeculae
5 innatarint, in imum deferantur. At venter ei, qui a periculo tutus est, reddit mollia, figurata, eodem fere tempore, quo secunda valetudine adsuevit, modo convenientia iis, quae adsumuntur. Peior cita alvus est: sed ne haec quidem terrere protinus debet, si matutinis temporibus coacta magis est, aut si procedente tempore paulatim contrahitur et rufa

---

^a Celsus uses *praecordia* to mean, (1) the lower chest in front of the heart; (2) the region over the diaphragm; (3) the upper abdomen below the ribs (hypochondria).

^b εἰλικρινής examined by the sun's light, tested and found genuine; and so pure, unmixed = Latin *sincerus*. This was used to denote the proper mingling of the four principles or elements (Proem. 14) to produce εὐκρασία, the equilibrium constituting the basis of health, which Celsus translated by

BOOK II. 3. 1–5

down; if the patient turns readily in bed, if he sleeps through the night, and keeps awake by day; if he breathes easily; if he does not toss about; if the skin around the navel and pubes is plump; if the parts below the ribs [a] on the two sides are uniformly soft, without any sensation of pain; for even although they are somewhat tumid, so long as they yield to pressure by the fingers, and are not tender, this illness, though it will continue for some time, yet will be safe. There is promise of freedom from anxiety when the body in general is uniformly soft and warm, and it sweats uniformly all over, and if with this sweating the touch of fever comes to an end. Among good signs are: sneezing, also a desire for food, whether maintained from the first, or even beginning after a distaste for food. Nor should a fever which ends on the same day cause alarm, nor indeed one which, although longer in disappearing, yet entirely quiets down before the next paroxysm, so that the body is rendered sound, or, as the Greeks call it, eilikrines.[b] But should any vomiting occur, it should be of bile and of phlegm mixed; any sediment in the urine should be white, slimy, and uniform, and so that even if small clouds, as it were, are swimming in it, they sink to the bottom. Again the belly of one who is safe from danger yields soft, formed motions, at much the same time as was customary in health, as well as proportionate to the food taken. A loose motion is worse; but not even this should cause alarm at once, if on the following morning the stool is rather more solid, or if each succeeding motion

integritas (soundness), III. 3–17. Galen uses the term ἀπυρεξία, absence of fever.

est neque foeditate odoris similem alvum sani hominis
6 excedit. Ac lumbricos quoque aliquos sub finem
morbi descendisse nihil nocet. Si inflatio in superioribus partibus dolorem tumoremque fecit, bonum signum est sonus ventris inde ad inferiores partes evolutus, magisque etiam, si sine difficultate cum stercore excessit.

4. Contra gravis morbi periculum est, ubi supinus aeger iacet porrectis manibus et cruribus; ubi residere volt in ipso acuti morbi impetu, praecipueque pulmonibus laborantibus; ubi nocturna vigilia premitur, etiamsi interdiu somnus accedit; ex quo tamen peior est, qui inter quartam horam et noctem
2 est, quam qui matutino tempore ad quartam. Pessimum tamen est, si somnus neque noctu neque interdiu accedit: id enim fere sine continuo dolore esse non potest. Neque vero signum bonum est etiam somno ultra debitum urgueri, peiusque, quo
3 magis se sopor interdiu noctuque continuat. Mali etiam morbi testimonium est vehementer et crebro spirare, a sexto die coepisse inhorrescere, pus expuere, vix excreare, dolorem habere continuum, difficulter morbum ferre, iactare bracchia et crura, sine voluntate lacrimare; habere umorem glutinosum dentibus inhaerentem, cutem circa umbilicum et pubem macram, praecordia inflammata, dolentia, dura, tumida, intenta, magisque si haec dextra parte quam sinistra sunt; periculosissimum tamen est, si

---

[a] Hipp. II. 22 (*Prog.* XI. 17).

becomes firmer, reddish, and smelling no worse than that of a man in health. There is no harm in passing off some round worms [a] towards the crisis of the malady. When flatulence causes pain and swelling in the upper part of the abdomen, it is a good sign when intestinal rumbling passes thence downwards towards the lower belly, and the more so, when without difficulty the wind escapes along with the faeces.

4. On the contrary there is danger of a severe disease: when the patient lies on his back with his arms and legs outstretched; when at the onset of an acute disease, especially in lung troubles, he wants to sit up; when he is worn down by insomnia even if he gets some sleep in the day-time, in which case to sleep between ten o'clock in the morning and night is worse than from early morning till ten o'clock. The worst, however, is if he gets sleep neither by day nor by night; for this generally cannot happen unless there is continuous pain. It is not a good sign, however, to be oppressed beyond measure by sleep, and it is the worse the more that somnolence continues day and night. It is also evidence of a serious malady: when the breathing is forcible and quick, when the patient begins to have shiverings from the sixth day, to spit up pus, to expectorate it with difficulty, to have continuous pain, to bear up against the disease with difficulty, to toss the arms and legs, to shed tears involuntarily, to have sticky humor adhering to the teeth, the skin about the navel and pubes wasted; the parts below the ribs inflamed, painful, hard, tumid, tense and this more on the right than on the left side; the greatest danger, however, is if in that region the

4 venae quoque ibi vehementer agitantur. Mali etiam morbi signum est nimis celeriter emacrescere; caput et pedes manusque calidas habere ventre et lateribus frigentibus, aut frigidas extremas partes acuto morbo urguente, aut post sudorem inhorrescere; aut post vomitum singultum esse vel rubere oculos; aut post cupiditatem cibi postve longas febres hunc fastidire; aut multum sudare, maximeque frigido 5 sudore, aut habere sudores non per totum corpus aequales, quique febrem non finiant, et febres eas, quae cotidie tempore eodem revertantur, quaeve semper pares accessiones habeant neque tertio quoque die leventur quaeque continuent, ut per accessiones increscant, tantum per decessiones molliantur, neque umquam integrum corpus dimittant. 6 Pessimum est, si ne levatur quidem febris, sed aeque concitata continuat. Periculosum etiam est post arcuatum morbum febrem oriri, utique si praecordia dextra parte dura manserunt. Ac dolentibus his nulla acuta febris leviter terrere nos debet; neque umquam in acuta febre aut a somno 7 non est terribilis nervorum distentio. Timere etiam ex somno mali morbi est, itemque in prima protinus febre mentem esse turbatam membrumve aliquod esse resolutum; ex quo casu quamvis vita redditur, tamen id fere membrum debilitatur. Vomitus etiam periculosus est si sincerus est nec [1] ei pituita vel bilis est mixta, peiorque, si viridis aut niger. 8 At mala urina est, in qua subsidunt subrubra et levia; deterior, in qua quasi folia quaedam tenuia atque alba; pessima ex his, si tamquam ex furfuribus

---

[1] Est nec ei *supplied by Marx*.

blood-vessels throb forcibly. It also indicates a serious malady: to become thinner too quickly; to have the head, feet and hands hot, and the belly and sides cold, or to have the extremities cold at the height of an acute disease, or to shiver after sweating; or after vomiting to hiccough or to get red eyes; or to have loss of appetite after eagerness for food or after prolonged fevers; or to sweat profusely, especially a cold sweat, or to have sweats unequally distributed over the body which do not put an end to the fever; and when those fevers which recur every day at the same hour, or which have always equal paroxysms, are not relieved on the third day, but continue; serious also are those fevers which, whilst they increase by paroxysms and are relieved by declining, yet never leave the body free. The worst is when the fever is not even relieved but continues uniformly at its height. It is likewise dangerous for a fever to supervene upon jaundice, especially if the parts below the ribs on the right side remain hard. In these sufferers every acute fever must make us seriously anxious; and never in acute fever or following on sleep is a spasm otherwise than terrifying. To lie in a fright on awaking from sleep is a sign of a serious malady; and also when, immediately upon the onset of a fever, there is mental disturbance, or any one of the limbs is paralysed; in which case, although there is a return of vitality, yet generally that limb is weakened. A vomit also is a danger-sign if purely of phlegm or of bile, unmixed, and it is the worse if green or black. It is a bad sign when the urinary sediment is reddish and slimy; worse if it is like flower-petals, thin and white; worst of all if there is an appearance as of fine clouds composed

factas nubeculas repraesentat. Diluta quoque atque
9 alba vitiosa est, sed in phreneticis maxime. Alvus
autem mala est ex toto suppressa; periculosa etiam,
quae inter febres fluens conquiescere hominem in
cubili non patitur, utique, si quod descendit est
perliquidum aut albidum aut pallidum aut spumans.
Praeter haec periculum ostendit id quod excernitur,
si est exiguum, glutinosum, leve, album, idemque
subpallidum; vel si est aut lividum aut biliosum aut
cruentum aut peioris odoris quam ex consuetudine.
Malum est etiam, quod post longas febres sincerum
est.

5. Post haec indicia votum est longum morbum
fieri: sic enim necesse est, nisi occidit. Neque vitae
alia spes in magnis malis est, quam ut impetum
morbi trahendo aliquis effugiat porrigaturque in
id tempus, quod curationi locum praestet. Protinus
tamen signa quaedam sunt, ex quibus colligere
possimus morbum, etsi non interemit, longius
2 tamen tempus habiturum: ubi frigidus sudor inter
febres non acutas circa caput tantum aut cervices
oritur, aut ubi febre non quiescente corpus insudat,
aut ubi corpus modo frigidum modo calidum est
et color alius ex alio fit, aut ubi, quod inter febres
aliqua parte abscessit, ad sanitatem non pervenit,
3 aut ubi aeger pro spatio parum emacrescit; item
si urina modo pura et liquida est, modo habet quaedam subsidentia, aut si[1] levia atque alba rubrave

---

[1] aut si *added by Marx.*

[a] For phrenesis cf. II. **1**, 15. III. **18**, 2, 3; a form of
insanity ('unsoundness') not connected with the brain by
Celsus and his predecessors. Cf. Hippocrates IV. 195. (*Aph.*
VII. 12.)

[b] Ἄκρητος, Hipp. IV. 194, note 3. (*Aph.* VII. 6.)

BOOK II. 4. 8–5. 3

of bran. Also thin and white urine is faulty, but above all in phrenetics.[a] Again it is bad for the motions to be totally suppressed; it is dangerous also during fevers when fluid stools allow the patient no rest in bed, and especially if the evacuation is quite liquid, whether it be whitish or greenish or frothy. In addition danger is indicated when the motion is scanty, viscid, slimy, white, the same when greenish yellow; or if it is either livid or bilious, or bloody, or if a worse odour than ordinary. It is bad after a prolonged fever when the stool is unmixed.[b]

5. After such signs the only thing to pray for is that the disease may be a long one, for so it must be unless it kills. Nor is there any other hope of life in grave illnesses except that the patient may avoid the attack of the disease by protracting it, and that it may be prolonged for sufficient time to afford opportunity for treatment. At the onset, however, there are certain signs from which it is possible to conclude that the disease, even if it be not fatal, nevertheless is going to last rather a long time: when in fevers which are not acute a cold sweat breaks out over the head and neck only, or when there is general sweating without the fever subsiding, or when the patient is at one time cold, at another time hot, or his colour changes from moment to moment, or when in the course of fever there is a congestion [c] in some part, which does not lead the way to recovery, or when the patient wastes a little for a considerable time,[d] again if the urine is at one time clear and limpid, at another time has some sediment which is slimy, white or red, or if there

[c] Cf. note [a], p. 88.

[d] Cf. Hipp. IV. 115 (*Aph.* II. 28), which Celsus seems to be translating here.

sunt, quae in ea subsidunt, aut si quasdam quasi miculas repraesentat, aut si bullulas excitat.

6. Sed inter haec quidem proposito metu spes tamen superest: ad ultima vero iam ventum esse testantur nares acutae, conlapsa tempora, oculi concavi, frigidae languidaeque aures et imis partibus leviter aversae, cutis circa frontem dura et intenta:
2 color aut niger aut perpallidus, multoque magis, si ita haec sunt, ut neque vigilia praecesserit neque ventris resolutio neque inedia. Ex quibus causis interdum haec species oritur, sed uno die finitur:
3 itaque diutius durans mortis index est. Si vero in morbo vetere iam triduo talis est, in propinquo mors est, magisque, si praeter haec oculi quoque lumen refugiunt et inlacrimant, quaeque in iis alba esse debent, rubescunt, atque in iisdem venulae pallent, pituitaque in iis innatans novissime angulis inhaerescit, alterque ex his minor est, iique aut vehementer subsederunt, aut facti tumidiores sunt, perque somnum palpebrae non committuntur, sed inter has ex albo oculorum aliquid apparet, neque
4 id fluens alvus expressit; eaedemque palpebrae pallent, et idem pallor labra et nares decolorat; eademque labra et nares oculique et palpebrae et supercilia aliquave ex his pervertuntur; isque propter inbecillitatem iam non audit, aut non videt.
5 Eadem mors denuntiatur, ubi aegri supini cubantis genua contracta sunt; ubi is deorsum ad pedes subinde delabitur; ubi brachia et crura nudat et

is in the sediment an appearance of bread crumbs, or it sends up bubbles.

6. But among the foregoing signs, though there are indeed grounds for fear, still there is hope left: however, that the last stage has now been reached is indicated by the nose becoming pointed, the temples sunken, the eyes hollowed, the ears cold and flaccid with the tips drooping slightly, the skin of the forehead hard and tight: the aspect is dusky or very pallid, and much more so when there has been no preceding insomnia, nor diarrhoea, nor loss of appetite. From which causes these appearances at times arise, but only last one day: and so when they last longer death is indicated. In the case of long-standing disease, when such signs have lasted for the third day, death is at hand, and the more so if besides this the eyes also shun the light and shed tears, and are reddened where they should be white, and the veins in them are pale, and phlegm floating in them comes to stick in the angles and one eye becomes smaller than the other, and either both are deep-sunken, or more tumid, and the eyelids are not closed in sleep, but some of the white of the eyes appears between them—always provided that this has not been occasioned by fluid motions; the same is the case when the eyelids become pale and a similar pallor renders colourless the lips and nostrils; so also when the lips and nostrils and eyes and eyelids and eyebrows or any one of them become distorted; and the patient owing to weakness either hears not or sees not. Death is likewise denoted: when the patient lies on his back with his knees bent; when he keeps on slipping down towards the foot of the bed; when he uncovers his arms and legs and tosses them

CELSUS

inaequaliter dispergit, neque iis calor subest; ubi hiat, ubi adsidue dormit; ubi is, qui mentis suae non est, neque id facere sanus solet, dentibus stridet; ubi ulcus, quod aut ante aut in ipso morbo natum est, aridum et aut pallidum aut lividum factum est.
6 Illa quoque mortis indicia sunt: ungues digitique pallidi, frigidus spiritus; aut si manibus quis in febre et acuto morbo vel insania pulmonisve dolore vel capitis in veste floccos legit fimbriasve deducit, vel in adiuncto pariete, si qua minuta eminent, carpit. Dolores etiam circa coxas et inferiores partes orti, si ad viscera transierunt, subitoque desierunt, mortem subesse testantur, magisque
7 si alia quoque signa accesserunt. Neque is servari potest, qui sine ullo tumore febricitans subito strangulatur, aut devorare salivam suam non potest; cuive in eodem febris corporisque habitu cervix convertitur sic, ut devorare aeque nihil possit; aut cui simul et continua febris et ultima corporis infirmitas est; aut cui febre non quiescente exterior pars friget, interior sic calet, ut etiam sitim faciat; aut qui febre aeque non quiescente simul et delirio et spirandi difficultate vexatur; aut qui epoto veratro exceptus distentione nervorum est; aut
8 qui ebrius ommutuit: is enim fere[1] nervorum distentione consumitur, nisi aut febris accessit, aut eo tempore, quo ebrietas solvi debet, loqui coepit. Mulier quoque gravida acuto morbo facile consumitur;

---

[1] fere *so Constantine for MSS.* febra adiecta, *which is deleted by Stangl followed by Marx.*

---

[a] Hipp. IV. 142. (*Aph.* IV. 35.)

about anyhow, whilst they lack warmth; when he gapes, when he continually falls asleep; when he whose mind is amiss grinds his teeth, which he did not do in health; when an ulceration, whether pre-existing or arising in the course of the illness, has become dry and either pallid or livid. The following are also indications of death: the nails and fingers pallid; the breath cold; or if the patient, in a fever or acute disease, or mad or with pain either in the lung or head, picks with his hands at the flock or pulls at the fringes of the bedclothes, or claws at anything small projecting from the adjacent wall. Pains about the hips and lower parts, which, after starting and spreading to the viscera, then suddenly subside, afford evidence of oncoming death, and the more so if there are other signs in addition. It is impossible for a patient to be saved, who, having fever without any swelling, is suddenly choked, or who cannot swallow his saliva; or who, in the same condition of fever and body, has the neck*a* twisted so that he can swallow nothing whatever; or who has continuous fever and is in the last stage of bodily weakness; or when, without the fever subsiding, the surface of the body becomes cold whilst the interior is so hot as even to produce thirst; or when, likewise without the fever subsiding, he is distressed at once by delirium and difficulty in breathing; or when, after a draught of hellebore, he is seized with spasm; or becoming drunk he loses his speech: for generally he is carried off in a spasm, unless either fever supervenes, or he begins to speak by the time that the intoxication should have passed off. A woman also when pregnant is easily carried off by an acute disease, as also a man

CELSUS

et is, cui somnus dolorem auget; et cui protinus in recenti morbo bilis atra vel infra vel supra se ostendit; cuive alterutro modo se prompsit, cum iam longo morbo corpus eius esset extenuatum
9 et adfectum. Sputum etiam biliosum et purulentum, sive separatim ista sive mixta proveniunt, interitus periculum ostendunt. Ac si circa septimum diem tale esse coepit, in proximum est, ut is circa quartum decimum diem decedat, nisi alia signa mitiora peiorave accesserint; quae quo leviora graviorave subsecuta sunt, eo vel seriorem mortem vel maturiorem
10 denuntiant. Frigidus quoque sudor in acuta febre pestifer est, atque in omni morbo vomitus, qui varius et multorum colorum est, praecipueque si malus in hoc odor est. Ac sanguinem quoque in
11 febre vomuisse pestiferum est. Urina vero rubra et tenuis in magna cruditate esse consuevit, et saepe, antequam spatio maturescat, hominem rapit: itaque si talis diutius permanet, periculum mortis ostendit. Pessima tamen est praecipueque mortifera nigra, crassa, mali odoris; atque in viris quidem et mulieribus talis deterrima est: in pueris vero quae tenuis
12 et diluta est. Alvus quoque varia pestifera est, quae strigmentum, sanguinem, bilem, viride aliquid, modo diversis temporibus, modo simul, et in mixtura quadam, discreta tamen repraesentat. Sed haec quidem potest paulo diutius trahere: in praecipiti vero iam esse denuntiat, quae livida eademque

---

a Celsus uses *pestifer* to translate the θανατώδης of Hippocrates, II, 14, 28. (*Prog.* IV, XIII), but he does not always give to it the meaning 'fatal.'

b Cf. Hipp. II, 26 (*Prog.* XII), ἔστ' ἂν δὲ λεπτὸν ᾖ τὸ οὖρον καὶ πυρρόν, ἄπεπτόν σημαίνει τὸ νόσημα εἶναι. According to

BOOK II. 6. 8-12

in whom sleep aggravates pain, and one in whom, at the very beginning of a fresh disorder, black bile presents itself, whether below or above; or after his body has become attenuated by a long illness and weakened, when such bile gains exit either way. Expectoration of bile or pus, whether they come up separately or mixed, discloses a danger of death. And when either commences about the seventh day, the patient will most likely die about the fourteenth day, unless other signs, better or worse, supervene; and according as these subsequent signs are the slighter or the graver, so they denote a later or earlier death. In an acute fever a cold sweat is noxious,[a] and so is a vomit in any malady when varied in composition and multicoloured, particularly so when malodorous. And to have vomited blood in a fever is also noxious. Now red and thin urine is usual in severe indigestion, and often, before there is time for it to mature, it carries the man off;[b] and so when such urine persists for a rather long while, danger of death is indicated. The worst and especially death-bringing urine, however, is that which is black, thick, malodorous; such urine is most to be dreaded both in men and in women; but in children urine which is thin and diluted. A motion also is noxious: when varied in composition, when it presents shreds, blood, bile, greenish matter, whether at different times, or simultaneously mixed together yet distinguishable. But although it is possible for the patient to bear up awhile against such symptoms, a speedy termination is denoted, when the motion is livid and also when it is either

Hippocrates it is the disease, not the urine, which is 'undigested.' Here *cruditas* may be the subject of *maturescat*.

CELSUS

vel nigra vel pallida vel pinguis est, utique si magna foeditas odoris accessit.

13 Illud interrogari me posse ab aliquo scio: si certa futurae mortis indicia sunt, quomodo interdum deserti a medicis convalescant?[1] quosdamque fama 14 prodiderit in ipsis funeribus revixisse. Quin etiam vir iure magni nominis Democritus ne finitae quidem vitae satis certas notas esse proposuit, quibus medici credidissent: adeo illud non reliquit, ut certa aliqua 15 signa futurae mortis essent. Adversus quos ne dicam illud quidem, quod in vicino saepe quaedam notae positae non bonos sed inperitos medicos decipiunt, quod Asclepiades funeri obvius intellexit [quendam] vivere qui efferebatur; nec protinus 16 crimen artis esse, si quod professoris sit. Illa tamen moderatius subiciam, coniecturalem artem esse medicinam, rationemque coniecturae talem esse, ut, cum saepius aliquando responderit, interdum tamen fallat. Non si quid itaque vix in millensimo corpore aliquando decipit, id notam non habet, cum per 17 innumerabiles homines respondeat. Idque non in iis tantum, quae pestifera sunt, dico, sed in iis quoque, quae salutaria; siquidem etiam spes interdum frustratur, et moritur aliquis, de quo medicus securus primo fuit: quaeque medendi causa reperta sunt, 18 nonnumquam in peius aliquid convertunt. Neque id evitare humana inbecillitas in tanta varietate cor-

---

[1] convalescant: *so one MSS. followed by Daremberg. Most of the MSS. read* convalescunt, *and Marx thinks some words have fallen out here such as* cum desperantibus medicis permultos convaluisse sit traditum: '*since it is on record that very many have recovered when the doctors have given them up.*'

BOOK II. 6. 12–18

black, or pallid, or fatty, especially if there is added an intensely fetid odour.

I know that on this point someone may question me:—if there are such sure signs of approaching death, how is it that patients who have been deserted by their medical attendants sometimes recover? And rumour has spread it about that some have revived whilst being carried out to burial. Democritus, indeed, a man justly renowned, even held that the signs of life having ended, upon which practitioners had relied, were not sufficiently sure; much more did he not admit that there could be any sure signs of approaching death. In answer to these I shall not even assert that some signs, stated as approximately certain, often deceive inexperienced practitioners, but not good ones; for instance Asclepiades, when he met the funeral procession, recognized that a man who was being carried out to burial was alive; and it is not primarily a fault of the art if there is a fault on the part of its professor. But I shall more modestly suggest that the art of medicine is conjectural, and such is the characteristic of a conjecture, that though it answers more frequently, yet it sometimes deceives. A sign therefore is not to be rejected if it is deceptive in scarcely one out of a thousand cases, since it holds good in countless patients. I state this, not merely in connexion with noxious signs, but as to salutary signs as well; seeing that hope is disappointed now and again, and that the patient dies whom the practitioner at first deemed safe; and further that measures proper for curing now and again make a change into something worse. Nor, in the face of such a variety of temperaments, can

porum potest. Sed tamen medicinae fides est, quae multo saepius perque multo plures aegros prodest. Neque tamen ignorare oportet in acutis morbis fallaces magis notas esse et salutis et mortis.

7. Sed cum proposuerim signa, quae in omni adversa valetudine communia esse consuerunt, eo quoque transibo, ut, quas aliquis in singulis morborum generibus habere possit notas, indicem. Quaedam autem sunt quae ante febres, quaedam quae inter eas, quid aut intus sit aut venturum sit, ostendunt. 2 Ante febres, si caput grave est, aut ex somno oculi caligant, aut frequentia sternumenta sunt, circa caput aliquis pituitae impetus timeri potest. Si sanguis aut calor abundat, proxumum est, ut aliqua parte profluvium sanguinis fiat. Si sine causa quis emacrescit, ne in malum habitum corpus eius recidat, metus est. Si praecordia dolent, aut inflatio gravis est, aut toto die non concocta fertur urina, cruditatem 3 esse manifestum est. Quibus diu color sine morbo regio malus est, ii vel capitis doloribus conflictantur, vel terram edunt. Qui diu habent faciem pallidam et tumidam, aut capite aut visceribus aut alvo laborant. Si in continua febre puero venter nihil reddidit, mutaturque ei color, neque somnus accedit, ploratque is adsidue, metuenda nervorum distentio 4 est. Frequens autem destillatio in corpore tenui longoque tabem timendam esse testatur. Ubi pluribus diebus non descendit alvus, docet aut

[a] In the corresponding passages of Hippocrates (Περὶ Ἐπικυήσιος, para. 10, Littré VIII. 487 and Κωακαὶ προγνώσεις, para. 333, Littré V. 687) reference is made to women's longings in early pregnancy to eat charcoal and earth. The statement in Celsus as to earth eating is more general, and is supported by recent discoveries in Africa of primitive tribes who eat earth composed of organic remains of palaeozoic molluscs.

human frailty avoid this. Nevertheless the medical art is to be relied upon, which more often, and in by far the greater number of patients, benefits the sick. It should not be ignored, however, that it is rather in acute diseases that signs, whether of recovery or of death, may be fallacious.

7. Now that I have set out the signs which are of general occurrence in any case of illness, I pass on to indicate signs which may be presented in particular sorts of diseases. There are, moreover, certain signs, some preceding, some in the course of fevers, which show what is, or what is about to become, the state of the internal parts. Before fever, if the head is heavy, or the eyes dimmed after sleep, or there are frequent sneezings, some attack of phlegm about the head is to be apprehended. If a man is full blooded or very hot, it is likely that there will be haemorrhage from some part. If a man without cause becomes thin, there is fear that his body may lapse into a dangerous state. If there is pain below the ribs or severe flatulence, or if for a whole day undigested urine is passed, there is clearly indigestion present. Persons whose colour is bad when they are not jaundiced are either sufferers from pains in the head or are earth eaters.[a] Those who for a long time have a pale or puffy face are sufferers from head, bowel or stomach trouble. If in the case of a child with constant fever no motion is passed, the colour is altered, and sleeplessness persists and constant crying, there is danger of spasms. Again running from the nose recurring often in a slender and tall man is a sign that consumption is to be apprehended. When for several days no motion passes, it shows that a sudden

CELSUS

subitam deiectionem aut febriculam instare. Ubi pedes turgent, longae deiectio⟨nes sunt;⟩ ubi dolor in imo ventre et coxis est, aqua inter cutem instat: 5 sed hoc morbi genus ab ilibus oriri solet. Idem propositum periculum est iis, quibus voluntas desidendi est, venter nihil reddit nisi et aegre et durum, tumor in pedibus est, idemque modo dextra modo sinistra parte ventris invicem oritur atque finitur: sed a iocinere id malum proficisci 6 videtur. Eiusdem morbi nota est, ubi circa umbilicum intestina torquentur (στρόφους Graeci nominant), coxaeque dolores manent, eaque neque tempore neque remediis solvuntur. Calor autem articulorum prout in pedibus manibusve aut alia qualibet parte sic est ut eo loco nervi contrahantur, aut si id membrum ex levi causa fatigatum aeque frigido calidoque offenditur, podagram cheiragramve, vel eius articuli, in quo id sentitur, morbum futurum 7 esse denuntiat. Quibus in pueritia sanguis ex naribus fluxit, deinde fluere desiit, vel capitis doloribus conflictentur necesse est, vel in articulis aliquas exulcerationes gravis habeant, vel aliquo morbo etiam debilitentur. Quibus feminis menstrua non proveniunt, necesse est capitis acerbissimi dolores 8 sint, vel quaelibet alia pars morbo infestetur. Eademque iis pericula sunt, quibus articulorum vitia, dolores tumoresque, sine podagra similibusque

---

[a] In the passage '*ubi pedes turgent . . . remediis solvuntur*' Celsus is describing three conditions under the heading '*aqua inter cutem*' or '*hydrops* (dropsy).'

1. General or partial oedema, treated by small incisions above the ankles, cf. III. 21. 11–13. Leukophlegmasia, III. 21. 2.

2. Ascites, treated by tapping, cf. VII. 15.

3. Tympanites, chronic peritonitis with effusion or dry

motion or a touch of fever is impending. Dropsy is impending, when with prolonged diarrhoea the feet swell; when there is pain in the lower belly and hips; but this class of disease is wont to arise from the flanks. There is danger, the same as just stated, to those in whom, when there is a desire for stool, the bowels yield nothing unless a forced hard motion; also in whom there is swelling in the feet, and a swelling in turn in the right and then the left half of the abdomen which rises and subsides: but this disease appears to begin from the liver. It is a sign of the same disease, when intestines in the umbilical region undergo twisting (the Greeks call it strophos), when pains in the hips persist, which are not dispersed either by time or by medicaments.[a] But when heat of joints, whether in the feet, hands, or any other part, is such that at that spot the sinews are contracted, or if that same limb, fatigued by a slight cause, is disturbed by heat and by cold alike, it denotes that there is about to set in pain[b] in feet or hands, or disease of that joint in which heat is felt. Children in whom there has been nose-bleeding, which then has ceased, are sure to be troubled by pains in the head, or they get some severe joint-ulcerations, or they also become debilitated by disease of some kind. Women in whom the menstrua are not forthcoming are sure to have the most acute pains in the head, or some part or other becomes subject to disease. There are similarly dangers for those in whom joint-disorders, pains and swellings, arise and subside

dropsy, treated by topical applications, cf. III. 21. 9, 10, and Hipp. IV. 136–7 (*Aph.* IV. 11).

[b] Celsus is not describing gout as now meant, though *podagra* later came to denote this disease, see I. 9, 1; IV. 31, 32 and Appendix, p. 463.

119

CELSUS

morbis, oriuntur et desinunt, utique, si saepe tempora iisdem dolent noctuque corpora insudant. Si frons prurit, lippitudinis metus est. Si mulier a partu vehementes dolores habet, neque alia praeterea signa mala sunt, circa vicensimum diem aut sanguis per nares erumpet, aut in inferioribus partibus
9 aliquid abscedet. Quicumque etiam dolorem ingentem circa tempora et frontem habebit, is alterutra ratione eum finiet, magisque si iuvenis erit, per sanguinis profusionem, si senior, per suppurationem. Febris autem, quae subito sine ratione, sine bonis signis
10 finita est, fere revertitur. Cui sanguine fauces et interdiu et noctu replentur, sic ut neque capitis dolores neque praecordiorum neque tussis neque vomitus neque febricula praecesserit, huius aut in naribus aut in faucibus ulcus reperietur. Si mulieri inguen et febricula orta est, neque causa apparet,
11 ulcus in vulva est. Urina autem crassa, ex qua quod desidet album est, significat circa articulos aut circa viscera dolorem metumque morbi esse. Eadem viridis aut viscerum dolorem tumoremque cum aliquo periculo subesse, aut certe corpus integrum non esse testatur. At si sanguis aut pus in urina est, vel
12 vesica vel renes exulcerati sunt. Si haec crassa carunculas quasdam exiguas quasi capillos habet, aut si bullat et male olet, et interdum quasi harenam, interdum quasi sanguinem trahit, dolent autem

---

[a] Hipp. IV. 38 (*Prog.* XVIII.).

BOOK II. 7. 8–12

without pain in the feet and such like diseases, especially if they have often pain in the temples and night sweats. Running from the eyes is to be apprehended when the forehead itches. If after childbirth a woman has severe pains, yet without other bad signs, about the twentieth day either blood will burst out from the nose, or there will be some congestion[a] in the lower parts. Indeed anyone getting great pain in the temples or forehead may be rid of it in one of these two ways, by haemorrhage especially if young, if older by suppuration. Fever, moreover, which suddenly, unaccountably and without good signs comes to an end, generally recurs. He will be found to have ulceration either in the nose or in the throat, whose throat, whether in the daytime, or by night, fills with blood, when this has been preceded neither by pains in the head, nor by pain over the heart, nor by coughing, nor by vomiting, nor by slight fever. In a woman, if without apparent cause an inguinal swelling has arisen with slight fever, there is ulceration in the womb. Again thick urine, the sediment from which is white, indicates that pain and disease are to be apprehended in the region of joints or viscera. Similar urine, when greenish, is a sign that there will be either visceral pain and swelling with some danger, or certainly that the patient is not free from fever. But if there is blood or pus in the urine, either the bladder or the kidneys have become ulcerated. The kidneys at any rate are the seat of disorder: if the urine is thick and contains bits of flesh like hairs; if it froths and is malodorous; if at one time it presents something like sand, at another time like blood; when the hips are painful,

coxae et quae inter has superque pubem sunt, et accedunt frequentes ructus, interdum vomitus biliosus, extremaeque partes frigescunt, urinae crebra cupiditas sed magna difficultas est, et quod inde excretum est, aquae simile vel rufum vel pallidum est, paulum tamen in eo levamenti est, alvus vero cum multo spiritu redditur, utique in renibus vitium
13 est. At si paulatim destillat, vel si sanguis per hanc editur, et in eo quaedam cruenta concreta sunt, idque ipsum cum difficultate redditur, et circa pubem inferiores partes dolent, in eadem vesica vitium est.
14 Calculosi vero his indiciis cognoscuntur: difficulter urina redditur paulatimque; interdum etiam sine voluntate destillat; eadem harenosa est; nonnumquam sanguis aut cruentum aut purulentum aliquid cum ea excernitur; eamque quidam promptius recti, quidam resupinati, maximeque ii, qui grandes calculos habent, quidam etiam inclinati reddunt,
15 colemque extendendo dolorem levant. Gravitatis quoque cuiusdam in ea parte sensus est; atque ea cursu omnique motu augentur. Quidam etiam, cum torquentur, pedes inter se, subinde mutatis vicibus, inplicant. Feminae vero oras naturalium suorum manibus admotis scabere coguntur: nonnumquam, si digitum admoverunt, ubi vesicae
16 cervicem is urguet, calculum sentiunt. At qui spumantem sanguinem excreant, iis in pulmone vitium est. Mulieri gravidae sine modo fusa alvus excutere partum potest. Eidem si lac ex mammis profluit, inbecillum est quod intus gerit: durae

---

[a] Defined, II. 1, 8; cf. also VII. 26, 1.
[b] VII. 26, 2 et seq.

as also the parts intermediate and above the pubes, and there are frequent eructations, now and again bilious vomiting, and the extremities become cold; when there is frequent desire to urinate but great urinary difficulty,[a] and when what is passed is like water, reddish or pallid, yet is followed by little relief, and much wind too is passed with a motion. But the bladder is the actual seat of the disorder: when urine is passed drop by drop, or when blood is emitted with it, and in the blood are some clots which are passed with difficulty, and when the lower parts in the region of the pubes are painful. Cases of stone[b] in the bladder are recognized by the following signs: urine is passed with difficulty and slowly, now and again even involuntarily, drop by drop, the urine being sandy; at times blood, or something blood-stained or purulent, is excreted with the urine; this some pass more readily standing, some whilst lying on the back and especially those with large calculi, some even pass urine bending forwards whilst they relieve the pain by drawing out the penis. There is in that part also a feeling of weight, increased by running, or by any kind of movement. Some also when in great pain interlock their feet, crossing alternatively the one over the other. Women again are forced to put their hands to their vulvar orifice and scratch; at times they feel the stone when they put a finger to the place where it is pressing upon the neck of the bladder. But there is a lung disease in those who spit up frothy blood. In a pregnant woman immoderate looseness of the bowels can drive out the foetus; in the same condition, what she is carrying is a weakling, if milk escapes from her breasts; firm breasts testify that it is

17 mammae sanum illud esse testantur. Frequens singultus et praeter consuetudinem continuus iecur inflammatum esse significat. Si tumores super ulcera subito esse desierunt, idque a tergo incidit, vel distentio nervorum vel rigor timeri potest: at si a priore parte id evenit, vel lateris acutus dolor vel
18 insania expectanda est: interdum etiam eiusmodi casum, quae tutissima inter haec est, profusio alvi sequitur. Si ora venarum, sanguinem solita fundere, subito suppressa sunt, aut aqua inter cutem aut
19 tabes sequitur. Eadem tabes subit, si in lateris dolore orta suppuratio intra quadraginta dies purgari non potuit. At si longa tristitia cum longo timore
20 et vigilia est, atrae bilis morbus subest. Quibus saepe ex naribus fluit sanguis, iis aut lienis tumet, aut capitis dolores sunt, quos sequitur, ut quaedam
21 ante oculos tamquam imagines obversentur. At quibus magni lienes sunt, iis gingivae malae sunt, et os olet, aut sanguis aliqua parte prorumpit; quorum si nihil evenit, necesse est in cruribus mala ulcera, et ex his nigrae cicatrices fiant. Quibus causa doloris neque sensus eius est, his mens labat. Si in ventrem sanguis confluxit, ibi in pus vertitur.
22 Si a coxis et inferioribus partibus dolor in pectus transit, neque ullum signum malum accessit, suppurationis eo loco periculum est. Quibus sine febre aliqua parte dolor aut prurigo cum rubore et calore est, ibi aliquid suppurat. Urina quoque, quae in homine sano parum liquida est, circa aures futuram aliquam suppurationem esse denuntiat.

---

[a] Cf. III. 18. 17, Hipp. IV. 184 (*Aph.* VI. 23).
[b] Hipp. *Prorrhetics* II. 35, 36.

healthy. It signifies that the liver is inflamed when there is hiccough both frequent and continuing longer than usual. When swellings which have supervened upon ulcerations subside suddenly, if situated in the back, either spasm or rigor may be apprehended; but if this happens in front, either acute pleural pain or madness is to be expected: at times also in such a case, diarrhoea follows, which is the safest thing. If a customary bleeding from haemorrhoids is suddenly suppressed, dropsy or phthisis follows. Phthisis likewise supervenes if, after beginning with pain in the side, suppuration cannot be cleared off within forty days. And the black bile [a] disease supervenes upon prolonged despondency with prolonged fear and sleeplessness. Those who often have bleeding from the nose, have swelling of the spleen, or pains in the head, and as a consequence some observe phantoms [b] before their eyes. But those in whom the spleens are enlarged, in these the gums are diseased, the mouth foul, or blood bursts out from some part. When none of these things happen, of necessity bad ulcers will be produced on the legs, and from these black scars. In those who, with a cause for pain, do not feel it, the mind is disordered. If blood flows into the abdomen it is there turned into pus. There is danger of suppuration in the chest when pain spreads there from the hips and lower parts, even although no other bad sign is added. When, without any fever, there is pain or itching in some part, with redness and heat, some suppuration is taking place there. Also urine which is not limpid enough for a man in health denotes that some parotid suppuration is about to set in.

CELSUS

23 Haec vero, cum sine febre quoque vel latentium vel futurarum rerum notas habeant, multo certiora sunt, ubi febris accessit, atque etiam aliorum mor- 24 borum tum signa nascuntur. Ergo protinus insania timenda est, ubi expeditior alicuius, quam sani fuit, sermo subitaque loquacitas orta est, et haec ipsa solito audacior; aut ubi raro quis et vehementer spirat, venasque concitatas habet praecordiis duris 25 et tumentibus. Oculorum quoque frequens motus, et in capitis dolore offusae oculis tenebrae, vel nullo dolore substante somnus ereptus, continuataque nocte et die vigilia, vel prostratum contra consuetudinem corpus in ventrem, sic ut ipsius alvi dolor id non coegerit, item robusto adhuc corpore insolitus dentium 26 stridor insaniae signa sunt. Si quid etiam abscessit, et antequam suppuraret manente adhuc febre subsedit, periculum adfert primum furoris, deinde interitus. Auris quoque dolor acutus cum febre continua vehementique saepe mentem turbat; ex eo casu iuveniores interdum intra septimum diem moriuntur, seniores tardius, quoniam neque aeque magnas febres experiuntur, neque aeque insaniunt: ita sustinent, dum is adfectus in pus vertatur. 27 Suffusae quoque sanguine mulieris mammae furorem venturum esse testantur. Quibus autem longae febres sunt, iis aut abscessus aliqui aut articulorum dolores erunt. Quorum faucibus in febre inliditur spiritus, instat his nervorum distentio. Si angina subito finita est, in pulmonem id malum transit;

---

[a] Note [b], p. 88.
[b] For the ear, cf. VI. 7. Hipp. II. 46 (*Prog.* XXII.).
[c] IV. 7. 10. 14.

Now these signs, though even in the absence of fever, they afford indications of latent or oncoming affections, do so with much more certainty when there is fever in addition; and then signs of other diseases besides may develop. Thus madness is to be apprehended immediately: when a patient speaks more hurriedly than he did when well, and of a sudden becomes loquacious, and that with more audacity than was his wont; or when he breathes slowly and forcibly, and has dilated blood-vessels, while the parts below the ribs are hard and swollen. Further signs of madness are: frequent movement of the eyes, and, in cases of headache, shadows passing before the eyes; or loss of sleep in the absence of pain, the wakefulness persisting night and day; or lying on the belly contrary to habit without being obliged to do so by abdominal pain; or, while the body is still vigorous, an unaccustomed grinding of the teeth. If also there has been congestion [a] which has subsided without the formation of pus, whilst fever persists, there is brought about danger first of delirium, then of death. Acute pain in the ear with continuous severe fever also often disturbs the mind; from which affection younger patients die at times within seven days; older ones later, for they experience neither such high fever, nor are equally delirious, hence they hold out until this condition [b] is converted into pus. The breasts of a woman, when they become suffused with blood, also indicate that delirium is about to supervene. But in those in whom fevers are prolonged, there will be an abscess somewhere or pains in the joints. When during fever the breathing in the throat becomes impeded, spasms are impending. If angina [c] subsides suddenly, the malady has passed

CELSUS

28 idque saepe intra septimum diem occidit. Quod nisi incidat, sequitur, ut aliqua parte suppuret. Deinde post alvi longam resolutionem tormina, post haec intestinorum levitas oritur; post nimias destillationes tabes, post lateris dolorem vitia pulmonum, post haec insania; post magnos fervores corporis nervorum rigor aut distentio; ubi caput vulneratum est, delirium; ubi vigilia torsit, nervorum distentio; ubi vehementer venae super ulcera moventur, sanguinis
29 profluvium. Suppuratio vero pluribus morbis excitatur: nam si longae febres sine dolore, sine manifesta causa remanent, in aliquam partem id malum incumbit, in iuvenioribus tamen: nam senioribus ex eiusmodi morbo quartana fere nascitur.
30 Eadem suppuratio fit, si praecordia dura, dolentia ante vicensimum diem hominem non sustulerunt, neque sanguis ex naribus fluxit, maximeque in adulescentibus, utique si inter principia aut oculorum caligo aut capitis dolores fuerunt: sed tum in in-
31 ferioribus partibus aliquid abscedit. Aut si praecordia tumorem mollem habent, neque habere intra sexaginta dies desinunt, haeretque per omne id tempus febris; sed tum in superioribus partibus fit abscessus; ac si inter ipsa viscera non fit, circa aures erumpit. Quomque omnis longus tumor ad suppurationem fere spectet, magis eo tendit is, qui in praecordiis quam is, qui in ventre est; is, qui supra

---

[a] II. 1, 8, p. 90; IV. 22. 23.
[b] I. 2, 3; III. 21. 3; IV. 5.
[c] Hipp. II. 48 (*Prog.* XXIV.), 4 *et seq.*
[d] III. 15. 16.
[e] Hipp. II. 18 (*Prog.* VII. 39).

into the lung; and it is then often fatal within seven days. If that does not happen, it follows that somewhere there is suppuration. Again after a prolonged looseness of the bowels there arise dysenteries, and after these intestinal lubricity;[a] phthisis after excessive runnings from the nose;[b] lung diseases after pain in the side; and from these madness; after ardent fevers rigor or spasm of sinews; after a head wound, delirium; when wakefulness tortures, spasms of sinews; when in wounds blood-vessels throb violently, haemorrhage. But suppuration is induced by many diseases; for if fever continues for a long while without pain and without evident cause, suppuration is developing in some part[c]—in younger patients, however; for generally in the elderly from a self-same malady a quartan[d] fever is developed. Suppuration is likewise being produced if the parts below the ribs are hard and painful, and have not carried off the patient by the twentieth day, or nose-bleeding has not occurred, and this chiefly in the case of adolescents, especially if from the commencement there has been dimness of vision, or headache; but in these cases something is abscessing in the lower parts of the abdomen.[e] Or if the parts below the ribs present a soft swelling which persists and does not subside within sixty days, and fever holds all that time; but in these cases an abscess is being produced in the upper parts of the abdomen. And if it is not produced in the actual viscera, it breaks out around the ears. Whilst every swelling of long standing is generally an expectant abscess, it tends more to this in the region in front of the heart, than in the abdomen, and in the abdomen rather above than below the

CELSUS

32 umbilicum quam is, qui infra est. Si lassitudinis etiam sensus in febre est, vel in maxillis vel in articulis aliquid abscedit. Interdum quoque urina tenuis et cruda sic diu fertur, ut alia salutaria signa sint, exque eo casu plerumque infra transversum septum, quod Διάφραγμα Graeci vocant, fit abcessus.
33 Dolor etiam pulmonis, si neque etiam per sputa neque per sanguinis detractionem neque per victus rationem finitus est, vomicas aliquas intus excitat aut circa vicesimum diem aut circa tricesimum aut circa quadragesimum, nonnumquam etiam circa
34 sexagensimum. Numerabimus autem ab eo die, quo primum febricitavit aliquis aut inhorruit aut gravitatem eius partis sensit. Sed hae vomicae modo a pulmone modo a contraria parte nascuntur. Quod suppurat, ab ea parte, quam adficit, dolorem inflammationemque concitat: ipsum calidius est et, si in partem sanam aliquis decubuit, onerare
35 eam ex pondere aliquo videtur. Omnis etiam suppuratio, quae nondum oculis patet, sic deprehendi potest: si febris non dimittit, eaque interdiu levior est, noctu increscit, multus sudor oritur, cupiditas tussiendi est, et paene nihil in tussi excreatur, oculi cavi sunt, malae rubent, venae sub lingua inalbescunt, in manibus fiunt adunci ungues, digiti maximeque summi calent, in pedibus tumores sunt, spiritus difficilius trahitur, cibi fastidium est, pustulae
36 toto corpore oriuntur. Quod si protinus initio dolor et tussis fuit et spiritus difficultas, vomica vel ante

---

a vomica, originally meaning the discharge from an abscess, came to be used of an abscess cavity (Greek ἀπόσκηψις) especially in the lung. Celsus uses it both to describe abscess

navel. Something is abscessing, either in the jaws or in the joints, if there is with the fever also a feeling of lassitude. At times too the urine remains thin and unconcocted for so long that other signs are salutary, and from this condition an abscess often occurs below the transverse membrane which the Greeks call diaphragma. Pain in the lung again, when not terminated by expectoration or by blood-letting, or by regulation of the diet, may excite some abscesses [a] in it about the twentieth, thirtieth, fortieth, occasionally sixtieth day. But we will count from the day when there is first fever or shivering or sense of weight in that part. These abscesses originate sometimes from the lung, sometimes from the opposite side.[b] Whichever side is affected the suppuration gives rise to pain and inflammation; it is hotter there, and if the patient lies on the sound side he seems to oppress it by some weight. Further, any suppuration, not yet evident to the eye, can be detected as follows: if the fever does not remit, but whilst diminishing by day, increases at night, if there is profuse sweating, a desire to cough, yet hardly anything is expectorated in coughing; the eyes are sunken, the cheeks flushed, the veins under the tongue pale; the finger nails become curved, the fingers hot, especially at their tips; there are swellings in the feet; there is greater difficulty in breathing, and distaste for food; pustules spring up all over the body. But if there was pain from the commencement with cough and difficult breathing, the abscess will burst before or

[a] of the lung or liver and empyema (pus in the pleural cavity); cf. IV. 13, 14, 15, and Hipp. II, 30–34 (*Prog.* XV., XVI).

[b] That is the side opposite to that where pain has first occurred.

vel circa vicesimum diem erumpet: si serius ista coeperint, necesse est quidem increscant, sed quo minus cito adfecerint, eo tardius solventur. Solent etiam in gravi morbo pedes cum digitis unguibusque nigrescere: quod si non est mors consecuta et reliquum corpus invaluit, pedes tamen decidunt.

8. Sequitur, ut in quoque morbi genere proprias notas explicem, quae vel spem vel periculum ostendant.

Ex vesica dolenti si purulenta urina processit, inque ea leve et album subsedit, metum detrahit.

2 In pulmonis morbo si sputo ipso levatur dolor, quamvis id purulentum est tamen aeger facile spirat, facile excreat, morbum ipsum non difficulter fert, potest ei secunda valetudo contingere. Neque inter initia terreri convenit, si protinus sputum mixtum est rufo quodam et sanguine, dummodo statim edatur.

3 Laterum dolores suppuratione facta, deinde intra quadragesimum diem purgata, finiuntur.

Si in iocinere vomica est, et ex ea fertur pus purum et album, salus facilis est: id enim malum in tunica est.

4 Ex suppurationibus vero eae tolerabiles sunt, quae in exteriorem partem feruntur et acuuntur. At ex iis, quae intus procedunt, eae leviores, quae contra se cutem non adficiunt, eamque et sine dolore esse et eiusdem coloris, cuius reliquae partes sunt, sinunt esse. Pus quoque, quacumque parte erumpit, si est leve, album, unius coloris, sine ullo metu est, et quo effuso febris protinus conquievit desieruntque urguere cibi fastidium et potionis desider-
5 ium. Si quando etiam suppuratio descendit in

---

a Hipp. II. 20. *Prog.* IX. 18.
b IV. 15. cf. Hipp. IV. 202 (*Aph.* VII. 45).

about the twentieth day; if these signs happen later they necessarily have to develop, but the less quickly they come to a head, the later the relief. In a grave disease the feet, toes and nails also tend to blacken; and when death does not follow, and the rest of the body recovers, nevertheless the feet fall off.[a]

8. It follows now that I have to explain the special signs which in any particular affection indicate either hope or danger.

When there is pain in the bladder, if purulent urine is discharged which has in it a sediment slimy and white, it allays apprehension.

In pulmonary disease a patient may possibly regain health, if expectoration, although purulent, relieves pain, so long as he breathes and expectorates freely, and bears the disease without difficulty. Nor is there cause for alarm at an early stage, if the expectoration is mixed with something reddish and with blood, so long as it is expectorated at once.

Pain in the side ends if the suppuration which has arisen is cleared off within forty days.[b]

If there is an abscess in the liver, and the pus let out is uniform and white, in that case recovery is easy, because the mischief is enclosed in a capsule.

Among suppurations too those are tolerable which point and discharge outwards. And of those which move inwards, those are the more favourable which do not affect the overlying skin, but leave it free from pain, and of the same colour as the surroundings. Pus indeed causes no fear, wherever it breaks out, when slimy and uniformly white, and if the fever subsides at once upon its discharge, and distaste for food and thirst cease to be troublesome. Also whenever suppuration descends into the legs,

crura, sputumque eiusdem factum pro rufo purulentum est, periculi minus est.

6 At in tabe eius, qui salvus futurus est, sputum esse debet album, aequale totum, eiusdemque coloris, sine pituita; eique etiam simile esse oportet, si quid in nares a capite destillat. Longe optimum est febrem omnino non esse; secundum est tantulam esse, ut neque cibum inpediat neque crebram sitim faciat. Alvus in hac valetudine ea tuta est, quae cotidie, quae coacta, quae convenientia iis, quae adsumuntur, reddit; corpus id, quod minime tenue maximeque lati pectoris atque saetosi est, cuiusque
7 cartilago exigua et carnosa est. Super tabem si mulieri suppressa quoque menstrua fuerunt, et circa pectus atque scapulas dolor mansit subitoque sanguis erupit, levari morbus solet: nam et tussis minuitur, et sitis atque febricula desinunt. Sed iisdem fere, nisi redit sanguis, vomica erumpit; quae quo cruentior, eo melior est.

8 Aqua autem inter cutem minime terribilis est, quae nullo antecedente morbo coepit; deinde, quae longo morbo supervenit, utique si firma viscera sunt, si spiritus facilis, si nullus dolor, si sine calore corpus est, aequaliterque in extremis partibus macrum est, si mollis venter, si nulla tussis, nulla sitis, si lingua
9 ne super somnum quidem inarescit; si cibi cupiditas est, si venter medicamentis movetur, si per se excernit mollia et figurata, si extenuatur; si urina

---

[a] Hipp. *Prorrhetics*, II. 7.
[b] III. 21.

and the patient's expectoration from being reddish becomes purulent, there is less danger.

But in phthisis,[a] he that is to recover should have his expectoration white, uniform in consistency and colour, unmixed with phlegm; and that which drips into the nose from the head should have similar characters. It is the best by far for there to be no fever; second best when the fever is so slight as not to impair the appetite or cause frequent thirst. In this affection the patient's state is favourable: when the bowels are moved once a day, the motions being formed and in amount corresponding to the food consumed; the body least attenuated, the chest most broad and hairy; its cartilages small, and covered with flesh. If supervening on phthisis, a woman's menses also become suppressed and pain is continuous over her chest and shoulders, a sudden eruption of blood customarily relieves the disease; for the cough becomes less, and the thirst and slight fever subside. But generally in these cases, unless the haemorrhage recurs, an abscess bursts, and the more blood comes from it the better.

Dropsy[b] is the least alarming when it has commenced without being preceded by any disease; next when it has supervened upon a long illness, certainly if the viscera are sound, if the breathing is easy, if there is no pain, if the body is not hot, and the extremities are wasted uniformly, if the abdomen is soft, if there is no cough, no thirst, if the tongue is not much parched even after sleep; if there is desire for food, if the bowels are moved by medicaments, if the motions when spontaneous are soft and formed, if the size of the abdomen is reduced; if the urine is altered both by a change of

CELSUS

et vini mutatione et epotis aliquibus medicamentis mutatur; si corpus sine lassitudine est et morbum facile sustinet: siquidem in quo omnia haec sunt, is ex toto tutus est; in quo plura ex his sunt, is in bona spe est.

10 Articulorum vero vitia, ut podagrae cheragraeque, si iuvenes temptarunt neque callum induxerunt, solvi possunt; maximeque torminibus leniuntur et quocumque modo venter fluit.

11 Item morbus comitialis ante pubertatem ortus non aegre finitur; et in quo ab una parte corporis venientis accessionis sensus incipit, optimum est a manibus pedibusve initium fieri, deinde a lateribus; pessimum inter haec a capite.

12 Atque in his quoque ea maxime prosunt, quae per deiectiones excernuntur. Ipsa autem deiectio sine ulla noxa est, quae sine febre est, si celeriter desinit, si contacto ventre nullus motus eius sentitur, si extremam alvum spiritus sequitur.

13 Ac ne tormina quidem periculosa sunt, si sanguis et strigmenta descendunt, dum febris ceteraeque accessiones huius morbi absint, adeo ut etiam gravida mulier non solum reservari possit, sed etiam partum reservare; prodestque in hoc morbo, si iam aetate aliquis processit.

14 Contra intestinorum levitas facilius a teneris aetatibus depellitur, utique si ferri urina et ali cibo corpus incipit.

Eadem aetas prodest et in coxae dolore et umerorum et omni resolutione nervorum; ex quibus

---

[a] III. 1. 6. 21. 7, 8.   [b] IV. 31. 32 and Appendix, p. 463.
[c] II. 8, 29–III. 23.   [d] Cf. IV. 26.
[e] Cf. IV. 22.

BOOK II. 8. 9-14

the wine and of certain medicinal draughts;[a] if there is no lassitude and the disorder is easily borne: a patient who presents all these signs is thoroughly safe, and that case is hopeful which exhibits the greater number of them.

Joint-disorders, too, such as foot and hand aches,[b] if they attack young people and have not induced callosities, can be resolved; for the most part they are removed by dysenteries and fluid motions, whatever the sort.

Epileptic[c] fits again are not difficult to bring to an end, when they have commenced before puberty, and whenever the sensation of the coming fit begins in some one part of the body. It is best for it to begin from the hands or feet, next from the flanks, worst of all from the head.

In such patients, also, the most favourable signs are when the disease can be discharged in the stools. Diarrhoea[d] is itself harmless, when there is no fever, if it is quickly over, if on touching the abdomen no movements are to be felt, if wind follows the last of the motion.

Even dysenteries[e] are not a danger although blood or shreds are passed, as long as fever and other accessories of this malady are absent, so that even a pregnant woman can not only be preserved herself, but the foetus preserved also. It is helpful in this malady if the patient's age is already mature.

Intestinal lubricity on the other hand is more easily got rid of in childhood, certainly if urine begins to be passed and the body to be nourished by the food.

The same age has the advantage in cases both of hip and shoulder pains, and of all forms of para-

coxa, si sine torpore est, si leviter friget, quamvis magnos dolores habet, tamen et facile et mature sanatur, resolutumque membrum, si nihilo minus alitur,[1] fieri sanum potest. Oris resolutio etiam alvo cita finitur; omnisque deiectio lippienti prodest.
15 At varix ortus vel per ora venarum subita profusio sanguinis vel tormina insaniam tollunt.

Umerorum dolores, qui ad scapulas vel manus tendunt, vomitu atrae bilis solvuntur; et quisquis dolor deorsum tendit, sanabilior est.

Singultus sternumento finitur.
16 Longas deiectiones supprimit vomitus.

Mulier sanguinem vomens profusis menstruis liberatur. Quae menstruis non purgatur, si sanguinem ex naribus fudit, omni periculo vacat. Quae locis laborat aut difficulter partum edit, sternumento levatur.

Aestiva quartana fere brevis est. Cui calor et tremor est, saluti delirium est. Lienosis bono tormina sunt.
17 Denique ipsa febris, quod maxime mirum videri potest, saepe praesidio est. Nam et praecordiorum dolores, si sine inflammatione sunt, finit; et iocineris dolori succurrit; et nervorum distentionem rigoremque, si postea coepit, ex toto tollit; et ex difficultate urinae morbum tenuioris intestini ortum, si urinam per calorem movet, levat.
18 At dolores capitis, quibus oculorum caligo et rubor cum quadam frontis prurigine accedunt, san-

[1] *Constantine for the MSS. aliter.*

---

[a] Hipp. *Prorrhetics*, II. 41.
[b] Hipp. IV. 184 (*Aph.* VI. 21).
[c] IV. **16**.

lysis; in such the hip^a may be cured easily and early, if it is not numbed, if slightly cool, even though the pains are severe, and a paralysed limb can be restored if its nutrition is not at all impaired. Paralysis of the face may be even ended by a quick motion; and any purging benefits runnings of the eyes. But madness^b is relieved rather by the formation of varicose veins or by a sudden effusion of blood from haemorrhoids or by dysentery.

Shoulder pains spreading to the shoulder-blades or hands are relieved by a vomit of black bile; and pain of any kind which moves downwards is the more curable.

Sneezing puts an end to hiccough.

Prolonged diarrhoeas are suppressed by vomiting.

In a woman a vomiting of blood is relieved by menstruation; when not cleared up by menstruation, nose-bleeding removes all danger. A woman in trouble with her womb or labour difficulty is relieved by sneezing.

Quartan fever in summer is mostly short. In a case of ardent fever with a tremor, delirium is salutary. For enlargement of the spleen^c dysenteries are good.

Then again fever itself is in the end often a protection, which may appear very strange. For it brings to an end pains over the heart if there is no inflammation; and it also relieves a painful liver; and if it begins after spasm and rigor, it gives entire relief; and it removes the disease of the small intestine arising from urinary difficulty, if by its heat it promotes urination.

Now pains in the head, accompanied by dimness of vision and redness of the eyes, along with some

CELSUS

guinis profusione vel fortuita vel etiam petita summoventur. Si capitis ac frontis dolores ex vento vel frigore aut aestu sunt, gravedine et sternumentis finiuntur.

19 Febrem autem ardentem, quam Graeci causoden vocant, subitus horror exsolvit. Si in febre aures obtunsae sunt, si sanguis naribus fluxit, aut venter resolutus est, illud malum desinit ex toto. Nihil plus adversus surditatem quam biliosa alvus potest.

20 Quibus in fistula urinae veluti minutiores abscessus, quos φύματα vocant, esse coeperunt, iis, ubi pus ea parte profluxit, sanitas redditur.

. . .[1] Ex quibus cum pleraque per se proveniant, scire licet inter ea quoque, quae ars adhibet, naturam plurimum posse.

21 Contra si vesica[2] cum febre continenti dolet, neque venter[3] quicquam reddit, malum atque mortiferum est; maximeque id periculum est pueris a septimo anno ad quartum decimum.

22 In pulmonis morbo, si sputum primis diebus non fuit, deinde a septimo die coepit et ultra septimum mansit, periculosum est; quantoque magis mixtos neque inter se diductos colores habet, tanto deterius. Et tamen nihil peius est quam sincerum id edi, sive rufum est sive cruentum sive album sive glutinosum

---

[1] *Marx proposes to fill the lacuna*: Haec sunt signa in quoque morbis genere propria quibus spes salutis ostenditur: "*These are the special signs in each kind of disease which hold out hope of recovery.*"

[2] *So Pantenus, followed by v.d. Linden for the* sive caput *of MSS.*

[3] venter *added by Targa.*

---

[a] Some words have fallen out here (see crit. note 1) probably to this effect: 'The following are the special

BOOK II. 8. 18–22

itching of the forehead, may be relieved by a haemorrhage, whether fortuitous or procured. Pains in the head and forehead due to wind or to cold or to heat are terminated by running from the nose and sneezings.

The ardent fever, however, which the Greeks call causodes, is got rid of by a sudden shivering. During a fever, if the ears have become dulled, that trouble is entirely removed by a flux of blood from the nose, or by loose motions from the bowel. Against deafness nothing can be more efficacious than a bilious stool. Those who have begun to suffer from the smaller kinds of abscesses in the urethra which they call phumata, get well when pus has come away from that part.[a]

. . . and since they mostly arise of themselves, we may know that even where the resources of art are applied, nature can do the most.

On the other hand, pain in the bladder with persistent fever, when nothing is passed by the bowel, is a fatal evil; the danger is greatest in boys from the seventh to the fourteenth year.[b]

In pulmonary disease, if there was no expectoration during the first days, if it then begins from the seventh day and persists beyond a further seven days, it is dangerous.[c] And the more the sputum has an undistinguishable admixture of colours, the worse it is. But nevertheless nothing is worse than for the expectoration emitted to be homogeneous, whether reddish, or clotted, or white, or glutinous, or pallid,

signs in each kind of disease which indicate hope of recovery.'

[b] From stone in the bladder, VII. 26, 2.

[c] Hipp. II. 30 ff. (*Prog.* XV. 10) of which this and the next sentence is a paraphrase.

CELSUS

sive pallidum sive spumans; nigrum tamen pessimum est. In eodem morbo periculosa sunt tussis, destillatio, etiam quod alias salutare habetur, sternumentum; periculosissimumque est, si haec secuta subita deiectio est.

Fere vero quae in pulmonis, eadem in lateris dolore et mitiora signa et asperiora esse consuerunt.

Ex iocinere si pus cruentum exit, mortiferum est.

23 At ex suppurationibus eae pessimae sunt, quae intus tendunt, sic ut exteriorem quoque cutem decolorent: ex iis deinde, quae in exteriorem partem prorumpunt, eae pessimae, quae maximae quaeque planissimae sunt. Quod si, ne rupta quidem vomica vel pure extrinsecus emisso, febris quievit, aut quamvis quierit, tamen repetit, item si sitis est, si cibi fastidium, si venter liquidus, si pus est lividum et pallidum, si nihil aeger excreat nisi pituitam spumantem, periculum certum est. Atque ex iis quidem suppurationibus, quas pulmonum morbi concitarunt, fere senes moriuntur: ex ceteris iuniores.

24 At in tabe sputum mixtum, purulentum, febris adsidua, quae et cibi tempora eripit et siti adfligit, in corpore tenui subesse periculum testantur. Si quis etiam in eo morbo diutius traxit, ubi capilli fluunt, ubi urina quaedam araneis similia subsidentia ostendit, atque in iis odor foedus est, maximeque ubi post haec orta deiectio est, protinus moritur, utique si tempus autumni est, quo fere qui cetera parte anni traxerunt, resolvuntur. Item pus ex-

---

* Contrasted with II. 8. 6, p. 134.

or frothy; worst of all, however, is the black. In this same disease the following are signs of danger: cough, catarrh, and even sneezing, which in other maladies is held salutary; and a sudden diarrhoea following upon the above is a most dangerous sign.

Generally too the same signs hold good for pain in the side as for that in the lung, both the more favourable as well as the graver signs.

If the pus discharged from the liver is bloody, it is a deadly sign.

Now of suppurations the worst are those which tend inwards, whilst also discolouring the overlying skin: of those again which burst externally, the worst are those which are largest and most widespread. But even after the abscess has ruptured, or the pus has been let outwards, there is danger for certain if the fever does not subside, or although it subsides, nevertheless recurs; or further if there is thirst, if distaste for food, if liquid motions, if the pus is livid and pallid, if the patient expectorates nothing but frothy phlegm. And of such suppurations, old people die mostly of those excited by lung diseases; younger people of other kinds.

But that in phthisis danger threatens a thin [a] man is signified as follows: the expectoration is purulent with admixtures, a persistent fever robs him of his appetite at meal-times and afflicts him with thirst. Death is at hand if, after the patient has dragged on for a long while, the hair falls out, the urine exhibits sediment like cobwebs and has a foul odour, and most of all when upon the above diarrhoea supervenes; especially if it is the autumn season, when patients who have lasted through the rest of the year are generally undone. Moreover, in this

puisse in hoc morbo, deinde ex toto spuere desisse
25 mortiferum est. Solent etiam in adulescentibus
ex eo morbo vomicae fistulaeque oriri; quae non
facile sanescunt, nisi si multa signa bonae valetudinis subsecuta sint. Ex reliquis vero minime
facile sanantur virgines aut eae mulieres, quibus
super tabem menstrua suppressa sunt. Cui vero
sano subitus dolor capitis ortus est, dein somnus
oppressit, sic ut stertat neque expergiscatur, intra
septimum diem pereundum est; magis si eum alvus
cita non antecesserit, si palpebrae dormientis non
coeunt, si album oculorum apparet. Quos tamen
ita mors sequitur, si id malum non est febre
discussum.

26 At aqua inter cutem, si ex acuto morbo coepit,
ad sanitatem raro perducitur, utique si contraria
iis, quae supra (§§ 8, 9) posita sunt, subsecuntur.
Aeque in ea quoque tussis spem tollit, item, si
sanguis sursum deorsumque erupit et aqua medium
corpus inplevit. Quibusdam etiam in hoc morbo
tumores oriuntur, deinde desinunt, deinde rursus
adsurgunt: hi tutiores quidem sunt, quam qui
supra conprehensi sunt, si adtendunt; sed fere
27 fiducia secundae valetudinis opprimuntur. Illud
iure aliquis mirabitur, quomodo quaedam simul et
adfligant nostra corpora, et parte aliqua tueantur:
nam sive aqua inter cutem quem implevit, sive in
magno abscessu multum puris coit, simul id omne
effudisse aeque mortiferum est, ac si quis sani
corporis vulnere factus exsanguis est.

[a] σύριγγες—Hipp. *Prorrhetics*, II. 7.
[b] Hipp. IV. 190 (*Aph.* VII. 51), II. 10 (*Prog.* II. 36).
[c] P. 135 paras. 8 and 9.
[d] Ascites. See further, III. 21, 3, 14; VII. 15.

disease, after pus has been expectorated, it is fatal for there to be an entire cessation of spitting. In the course of phthisis, even in adolescents, abscesses followed by fistulae [a] arise in the lung; and unless numerous signs of convalescence follow, they do not readily heal. But as regards others, the least easily cured are girls, or those women in whom suppression of menses has supervened upon the phthisis. When again in a man who has been healthy there arises suddenly pain in the head, next he is so overcome by sleep that he snores and cannot be awaked, he will die by the seventh day; the more so, if a loose motion has not preceded, if the eyelids of the sleeper are unclosed and the whites of the eyes show.[b] And in these cases death follows except if the malady has been dispersed by fever.

Again dropsy, if caused by an acute disease, is seldom conducted to a cure, at any rate when signs supervene the reverse of those noted above.[c] Likewise too in this disease a cough takes away hope as is also the case if there is an outburst of blood whether upwards or downwards and water fills the middle [d] of the body. In some also in this disease swellings arise, then subside, and again recur: such patients are in a somewhat safer state than those mentioned above, if they give attention; but generally they are undone by over-confidence in their health. Here we may wonder with good reason why there should be occurrences which cause our bodies harm, and yet at the same time in a measure are beneficial: for whether it is dropsical fluid which has filled a patient up, or whether it is a quantity of pus which has collected in a large abscess, evacuation all at once is as fatal to him, as if a healthy man loses blood by a wound.

28 Articulis vero qui sic dolent, ut super eos ex callo quaedam tubercula innata sint, numquam liberantur: quaeque eorum vitia vel in senectute coeperunt, vel ad senectutem ab adulescentia pervenerunt, ut aliquando leniri possunt, sic numquam ex toto finiuntur.

29 Morbus quoque comitialis post annum XXV ortus aegre curatur, multoque aegrius is, qui post XL annum coepit, adeo ut in ea aetate aliquid in natura spei, vix quicquam in medicina sit. In eodem morbo si simul totum corpus adficitur, neque ante in partibus aliquis venientis mali sensus est, sed homo inproviso concidit, cuiuscumque is aetatis est, vix sanescit: si vero aut mens laesa est, aut nervorum facta resolutio, medicinae locus non est.

30 Deiectionibus quoque si febris accessit, si inflammatio iocineris aut praecordiorum aut ventris, si inmodica sitis, si longius tempus, si alvus varia, si cum dolore est, etiam periculum mortis subest, maximeque si inter haec tormina vera[1] esse coeperunt; isque morbus maxime pueros absumit usque ad annum decimum: ceterae aetates facilius sustinent. Mulier quoque gravida eiusmodi casu rapi potest; atque, etiamsi ipsa convaluit, tamen partum
31 perdit. Quin etiam tormina ab atra bile orsa mortifera sunt, aut si sub his extenuato iam corpore subito nigra alvus profluxit.

32 At intestinorum levitas periculosior est, si frequens deiectio est, si venter omnibus horis et cum

[1] *So Marx for the* vetera *of the MSS.; the word is deleted by Targa.*

---

[a] Epileptic and apoplectic. Hipp. *Prorrhetics*, II. 9.
[b] φλεγμονὴ γαστρός—Hipp. *Prorrhetics*, II. 22.

BOOK II. 8. 28-32

Those too who suffer in their joints, so that growths of hard stuff are formed upon them, are never relieved entirely: all these damages, whether they have begun in old age, or have lasted from youth up to old age, although there is a possibility of some alleviation, are never entirely cured.

Also fits [a] which have arisen after the twenty-fifth year are hard to relieve, much harder when they begin after the fortieth; hence at this age, whilst there may be some hope from nature, there is scarcely any from the Art of Medicine. In this affection, if the whole body is affected all together, and there has not been beforehand in any part some feeling of an oncoming ill, but the patient falls down unexpectedly, he scarcely ever gets well, be his age what it may: further, if either the mind is diseased, or paralysis has been set up, there is no opportunity for the Art of Medicine.

Again in cases of diarrhoea, danger of death is at hand: if there is fever in addition, if there is inflammation of the liver or of the parts over the heart or of the stomach,[b] if excessive thirst, if the affection is prolonged; if the stools are varied and passed with pain, and especially if with these signs true dysenteries set in; and this disease carries off mostly children up to the age of ten; other ages bear it more easily. Also a pregnant woman can be swept away by such an event, and even if she herself recovers, yet she loses the child. Dysenteries are fatal, moreover, when originated by black bile, or if a black motion suddenly issues from a body already wasted by dysentery.

Now intestinal lubricity is the more dangerous, if there is a frequent motion, if there is a flux from

CELSUS

sono et sine hoc profluit; si similiter noctu et interdiu, si, quod excernitur, aut crudum est aut nigrum et praeter id etiam leve et mali odoris; si sitis urget, si post potionem urina non redditur (quod evenit, quia tunc liquor omnis non in vesicam sed intestina 33 descendit); si os exulceratur, rubet facies et quasi maculis quibusdam colorum omnium distinguitur; si venter est quasi fermentatus, pinguis atque rugosus, si cibi et . . .[1] cupiditas non est; inter quae cum evidens mors sit, multo evidentior est, si iam longum quoque id vitium est, id maxime etiam, si in corpore senili est.

34 Si vero in tenuiore intestino morbus est, vomitus, singultus, nervorum distentio, delirium mala sunt.

At in morbo arquato durum fieri iecur perniciosissimum est.

Quos lienis male habet, si tormina prenderunt, deinde inversa sunt vel in aquam inter cutem vel intestinorum levitatem, vix ulla medicina periculo subtrahit.

35 Morbus intestini tenuioris[2] nisi resolutus est, intra septimum diem occidit.

Mulier ex partu si cum febre vehementibus etiam

---

[1] *Daremberg inserted* et ambulationis *"and exercise." Marx suggests that more has fallen out—comparing the corresponding passage of Hippocrates. Prorrhet. II. 23.*

[2] *Marx, following Constantine and v.d. Linden would insert* in urinae difficultate febre; *cf. Hipp. IV. 188 (Aph. VI. 44) of which this passage seems to be a translation.*

---

[a] Hipp. *Prorrhetics*, II. 23—ῥυπαρὰς καὶ ῥυτιδώδεας, and it has been questioned whether Celsus read λιπαράς, fatty, for

the bowel at all hours, with or without noise, if the same condition continues by night and in the day-time, if what is passed is either undigested or black, and besides that also slimy and foul; if there is urgent thirst, if urine is not passed after a drink (which happens because then all fluid passes down, not into the bladder, but into the intestine); if the mouth becomes ulcerated, the face reddened and marked as if by kinds of spots of all colours; if the belly is as though in a state of fermentation, fatty and wrinkled,[a] also if there is no desire for food . . . ; while death is imminent in these circumstances, it is much more imminent if also this disease has already lasted a long while, especially if it be in an old patient.

If again there is disease in the smaller intestine, vomiting, hiccough, spasm, delirium are bad signs.

In jaundice again it is most pernicious[b] for the liver to become hard.

If dysentery has seized upon those with disease of the spleen which has then turned into dropsy or into leientery,[c] scarcely any medical treatment can save them from danger.

The disease of the smaller intestine,[d] unless resolved, kills within seven days.

A woman after childbirth is in danger of death, if

---

[a] ῥυπαράς, dirty—' dirty and wrinkled ' makes better medical sense.

[b] Perniciosus here means, not fatal, but much the same as pestifer, noxious, II. 6, 10. See Cicero, *Laws*, II. 5, multa perniciose, multa pestifera.

[c] IV. 23.

[d] Translating Constantine's emendation (see critical note) the passage runs " disease of the smaller intestine arising from difficulty in passing urine, unless resolved by fever, kills, etc." See above para. 17, also II. 1. 8.

et adsiduis capitis doloribus premitur, in periculo mortis est.

Si dolor atque inflammatio est in iis partibus, quibus viscera continentur, frequenter spirare signum malum est.

36 Si sine causa longus dolor capitis est, et in cervices ac scapulas transit, rursusque in caput revertitur, aut a capite ad cervices scapulasque pervenit, perniciosus est, nisi vomicam aliquam excitavit, sic ut pus extussiretur, aut nisi sanguis aliqua parte prorupit, aut nisi in capite multa porrigo totove corpore
37 pustulae ortae sunt. Aeque magnum malum est, ubi torpor atque prurigo pervagantur, modo per totum caput, modo in parte, aut sensus alicuius ibi quasi frigoris est, eaque ad summam quoque linguam perveniunt. Et cum in iisdem abscessibus auxilium sit, eo difficilior sanitas est, quo minus saepe sub his malis illi subsecuntur.

38 In coxae vero doloribus si vehemens torpor est, frigescitque crus et coxa, alvus nisi coacta non reddit, idque quod excernitur muccosum est, iamque aetas eius hominis XL annum excessit, is morbus erit longissimus minimeque annuus, neque finiri poterit nisi aut vere aut autumno.

39 Difficilis aeque curatio est in eadem aetate, ubi umerorum dolor vel ad manus pervenit vel ad scapulas tendit torporemque et dolorem creat, neque bilis vomitu levatur.

---

[a] This refers to parts above the diaphragm. Hipp. II. 14 (*Prog.* V. 1).

[b] Hipp. *Prorrhetics*, II. 30.

[c] Indicating an elimination of the material of diseases, cf. VI. 2. 2.

[d] Cf. IV. 29, p. 453. He is here describing osteoarthritis and sciatica.

also oppressed by violent and persistent pain in the head along with fever.

To breathe rapidly is a bad sign if there is pain and inflammation in those parts which contain viscera.[a]

A prolonged pain [b] in the head, if without cause it shifts to the neck and shoulders, and again returns to the head, or if it spreads from the head to reach the neck and shoulders, is most pernicious, unless it induces some abscess so that pus is coughed up, or unless there is an outburst of blood from some part, or unless there is upon the head an eruption [c] of much scurf or of pimples all over the body. Equally severe is this malady when a numbness or an itching wanders, now all over the head, now over part of it, or there is felt there a sensation as of something cold, and when these symptoms extend to the tip of the tongue. And since the abscesses described above are beneficial, recovery is more difficult, in proportion as they supervene less often upon such maladies.

When there are pains [d] in the hips, if there is great numbness, and both the leg and hip become cold, if there is no movement of the bowel except forced, and the stool passed is mucous,[e] and if the patient is already over forty, there will be a very prolonged illness, lasting at least a year, nor will it possibly come to an end except either in spring or in autumn.

Treatment is likewise difficult at that age when pain in the shoulders either spreads to the hands or extending to the shoulder-blades gives rise to numbness and pain there, which is not relieved by a vomit of bile.[f]

[e] Hipp. *Prorrhetics*, II. 41.
[f] Hepatic Disease; cf. IV. 15, and Hipp. *Prorrhetics*, II. 40.

CELSUS

40 Quacumque vero corporis parte membrum aliquod resolutum est, si neque movetur et emacrescit, in pristinum habitum non revertitur, eoque minus, quo vetustius id vitium est, et quo magis in corpore senili est. Omnique resolutioni nervorum ad medicinam non idonea tempora sunt hiemps et autumnus: aliquid sperari potest vere et aestate; is morbus mediocris vix sanatur, vehemens sanari non potest.

Omnis etiam dolor minus medicinae patet, qui sursum procedit.

41 Mulieri gravidae si subito mammae emacuerunt, abortus periculum est. Quae neque peperit neque gravida est, si lac habet, a menstruis defecta est.

42 Quartana aestiva brevis, autumnalis fere longa est maximeque quae coepit hieme adpropinquante. Si sanguis profluit, dein secuta est dementia cum distentione nervorum, periculum mortis est, itemque si medicamentis purgatum et adhuc inanem nervorum distentio oppressit, item si in magno alvi[1] dolore extremae partes frigent.

43 Neque is ad vitam redit, qui ex suspendio spumante ore detractus est.

Alvus nigra, sanguini atro similis, repentina, sive cum febre sive etiam sine hac est, perniciosa est.

---

[1] alvi *added by Marx.*

---

[a] Hipp. *Prorrhetics*, II. 39.

[b] 'Nervus' was used by Celsus, as also later by Galen, for fibrous tissues and membranes, which were regarded as the vitally active parts; the soft material in the nervous system and in muscles was looked upon as a sort of padding, and named 'caro,' flesh. Hence, in translating, the term 'sinew' has to be employed to include 'tendon' and 'ligament' as well as nerves.

BOOK II. 8. 40–43

Whatever too the part of the body, any limb which becomes paralysed if it is not moved and wastes, will not be restored to its former state, and the less so the longer the paralysis [a] has been, and the older the patient. And for the cure of all cases of paralysis,[b] winter and autumn are not favourable seasons; there is possibly hope in spring and summer; even when mild this disease is scarcely curable, a severe attack cannot be cured.[c]

All pain also becomes less amenable to treatment as it spreads upwards.[d]

In a pregnant woman,[e] if the breasts suddenly shrivel up, there is danger of abortion. A woman has a defective menstruation who has milk in her breasts, not having just borne a child, or being pregnant.

Quartan fever,[f] whilst brief in summer, is generally prolonged in autumn, and especially so when beginning at the approach of winter. There is danger of death if haemorrhage is followed by dementia and by spasm; the same is the case when, after purgation by medicaments, and, with the bowel still empty, there is an attack of spasm, as also if with great pain in the bowel the extremities become cold.

He does not return to life [g] who has been taken down from hanging with foam round the mouth.

A black stool resembling black blood, passed suddenly, whether accompanied by fever, or even without fever, is dangerous.

---

[c] III. 27, Hipp. IV. 118 (*Aph*. II. 42).
[d] Hipp. *Prorrhetics*, II. 41.
[e] Hipp. IV. 166, 168 (*Aph*. V. 37, 39).
[f] Chronic malaria of Mediterranean, cf. III. 15, 16, pp. 283 ff.
[g] Hipp. IV. 118 (*Aph* II. 43).

CELSUS

9. Cognitis indiciis, quae nos vel spe consolentur vel metu terreant, ad curationes morborum transeundum est. Ex his quaedam communes sunt, quaedam propriae. Communes, quae pluribus opitulantur morbis; propriae, quae singulis. Ante de communibus dicam (X—XXXIII): ex quibus tamen quaedam non aegros solum sed sanos quoque sustinent, quaedam in adversa tantum valetudine adhibentur.

2 Omne vero auxilium corporis aut demit aliquam materiam aut adicit, aut evocat aut reprimit, aut refrigerat aut calefacit, simulque aut durat aut mollit: quaedam non uno modo tantum sed etiam duobus inter se non contrariis adiuvant. Demitur materia sanguinis detractione, cucurbitula, deiectione, vomitu, frictione, gestatione omnique exercitatione corporis, abstinentia, sudore; de quibus protinus dicam. (X—XVII.)

10. Sanguinem incisa vena mitti novum non est: sed nullum paene esse morbum, in quo non mittatur, novum est. Item mitti iunioribus feminis uterum non gerentibus vetus est: in pueris vero idem experiri et in senioribus et in gravidis quoque mulieribus vetus non est: siquidem antiqui primam ultimamque aetatem sustinere non posse hoc auxilii genus iudicabant, persuaserantque sibi mulierem gravidam, quae ita curata esset, abortum esse
2 facturam. Postea vero usus ostendit nihil in his esse perpetuum, aliasque potius observationes adhibendas esse, ad quas derigi curantis consilium

---

[a] Chs. 10–33. [b] Books III and IV.
[c] Chs. 19–33. [d] Chs. 10–17.

BOOK II. 9. 1–10. 2

9. Having recognized the indications which either console us with hope, or terrify us with fear, we must pass to the methods of treating Diseases. Of these some are general aids,[a] some special. General[b] Aids are those which are beneficial in most diseases, Special Aids in particular ones. I shall speak first of the general, some of which,[c] however, sustain not alone the sick but also those in health; some are applied against illness only.[d]

Now every corporeal aid either diminishes substance or adds to it, either draws it out or represses it, either cools or warms, either hardens or softens;[e] some act, not merely in one way, but even in two ways, not contrary the one to the other. Substance is withdrawn by blood-letting, cupping, purging, vomiting, rubbing, rocking, and by bodily exercises of all kinds, by abstinence, by sweating; of these I shall now speak.[d]

10. To let blood by incising a vein is no novelty; what is novel is that there should be scarcely any malady in which blood may not be let. Again, to let blood in young women who are not pregnant is an old practice; but it is not an old practice for the same to be tried in children and in the elderly and also in pregnant women: for indeed the ancients were of opinion that the first and last years could not sustain this kind of treatment, and they were persuaded that a pregnant woman, so treated, would abort.[f] Practice subsequently showed indeed that in these matters there is no unvarying rule, and that other observations are rather to be made, to which the consideration of the practitioner ought to be directed. For

[e] Ch. 33.
Hipp. IV. 166, *Aph.* V. 31.

155

debeat. Interest enim, non quae aetas sit, neque quid in corpore intus geratur, sed quae vires sint. Ergo si iuvenis inbecillus est, aut si mulier, quae gravida non est, parum valet, male sanguis emittitur: emoritur enim vis, si qua supererat, hoc 3 modo erepta. At firmus puer et robustus senex et gravida mulier valens tuto curatur. Maxime tamen in his medicus inperitus falli potest, quia fere minus roboris illis aetatibus subest; mulierique praegnati, post curationem quoque, viribus opus est, non tantum ad se, sed etiam ad partum sus- 4 tinendum. Non quicquid autem intentionem animi et prudentiam exigit protinus faciendum[1] est, cum praecipua in hoc ars sit, quae non annos numeret, neque conceptionem solam videat, sed vires aestimet, et eo colligat, possit necne superesse, quod vel puerum vel senem vel in una muliere duo corpora 5 simul sustineat. Interest etiam inter valens corpus et obesum, inter tenue et infirmum: tenuioribus magis sanguis, plenioribus magis caro abundat. Facilius itaque illi detractionem eiusmodi sustinent: celeriusque ea, si nimium est pinguis, aliquis adfligitur; ideoque vis corporis melius ex venis quam ex ipsa specie aestimatur. Neque solum haec consideranda sunt, sed etiam morbi genus quod sit, utrum superans an deficiens materia laeserit, 6 corruptum corpus sit an integrum. Nam si materia

[1] *So the best MS., but Daremberg with some MSS. authority reads* eiiciendum, *' not that we should reject anything '—which makes the argument easier to follow.*

---

[a] Throughout this passage Celsus distinguishes the *materies corporis* from the *materies morbi*.

BOOK II. 10. 2–6

it matters not what is the age, nor whether there is pregnancy, but what may be the patient's strength. So, then, if a youth is weakly, or a woman, although not pregnant, has little strength, it is bad to let blood; for any remaining strength dies out if it is thus stripped away. But a strong child, or a robust old man, or a pregnant woman in good health, may be so treated with safety. It is mostly, however, in such cases that an inexperienced practitioner can be deceived, because at the above ages there is usually a less degree of strength; and a pregnant woman has need also, after the bloodletting, of forces to sustain, not merely herself, but also her unborn child. Not that we should be in a hurry to do anything that demands anxious attention and care; for in that very point lies the art of medicine, which does not count years, or regard only the pregnancy, but calculates the strength of the patient, and infers from that whether possibly or no there is a superfluity, enough to sustain either a child or an old man or simultaneously two beings within one woman. There is a difference between a strong and an obese body, between a thin and an infirm one: thinner bodies have more blood, those of fuller habit more flesh. The more easily, therefore, do the former sustain this sort of depletion; and the more quickly is he who is over-fat distressed by it; hence it is that the body's strength may be estimated better by its blood-vessels than by its actual appearance. And the foregoing are not the sole considerations, but there is also the kind of disease, whether a superabundance or a deficiency of bodily material[a] has done the harm, whether the body is corrupted or sound. For if the material

vel deest vel integra est, istud alienum est: at si vel copia sui male habet, vel corrupta est, nullo modo melius succurritur. Ergo vehemens febris, ubi rubet corpus, venaeque plenae tument, sanguinis detractionem requirit; item viscerum morbi, nervorum et resolutio et rigor et distentio, quicquid denique fauces difficultate spiritus strangulat, quicquid supprimit subito vocem, quisquis intolerabilis dolor est, et quacumque de causa ruptum aliquid
7 intus atque collisum est; item malus corporis habitus omnesque acuti morbi, qui modo, ut supra dixi, non infirmitate sed onere nocent. Fieri tamen potest, ut morbus quidem id desideret, corpus autem vix pati posse videatur: sed si nullum tamen appareat aliud auxilium, periturusque sit qui laborat, nisi temeraria quoque via fuerit adiutus, in hoc statu boni medici est ostendere, quam nulla spes sit sine sanguinis detractione, faterique, quantus in hac ipsa metus sit, et tum demum, si exigetur,
8 sanguinem mittere. De quo dubitari in eiusmodi re non oportet: satius est enim anceps auxilium experiri quam nullum; idque maxime fieri debet, ubi nervi resoluti sunt; ubi subito aliquis ommutuit; ubi angina strangulatur; ubi prioris febris accessio paene confecit, paremque subsequi verisimile est neque eam videntur sustinere aegri vires posse.
9 Cum sit autem minime crudo sanguis mittendus,

---

[a] Para. 5.   [b] Cf. IV. 7, p. 380.

of the body is either deficient, or is sound, blood-letting is unsuitable; but if the harm is its copiousness, or the material has become corrupted, there is no better remedy. Therefore severe fever, when the bodily surface is reddened, and the blood-vessels full and swollen, requires withdrawal of blood; so too diseases of the viscera, also paralysis and rigor and spasm of sinews, in fact whatever strangulates the throat by causing difficulty of breathing, whatever suppresses the voice suddenly, whenever there is intolerable pain, and whenever there is from any cause rupture and contusion of internal organs; so also a bad habit of body and all acute diseases, provided, as I have stated above,[a] they are doing harm, not by weakness, but by overloading. But it may happen that some disease demands blood-letting, although the body seems scarcely able to bear it; if, however, there appears to be no other remedy, and if the patient is likely to die unless he be helped even at some risk—that being the position, it is the part of a good practitioner to show that without the withdrawal of blood there is no hope, and to confess how much fear there may be in that step, and then at length, if the attempt is demanded, to let blood. In such a case there should be no hesitation about it; for it is better to try a double-edged remedy than none at all; and in particular it should be done: when there are paralyses; when a man becomes speechless suddenly; when angïna[b] causes choking; when the preceding paroxysm of a fever has been almost fatal, and it is very probable that a like paroxysm is about to set in which it seems impossible for the patient's strength to sustain. Further although it is least proper to let blood whilst food is

tamen ne id quidem perpetuum est: neque enim semper concoctionem res expectat. Ergo si ex superiore parte aliquis decidit, si contusus est, si ex aliquo subito casu sanguinem vomit, quamvis paulo ante sumpsit cibum, tamen protinus ei demenda materia est, ne, si subsederit, corpus adfligat; idemque etiam in aliis casibus repentinis, qui strangu-
10 labunt, dictum erit. At si morbi ratio patietur, tum demum nulla cruditatis suspicione remanente id fiet; ideoque ei rei videtur aptissimus adversae valetudinis dies secundus aut tertius. Sed ut aliquando etiam primo die sanguinem mittere necesse est, sic numquam utile post diem quartum est, cum iam spatio ipsa[1] materia et exhausta est et corpus conrupit, ut detractio inbecillum id facere
11 possit, non possit integrum. Quod si vehemens febris urget, in ipso impetu eius sanguinem mittere hominem iugulare est; expectanda ergo remissio est: si non decrescit, sed crescere desiit, neque speratur remissio, tum quoque, quamvis peior, sola
12 tamen occasio non omittenda est. Fere etiam ista medicina, ubi necessaria est, in biduum dividenda est: satius est enim in primo levare aegrum, deinde perpurgare, quam simul omni vi effusa fortasse praecipitare. Quod si in pure quoque aquaque, quae inter cutem est, ita respondet, quanto magis necesse est in sanguine respondeat. Mitti vero is debet, si totius corporis causa fit, ex brachio; si partis alicuius,

---

[1] *Daremberg* ipso *with one MS.*

[a] See p. 144, note [a].

undigested, yet that is not an invariable precept; for the case will not always wait for digestion. Thus if a man falls from a height, if there is contusion, or something else happening suddenly has caused vomiting of blood, although food may have been taken but a short while before, yet at once the bodily material should be depleted, lest, if it forms a congestion, it should harm the body; and the same rule will hold good also in other sudden accidents which cause suffocation. But if the character of the affection permits, it should be done then only when there remains no suspicion of undigested food; and therefore the second or third day of the illness may seem the most fitting for the procedure. But whilst there is sometimes a necessity for blood-letting even on the first day, it is never of service after the fourth day, for within that interval the material itself has both been sucked up and corrupted the body, so that then depletion can make it weak but cannot make it sound. But if there is the oppression of a vehement fever, to let blood during the actual paroxysm is to cut the man's throat; the remission is therefore to be awaited: if the fever does not decrease, but merely stops increasing, and there is no hope of remission, then also the opportunity, bad as it is, as it is the only one, should not be missed. When the measure is necessary it should generally be divided between two days; on the first it is better to relieve, and later to deplete the patient, rather than perchance to precipitate his end by dissipating his strength all at once. But if this answers in the case of pus, or of the water in dropsy,[a] all the more necessarily should it answer in the letting out of blood. If the cause affects the body as a whole, blood should be let from the arm;

CELSUS

ex ea ipsa parte aut certe quam proxima, quia non ubique mitti potest, sed in temporibus, in brachiis,
13 iuxta talos. Neque ignoro quosdam dicere quam longissime sanguinem inde, ubi laedit, esse mittendum: sic enim averti materiae cursum: at illo modo in id ipsum, quod gravat, evocari. Sed id falsum est: proximum enim locum primum exhaurit, ex ulterioribus autem eatenus sanguis sequitur, quatenus emittitur; ubi is suppressus est, quia non trahitur,
14 ne venit quidem. Videtur tamen usus ipse docuisse, si caput fractum est, ex brachio potius sanguinem esse mittendum; si quod in umero vitium est, ex altero brachio: credo quia, si quid parum cesserit, opportuniores hic eae partes iniuriae sunt, quae iam male habent. Avertitur quoque interdum sanguis, ubi alia parte prorumpens alia emittitur. Desinit enim fluere qua nolumus, inde obiectis quae prohibeant,
15 alia dato itinere. Mittere autem sanguinem cum sit expeditissimum usum habenti, tum ignaro difficillimum est: iuncta enim est venae arteria, his nervi. Ita, si nervum scalpellus attingit, sequitur nervorum distentio, eaque hominem crudeliter consumit. At arteria incisa neque coit neque sanescit; interdum etiam, ut sanguis vehementer erumpat, efficit.
16 Ipsius quoque venae, si forte praecisa est, capita

---

[a] In the controversy as to whether blood should be let as near or as far as possible from the seat of the disease Celsus on the whole takes the former view but his attitude is not prejudiced.

[b] e.g. in the arm, the basilic vein, the brachial artery, also the median nerve, and the tendons of the biceps and brachialis anticus muscles.

BOOK II. 10. 12–16

if some part, then actually from that part, or at any rate from a spot as near as may be, for it is not possible to let blood from everywhere, but only from the temples, arms and near the ankles. Nor am I ignorant that some say blood should be let from a place the furthest[a] away from the damaged part, for that thus the course of the material of the disease is diverted, but that otherwise it is drawn into the very part which is damaged. Yet this is erroneous, for blood-letting draws blood out of the nearest place first, and thereupon blood from more distant parts follows so long as the letting out of blood is continued; when put a stop to, no more blood comes to the part diseased, because it is no longer drawn to the opened vein. Practice itself, however, seems to have taught that for a broken head blood should be let preferably from the arm; when the pain is situated in one upper limb, then from the arm opposite; I believe because, if anything goes wrong, those parts are more liable to take harm which are already in a bad state. Blood is also at times diverted when, having burst out at one place, it is let out at another. For bleeding from a place where it is not desired ceases after something is applied to stop it there, when the blood is given another exit. Now blood-letting, whilst it may be very speedily done by one practised in it, yet for one without experience is very difficult, for to the vein is joined an artery, and to both sinews.[b] Hence should the scalpel strike a sinew, spasm follows, and this makes a cruel end to the patient. Again, when an artery is cut into, it neither coalesces nor heals; it even sometimes happens that a violent outburst of blood results. As to the actual vein, when completely divided by a forceful cut, its two ends

comprimuntur, neque sanguinem emittunt. At si timide scalpellus demittitur, summam cutem lacerat neque venam incidit: nonnumquam etiam ea latet neque facile reperitur. Ita multae res id difficile inscio faciunt, quod perito facillimum est. Incidenda ad medium vena est. Ex qua cum sanguis erumpit, 17 colorem eius habitumque oportet attendere. Nam si is crassus et niger est, vitiosus est, ideoque utiliter effunditur: si rubet et perlucet, integer est; eaque missio sanguinis adeo non prodest, ut etiam noceat; protinusque is supprimendus est. Sed id evenire non potest sub eo medico, qui scit, ex quali corpore 18 sanguis mittendus sit. Illud magis fieri solet, ut aeque niger adsidue ac primo die profluat; quod quamvis ita est, tamen si iam satis fluxit, supprimendus est, semperque ante finis faciendus est, quam anima deficiat, deligandumque brachium superinposito expresso ex aqua frigida penicillo, et postero die averso medio digito vena ferienda, ut recens coitus eius resolvatur iterumque sanguinem fundat. 19 Sive autem primo sive secundo die sanguis, qui crassus et niger initio fluxerat, et rubere et perlucere coepit, satis materiae detractum est, atque quod superest sincerum est; ideoque protinus brachium deligandum habendumque ita est, donec valens cicatricula sit; quae celerrime in vena confirmatur.

11. Cucurbitularum duo vero genera sunt, aeneum et corneum. Aenea altera parte patet, altera clausa est: altera cornea parte aeque patens altera foramen

BOOK II. 10. 16–11. 1

are pressed together, and do not let out the blood. Yet if the scalpel is entered timidly, it lacerates the skin but does not enter the vein; at times, indeed, the vein is concealed and not readily found. Thus many things make difficult to one who is unskilled what to one experienced is very easy. The vein ought to be cut half through. As the blood streams out its colour and character should be noted. For when the blood is thick and black, it is vitiated, and therefore shed with advantage, if red and translucent it is sound, and that blood-letting, so far from being beneficial, is even harmful; and the blood should be stopped at once. But this cannot happen under that practitioner who knows from what sort of body blood should be let. It more often happens that the flow of blood continues as black as on the first day; although this be so, nevertheless, if enough has flowed out, blood-letting should be stopped, and always an end should be put to it before the patient faints, and the arm should be bandaged after superimposing a pad squeezed out of cold water, and the next day the vein is to be flicked open by the tip of the middle finger so that, its recent coalescence being undone, it may again let out blood. Whether it be on the first or on the second day that the blood, which has at first flowed out thick and black, begins to become red and translucent, a sufficient quantity has been withdrawn, and the rest of the blood is pure; and so at once the arm should be bandaged and kept so until the little scar is strong, and this, in a vein, becomes firm very quickly.

11. Now there are two kinds of cups, one made of bronze, the other of horn. The bronze cup is open at one end, closed at the other; the horn one, likewise

CELSUS

habet exiguum. In aeneam linamentum ardens coicitur, ac sic os eius corpori aptatur inprimiturque, 2 donec inhaereat. Cornea corpori per se inponitur, deinde, ubi ea parte, qua exiguum foramen est, ore spiritus adductus est, superque cera cavum id clausum est, aeque inhaerescit. Utraque non ex his tantum materiae generibus, sed etiam ex quolibet alio recte fit: ac si cetera defecerunt, caliculus quoque aut pultarius 3 oris compressioris ei rei commode aptatur. Ubi inhaesit, si concisa ante scalpello cutis est, sanguinem extrahit, si integra est, spiritum. Ergo ubi materia, quae intus est, laedit, illo modo, ubi inflatio, hoc inponi solet. Usus autem cucurbitulae praecipuus est, ubi non in toto corpore sed in parte aliqua vitium est, quam exhauriri ad confirmandam valetudinem 4 satis est. Idque ipsum testimonium est etiam scalpello sanguinem, ubi membro succuritur, ab ea potissimum parte, quae iam laesa est, esse mittendum, quod nemo cucurbitulam diversae parti inponit, nisi cum profusionem sanguinis eo avertit, sed ei ipsi, quae dolet quaeque liberanda est. Opus etiam esse cucurbitula potest in morbis longis, quamvis iam eis spatium aliquod accessit, sive corrupta materia 5 sive spiritu male habente: in acutis quoque quibusdam, si et levari corpus debet et ex vena sanguinem mitti vires non patiuntur; idque auxilium ut minus vehemens, ita magis tutum neque umquam pericu-

---

[a] The dry cup produced a subcutaneous oedema, termed emphysema because supposed to consist partly of flatus, φύσα, derived from the pneuma. See note p. 9.

BOOK II. 11. 1-5

at one end open, has at the other a small hole. Into the bronze cup is put burning lint, and in this state its mouth is applied and pressed to the body until it adheres. The horn cup is applied as it is to the body, and when the air is withdrawn by the mouth through the small hole at the end, and after the hole has been closed by applying wax over it, the horn cup likewise adheres. Either form of cup may be made, not only of the above materials, but also of anything else suitable; when others are lacking, a small drinking-cup or porridge bowl with a narrowish mouth may be adapted conveniently for the purpose. If the skin upon which the cup is to be stuck is cut beforehand with a scalpel, the cup extracts blood; when the skin is intact, wind.[a] Therefore when it is some matter inside which is doing the harm, the former method of cupping should be employed, when it is flatulency, then the latter. Now the use of a cup is the rule for a disease, not of the body as a whole, but of some part, the sucking out of which suffices for the re-establishment of health. And this same fact is a proof that with a scalpel, when a part is being relieved, blood must be let from that very part where the injury already exists; since unless it be to divert haemorrhage in that direction, nobody applies a cup to a part distant from the disease, but to that which is actually affected and has to be relieved. Further there may be need for cupping in chronic maladies, although already of somewhat long duration, if there is corrupted material or an unhealthy condition of wind; in certain acute cases also, if the body ought to be depleted and at the same time the patient's strength does not admit of cutting a vein; and cupping, as it is a less severe remedy, so it is a safer

CELSUS

losum est, etiamsi in medio febris impetu, etiamsi in
6 cruditate adhibetur. Ideoque ubi sanguinem mitti
opus est, si incisa vena praeceps periculum est, aut
si in parte corporis etiamnum vitium est, huc potius
confugiendum est, cum eo tamen, ut sciamus hic
ut nullum periculum, ita levius praesidium esse, nec
posse vehementi malo nisi aeque vehemens auxilium
succurrere.

12. 1. Deiectionem autem antiqui variis medicamentis crebraque alvi ductione in omnibus paene morbis moliebantur; dabantque aut nigrum veratrum aut filiculam aut squamam aeris, quam λεπίδα χαλκοῦ Graeci vocant, aut lactucae marinae lac, cuius gutta pani adiecta abunde purgat, aut lac vel asininum aut bubulum, vel caprinum, eique salis paulum adiciebant, decoquebantque id et sublatis is, quae coierant, quod quasi serum supererat, bibere
B cogebant. Sed medicamenta stomachum fere laedunt: alvus si vehementius fluit aut saepius ducitur, hominem infirmat. Ergo numquam in adversa valetudine medicamentum eius rei causa recte datur, nisi ubi is morbus sine febre est, ut cum veratrum nigrum aut atra bile vexatis aut cum tristitia insanientibus aut iis, quorum nervi parte aliqua resoluti
C sunt, datur. At ubi febres sunt, satius est eius rei causa cibos potionesque adsumere, qui simul et alant et ventrem molliant; suntque valetudinis genera, quibus ex lacte purgatio convenit.

2. Plerumque vero alvus potius ducenda est; quod ab Asclepiade quoque sic temperatum, ut tamen

one; nor is it ever dangerous, even if adopted in the midst of the attack of a fever, or even with food undigested. Therefore, when blood-letting is needed, if cutting a vein is an instant danger, or if the mischief is still localised, recourse is to be had rather to cupping, not forgetting that whilst we recognize an absence of danger, its efficacy is thus the less, and it is impossible to remedy a severe malady unless by a remedy likewise severe.

12. Now purging was promoted by the ancients in almost all diseases by various medicaments, and by frequent clysters; they administered either black hellebore root, or polypody fern root, or the copper scales which the Greeks call lepida chalkou, or the milky juice of seaside spurge, of which one drop on bread purges freely, or milk, whether from an ass or cow, or goat, to which a little salt was added, which they boiled, and having removed the solidified skin, they obliged their patients to drink the whey-like remainder. But medicaments generally irritate the stomach; a motion when excessively liquid, or a clyster often repeated, weakens the patient. Never, therefore, in illness is a medicament which causes such a motion rightly given, unless when that malady is without fever, as when black hellebore root is given either to those with black bile and to those suffering from insanity with melancholy, or to those who have their sinews in some part paralysed. But in the presence of fevers, it suffices for the purpose of a purge to take such food and drink as both nourish and at the same time soften the belly; and there are sorts of illness in which purgation by milk is suitable.

Still, for the most part the bowel preferably is to be clystered; the practice was limited by Asclepiades

servatum sit, video plerumque saeculo nostro praeteriri. Est autem ea moderatio, quam is secutus videtur, aptissima, ut neque saepe ea medicina temptetur, et tamen semel, summum vel bis non omittatur:
B si caput grave est; si oculi caligant; si morbus maioris intestini est, quod Graeci colum nominant; si in imo ventre aut si in coxa dolores sunt; si in stomachum quaedam biliosa concurrunt, vel etiam pituita eo se umorve aliquis aquae similis confert; si spiritus difficilius redditur; si nihil per se venter excernit, utique si iuxta quoque stercus est et intus remanet, aut si stercoris odorem nihil deiciens aeger ex spiritu suo sentit, aut si corruptum est quod excernitur; aut si prima inedia febrem non sustulit; aut si sanguinem mitti, cum opus sit, vires non patiuntur tempusve eius rei praeterit, aut si multum ante
C morbum aliquis potavit; aut si is, qui saepe vel sponte vel casu purgatus est, subito habet alvum suppressam. Servanda vero illa sunt, ne ante tertium diem ducatur; ne ulla cruditate substante; ne in corpore infirmo diuque in adversa valetudine exhausto, neve in eo, cui satis alvus cottidie reddit quive eam liquidam habebit; ne in ipso accessionis impetu, quia quod tum infusum est, alvo continetur, regestumque in caput multo gravius periculum
D efficit. Pridie vero abstineri debet aeger, ut aptus tali curationi sit, eodem die ante aliquot horas aquam

though still kept, but I see that in our time it is usually neglected. But the limitation which he seems to have adopted is most fitting: that this remedy should not be tried often, and yet we should not omit to use it once, or at most twice: if the head is heavy; if the eyes are dim; if the disease is in the larger intestine, which the Greeks call colon; if there are pains in the lower belly or in the hips; if bilious fluid collects in the stomach, or even phlegm or other water-like humor forms there; if wind is passed with undue difficulty; if there is no spontaneous motion, and especially if the faeces remain inside although close to the anus, or if the patient who fails to pass anything perceives a foul odour in his breath, or if the motions have become corrupted; or if abstinence does not at once get rid of the fever; or if the patient's strength does not allow of bloodletting when it is needed, or the time for that measure has passed; or if previous to the malady the patient has been drinking freely; or if a patient who has been purged repeatedly, whether that has been intentional or casual, has suddenly a suppression of motions. However, the following rules are to be observed: that the clyster is not to be administered before the third day, nor whilst there is any undigested food; nor in a case of weakness due to exhaustion by a long illness; nor to a patient who has daily a sufficient motion, nor to one whose stools are liquid; nor during the acme of the paroxysm of a fever, for what is then injected is retained in the bowel and mounting up into the head brings about a much graver danger. On the day too before the clyster the patient ought to fast, in order to fit himself for such a treatment, and on the actual day, some hours

calidam bibere, ut superiores eius partes madescant; tum inmittenda in alvum est, si levi medicina contenti sumus, pura aqua, si paulo valentiore, mulsa; si leni, ea, in qua faenum Graecum vel tisana vel
E malva decocta sit, [si reprimendi causa, ex verbenis][1] acris autem est marina aqua vel alia sale adiecto; atque utraque decocta commodior est. Acrior fit adiecto vel oleo vel nitro vel melle: quoque acrior est, eo plus extrahit, sed minus facile sustinetur. Idque, quod infunditur, neque frigidum esse oportet neque calidum, ne alterutro modo laedat. Cum infusum est, quantum fieri potest, continere se in lecticulo debet aeger, nec primae cupiditati deiectionis protinus cedere: ubi necesse est, tum demum
F desidere. Fereque eo modo dempta materia, superioribus partibus levatis, morbum ipsum mollit. Cum vero, quotiens res coegit, desidendo aliquis se exhausit, paulisper debet conquiescere; et ne vires deficiant, utique eo die cibum adsumere; qui plenior an exiguus sit, ex ratione eius accessionis, quae expectabitur aut in metu non erit, aestimari oportebit.

13. At vomitus ut in secunda quoque valetudine saepe necessarius biliosis est, sic etiam in iis morbis, quos bilis concitavit. Ergo omnibus, qui ante febres horrore et tremore vexantur, omnibus, qui cholera laborant, omnibus etiam cum quadam hilaritate insanientibus, et comitiali quoque morbo oppressis necessarius est. Sed si acutus morbus est, sicut in

[1] *Targa deleted these words as a gloss.*

---

[a] Cf. II. **1**, 21, and also IV. **18**, cholera nostras. Celsus did not refer to Asiatic cholera.

[b] ἱλαρότης, associated with yellow bile, note p. 8, and a symptom of mania, III. **18**, 3 *et seq*.

beforehand, he should drink warm water to moisten his upper parts; there should then be introduced into the bowel simply water when we are content with a gentle remedy, or hydromel as one a little stronger; or as a soothing enema a decoction of fenugreek, or of pearl barley, or of mallow, [or as an astringent clyster a decoction of vervains], but a drastic one is sea-water or ordinary water with salt added; and the better in both instances for boiling. A clyster is made more drastic by the addition of olive oil, or soda, or honey: the more drastic the clyster, the more it extracts, but the less easily is it borne. The fluid injected should be neither cold nor hot, lest either way it should do harm. Following upon the injection the patient ought to keep in bed as long as he can, and not give way to his first desire to defaecate; then go to stool only when he must. In this way generally when the material is extracted, and the upper parts relieved, the disease itself is mollified. But when the patient has become exhausted owing to forced calls to stool, he ought for a while to keep quiet; and lest his strength fail, he should certainly take food that day, but whether it should be abundant or scanty, should be regulated according to the strength of the paroxysm anticipated, or the absence of such apprehension.

13. Again, a vomit, as it is often quite a necessity for one who in health is bilious, so is it also in those diseases which bile has occasioned. It is the more necessary, therefore, for all who are troubled by shivering and trembling before fevers, for all suffering from cholera,[a] even for all suffering from insanity accompanied by a kind of hilarity,[b] and also for those afflicted by epilepsy. But if the disease is an acute one,

2 cholera, si febris est, ut inter horrores, asperioribus medicamentis opus non est, sicut in deiectionibus quoque supra (I. 2, 1 C) dictum est; satisque est ea vomitus causa sumi, quae sanis quoque sumenda esse proposui (I. 3, 22). At ubi longi valentesque morbi sine febre sunt, ut comitialis, ut insania, veratro
3 quoque albo utendum est. Id neque hieme neque aestate recte datur, optime vere, tolerabiliter autumno. Quisquis daturus erit, id agere ante debebit, ut accepturi corpus umidius sit. Illud scire oportet, omne eiusmodi medicamentum, quod potui datur, non semper aegris prodesse, semper sanis nocere.

14. De frictione vero adeo multa Asclepiades tamquam inventor eius posuit in eo volumine, quod communium auxiliorum inscripsit, ut,[1] cum trium faceret tantum mentionem, huius et aquae et gestationis, tamen maximam partem in hac consumpserit. Oportet autem neque recentiores viros in iis fraudare, quae vel reppererunt vel recte secuti sunt, et tamen ea, quae apud antiquiores aliquos posita
2 sunt, auctoribus suis reddere. Neque dubitari potest, quin latius quidem et dilucidius, ubi et quomodo frictione utendum esset, Asclepiades praeceperit, nihil tamen reppererit, quod non a vetustissimo auctore Hippocrate paucis verbis comprehensum sit, qui dixit frictione, si vehemens sit, durari corpus, si lenis, molliri:

[1] ut *added by v. d. Linden.*

---

[a] I. 3, 22.   [b] Hipp. IV. 136 (*Aph.* IV. 13).

BOOK II. 13. 1–14. 2

as in the case of cholera, if there is fever, during the shivering fits, then the sharper medicaments are out of place, as mentioned also above in relation to purgations, and for the purpose of a vomit it is sufficient to take the emetics which I have prescribed to be taken by those in health.[a] But when there are chronic and violent diseases without fever, such as epilepsy and insanity, white hellebore root[b] should also be used. But it is not right to give it either in winter or in summer; the spring is the best time, and autumn tolerably good. Whoever is going to administer it ought to take care beforehand that the body of the prospective recipient is rendered more humid. This should be recognized, that all such medicaments given as a drink do not always benefit the sick, and are always harmful to those in health.

14. Now concerning rubbing Asclepiades as if he were the inventor of the practice has treated it in his volume, entitled "Common Aids," at such great length, that, though making mention only of three such aids, namely, Rubbing, Water-drinking, and Rocking, yet he has taken up the greatest part with the first-named subject. Now on such matters recent writers ought to have credit where they have made discoveries, or where they have rightly followed others; yet we must not omit to attribute to their true authors teaching found among the more ancient writers. And it cannot be disputed that Asclepiades has taught when and how rubbing should be practised, with a wider application, and in a clearer way, although he has discovered nothing which had not been comprised in a few words by that most ancient writer Hippocrates, who said that rubbing, if strenuous, hardens the body, if gentle,

CELSUS

si multa, minui, si modica, inpleri. Sequitur ergo, ut tum utendum sit, cum aut adstringendum corpus sit, quod hebes est, aut molliendum, quod induruit, aut digerendum in eo, quod copia nocet, aut alendum 3 id, quod tenue et infirmum est. Quas tamen species si quis curiosius aestimet (quod iam ad medicum non pertinet), facile intelleget omnes ex una causa pendere, quae demit. Nam et adstringitur aliquid eo dempto, quod interpositum, ut id laxaretur, effecerat, et mollitur eo detracto, quod duritiem creabat, et inpletur non ipsa frictione sed eo cibo, qui postea usque ad cutem digestione quadam relaxa- 4 tam penetrat. Diversarum vero rerum in modo causa est.

Inter unctionem autem et frictionem multum interest. Ungui enim leviterque pertractari corpus etiam in acutis et recentibus morbis oportet, in remissione tamen et ante cibum. Longa vero frictione uti neque in acutis morbis neque increscentibus convenit, praeterquam cum phreneticis 5 somnus ea quaeritur. Amat autem hoc auxilium valetudo longa et iam a primo inpetu inclinata. Neque ignoro quosdam dicere omne auxilium necessarium esse increscentibus morbis, non cum iam per se finiuntur. Quod non ita se habet. Potest enim morbus, etiam qui per se finem habiturus est, ce- 6 lerius tamen adhibito auxilio pelli. Quod duabus de causis necessarium est, et ut quam primum bona

---

[a] Hipp. III. 76. (*Surgery*, XVII.)
[b] Massage relieves the result of a severe bruising by causing the disappearance of induration and discolouration; the skin becomes movable upon the underlying parts, and regains warmth and natural colours owing to the restored circulation of the blood.

BOOK II. 14. 2-6

relaxes; if much, it diminishes, if moderate, fills out.[a] It follows, therefore, that in the following cases rubbing should be employed, when either a feeble body has to be toned up, or one indurated has to be softened, or a harmful superfluity is to be dispersed, or a thin and infirm body has to be nourished. Yet when examined with attention (although this no longer concerns the medical man) the various species of rubbing may be easily recognized as all dependent on causing one thing, depletion. For an object is toned up when that is removed, which, by its presence was the cause of the laxness; and is softened when that which has been producing induration is abstracted; and it is filled up, not by the rubbing itself, but by the nutriment, which subsequently penetrates by some sort of dispersal to the very skin itself after it has become relaxed.[b] The cause of the different results lies in the degree.

Now there is a great difference between anointing and rubbing. For it is desirable that even in acute and recent diseases the body should be anointed and then gently stroked, but only during remissions and before food. But prolonged rubbing is unsuitable in acute and increasing diseases, unless it be in madness to procure sleep. Yet a prolonged illness and one declining from its primary vehemence loves this aid. I am quite aware that some say that the need for any aid is during the increase of diseases, not when diseases are tending to end of themselves. But this is not the case. For a disorder, even although it will end of itself, may be expelled yet more speedily by adopting the aid. An aid is necessary on two accounts, both that health may be regained at the earliest possible

valetudo contingat, et ne morbus, qui remanet, iterum, quamvis levi de causa, exasperetur. Potest morbus minus gravis esse quam fuerit, neque ideo tamen solvi, sed reliquiis quibusdam inhaerere, quas 7 admotum aliquod auxilium discutit. Sed ut levata quoque adversa valetudine recte frictio adhibetur, sic numquam adhibenda est febre increscente: verum, si fieri potest, cum ex toto corpus ea vacabit; si minus, certe cum ea remiserit. Eadem autem modo in totis corporibus esse debet, ut cum infirmus aliquis implendus est, modo in partibus, aut quia ipsius eius membri inbecillitas id requirit aut quia 8 alterius. Nam et capitis longos dolores ipsius frictio levat, non in inpetu tamen doloris, et membrum aliquod resolutum ipsius frictione confirmatur. Longe tamen saepius aliud perfricandum est, cum aliud dolet; maximeque cum a summis aut a mediis partibus corporis evocare materiam volumus, ideoque 9 extremas partes perfricamus. Neque audiendi sunt qui numero finiunt, quotiens aliquis perfricandus sit: id enim ex viribus hominis colligendum est; et si is perinfirmus est, potest satis esse quinquagies, si robustior, potest ducenties esse faciendum; inter utrumque deinde, prout vires sunt. Quo fit, ut etiam minus saepe in muliere quam in viro, minus saepe in puero vel sene quam iuvene manus dimo-10 vendae sint. Denique si certa membra perfricantur,

---

^a IV. 2, 8.

moment, and that what remains of the disease may not again become exacerbated from however slight a cause. Possibly the disease may have become less grave than it had been, yet is not completely got rid of, but some remnants of it persist, which the application of a remedy disperses. But while rubbing is rightly applied after a disorder has been lessened, yet it should never be applied whilst a fever is increasing: but if possible after the fever has entirely left the body, or if not, at least when it has remitted. Sometimes, moreover, rubbing should be applied to the body all over, as when a thin man ought to put on flesh; sometimes to a part only, either because weakness of the limb actually rubbed demands it, or that of some other part. For both prolonged headaches [a] are relieved by rubbing of the head, although not at the height of the pain, and any partially paralysed limb is strengthened by being itself rubbed. Much more often, however, some other part is to be rubbed than that which is the seat of the pain; and especially when we want to withdraw material from the head or trunk, and therefore rub the arms and legs. Neither should we listen to those who would fix numerically how many times a patient is to be stroked; for that is to be regulated by his strength; and if he is very infirm fifty strokes may possibly be enough, if more robust possibly two hundred may be made; then an intermediate number according to his strength. Hence it is that the hand is to be passed even fewer times over a woman than over a man, fewer over a child or old man, than over a young adult. Finally, if particular limbs are rubbed, many strokes are re-

multa valentique frictione opus est: nam neque totum corpus infirmari cito per partem potest, et opus est quam plurimum materiae digeri, sive id ipsum membrum sive per id aliud levamus. At ubi totius corporis inbecillitas hanc curationem per totum id exigit, brevior esse debet et lenior, ut tantummodo summam cutem emolliat, quo facilius capax ex
11 recenti cibo novae materiae fiat. In malis iam aegrum esse, ubi exterior pars corporis friget, interior cum siti calet, supra (II. 6, 7) posui. Sed tunc quoque unicum in frictione praesidium est; quae si calorem in cutem evocavit, potest alicui medicinae locum facere.

15. Gestatio quoque longis et iam inclinatis morbis aptissima est; utilisque est et iis corporibus, quae iam ex toto febre carent sed adhuc exerceri per se non possunt, et iis, quibus lentae morborum reliquiae remanent neque aliter eliduntur. Asclepiades etiam in recenti vehementique praecipueque ardente febre ad discutiendam eam gestatione dixit
2 utendum. Sed id periculose fit, meliusque quiete eiusmodi impetus sustinetur. Si quis tamen experiri volet, sic experiatur: si lingua non erit aspera, si nullus tumor, nulla durities, nullus dolor visceribus aut capiti aut praecordiis suberit. Et ex toto numquam gestari corpus dolens debet, sive id in toto sive in parte est, nisi tamen solis nervis dolentibus, neque

---

[a] Such as is used for babies; Greek σεισμός, cf. Plato *Laws*, VII. 789, C–E. Similar results were attained by horse-riding (*equitatio*) which Celsus prescribes for sufferers from looseness of the bowels (IV. 26, 5) and for convalescents (IV. 32, 1).

quired and forcible rubbing; both because the body cannot be as a whole quickly rendered weak through a part, and it is necessary that as much as possible of the diseased matter should be dispersed, whether our aim is to relieve the limb actually rubbed, or through it another limb. When, however, general bodily weakness requires that the rubbing should be applied all over, it should be shorter and more gentle, just to the extent of softening the skin, so that the body may be more easily capable of forming new material from food recently consumed. As I have stated above (II. 6, 7), a patient is already in a bad way, when the exterior of the body is cold, whilst his interior is hot and there is thirst. But even then rubbing is the only remedy; if it draws the heat outwards into the skin, it makes possible an opportunity for other treatment.

15. Rocking [a] also is very suitable for chronic maladies which are already abating; it is also of service both for those who are now entirely free from fever, but cannot as yet themselves take exercise, and also for those in whom persist sluggish remnants of maladies, not otherwise to be got rid of. Asclepiades has stated that use is to be made of rocking even for dispersal of a recent and severe fever, especially an ardent fever. But that gives rise to danger; an attack of that sort is better sustained by keeping quiet. If anyone, however, wants to give it a trial, let him try it when the tongue is not furred, when there is no swelling, no induration, no pain, either in the viscera or head or about the heart. And on the whole a body in pain should never be rocked, whether the pain be general or local, except, however, when sinews alone are in

umquam increscente febre sed in remissione eius.
3 Genera autem gestationis plura sunt adhibendaque sunt et pro viribus cuiusque et pro opibus, ne aut inbecillum hominem nimis digerant, aut humili desint. Lenissima est navi vel in portu vel in flumine, vehementior vel in alto mari navi vel lectica, etiamnum acrior vehiculo: atque haec ipsa et intendi et leniri
4 possunt. Si nihil eorum est, suspendi lectus debet et moveri: si ne id quidemst, at certe uni pedi subiciendum fulmentum est, atque ita lectus huc et illuc manu inpellendus.

Et levia quidem genera exercitationis infirmis conveniunt, valentiora vero iis, qui iam pluribus diebus febre liberati sunt, aut iis, qui gravium morborum initia sic sentiunt, ut adhuc febre vacent (quod et in tabe et in stomachi vitiis, et cum aqua cutem subit, et interdum in regio fit), aut ubi quidam morbi, qualis comitialis, qualis insania est, sine febre quamvis
5 diu manent. In quibus adfectibus ea quoque genera exercitationum necessaria sunt, quae conprehendimus eo loco (I. 2, 6), quo, quemadmodum sani neque firmi homines se gererent, praecepimus.

16. Abstinentiae vero duo genera sunt, alterum ubi nihil adsumit aeger, alterum ubi non nisi quod oportet. Initia morborum primum famem sitimque desiderant, ipsi deinde morbi moderationem, ut neque aliud quam expedit neque eius ipsius nimium sumatur: neque enim convenit iuxta inediam

---

[a] Seneca, *Ep. Mor.* LV. (L.C.L. vol. I, p. 364).
[b] αἰώρα, a swing, or hammock.
[c] ὑπόβαθρον Xenophon *Memorabilia*, mem. II. 1, 30.
[d] Cf. I. 2. 6.

pain, and never during the rise of a fever, but only during its remissions. But there are many sorts of rocking, and they are to be regulated both by the patient's strength and by his resources, lest either a weak patient undergo overmuch depletion, or a poor man come short. The gentlest rocking is that on board ship either in harbour or in a river, more severe is that aboard ship on the high seas, or in a litter,[a] even severer still in a carriage: but each of these can either be intensified or mitigated. Failing any of the above, the bed should be so slung[b] as to be swayed; if not even that, at any rate a rocker[c] should be put under its foot so that the bed may be moved from side to side by hand.

And this sort of exercise of the lighter kinds suits the infirm, the stronger kinds again those who have already become free from fever for several days, or those who, whilst feeling the commencement of grave disorders, as yet are free from fever (which happens in the case both of phthisis and of stomach disease, and of dropsy, also, at times, of jaundice), or when certain diseases such as epilepsy and madness persist without fever, for however long. In which affections also those kinds of exercises are necessary, which we have included in the passage where we prescribed what healthy, yet not strong, men should carry out.[d]

16. Now Abstinence is of two kinds, in one of which the patient takes nothing at all, in the other only what he must. The beginnings of diseases require at first hunger and thirst, the actual diseases then moderation, so that nothing but what is expedient, and not too much of that, may be consumed; for it is not at all proper to have surfeit at once after a fast.

protinus satietatem esse. Quod si sanis quoque corporibus inutile est, ubi aliqua necessitas famem fecit, quanto inutilius est etiam in corpore aegro? 2 Neque ulla res magis adiuvat laborantem quam tempestiva abstinentia. Intemperantes homines apud nos ipsi cibi . . .[1] tempora curantibus dantur: rursus alii tempora medicis pro dono remittunt, sibi ipsis modum vindicant. Liberaliter agere se credunt, qui cetera illorum arbitrio relinquant, in genere cibi liberi sunt: quasi quaeratur quid medico liceat, non quid aegro salutare sit, cui vehementer nocet, quotiens in eius, quod adsumitur, vel tempore vel modo vel genere peccatur.

17. Sudor etiam duobus modis elicitur, aut sicco calore aut balneo. Siccus calor est et harenae calidae et Laconici et clibani et quarundam naturalium sudationum, ubi terra profusus calidus vapor aedificio includitur, sicut super Baias in murtetis habemus. Praeter haec sole quoque et exercitatione movetur. Utiliaque haec genera sunt, quotiens umor intus nocet, isque digerendus est. Ac nervorum quoque quae- 2 dam vitia sic optime curantur. Sed cetera infirmis possunt convenire: sol et exercitatio tantum robustioribus, qui tamen sine febre vel inter initia

---

[1] *There is a lacuna here. Marx adds*: Modum sibi constituunt nec magis cibi: '*determine the amount of food for themselves, and do not even leave the times to the doctors to settle*,' *v. d. Linden with some MSS. reads*: dant sibi cibi tempora, modum curantibus dant: '*dictate the times for their meals, and leave the quantity to their doctors.*'

---

[a] The text as it stands cannot be translated; for suggested emendations, see Critical Note.
[b] Laconicum: first used by the Lacedaemonians, see Vitruv. V. 10.

BOOK II. 16. 1-17. 2

But if this be not good even in healthy bodies, when some necessity has imposed fasting, how much worse it is in a sick body! To a sufferer nothing is more advantageous than a timely abstinence. Among us intemperate men with regard to their food themselves . . . the times are left to the doctors [a]; again others make a present of the times to their medical men, but reserve to themselves as to quantity. They think that they are generous, when they leave them to decide as to all else, and keep free as to the kind of food; as if it were a question of what may be yielded to the doctor, not what may be good for the patient, who suffers grievous harm, as often as he transgresses in what he consumes, whether as to the time of the meal, its quantity or its quality.

17. Sweating also is elicited in two ways, either by dry heat, or by the bath. The dry is the heat of hot sand, of the Laconian sweating-room,[b] and of the dry oven,[c] and of some natural sweating places, where hot vapour exhaling from the ground is confined within a building, as we have it in the myrtle groves above Baiae. Besides these it is also derived from the sun and through exercise. These treatments are useful whenever humor is doing harm inside, and has to be dispersed. And also some diseases of sinews are best treated thus. But the other treatments may suit the infirm: sun and exercise only the more robust, who must, however, be free from fever, whether only at the com-

[c] Clibanus, a small oven of burnt clay, which was heated like a bread-oven, into which the affected limb was put: see III. 21, 6, p. 314 note.

morborum vel etiam gravibus morbis tenentur. Cavendum autem est, ne quid horum vel in febre vel in cruditate temptetur. At balnei duplex usus est: nam modo discussis febribus initium cibi plenioris vinique firmioris [valetudinis] facit, modo ipsam febrem tollit; fereque adhibetur, ubi summam cutem relaxari evocarique corruptum umorem et
3 habitum corporis mutari expedit. Antiqui timidius eo utebantur, Asclepiades audacius. Neque terrere autem ea res, si tempestiva est, debet: ante tempus nocet. Quisquis febre liberatus est, simulatque ea uno die non accessit, eo, qui proximus est post tempus accessionis, tuto lavari potest. At si circumitum habere ea febris solita est, sic ut tertio quartove die revertatur, quandocumque non
4 accessit, balneum tutum est. Manentibus vero adhuc febribus, si eae sunt quae lentae lenesque iam diu male habent, recte medicina ista temptatur, cum eo tamen, ne praecordia dura sint neve ea tumeant, neve lingua aspera sit, neve aut in medio corpore aut in capite dolor ullus sit, neve tum febris increscat. Atque in iis quidem febribus, quae certum circuitum habent, duo balnei tempora sunt, alterum ante
5 horrorem, alterum febre finita: in iis vero, qui lentis febriculis diu detinentur, cum aut ex toto recessit accessio, aut si id non solet, certe lenita est, iamque

---

[a] Note, p. 52.   [b] Cf. III. 3, *et. seq.*

mencement of a disease, or when actually in the grasp of a grave malady. But care must be taken that none of the above are tried either during fever or with food undigested. Now the bath[a] is of double service: for at one time after fevers have been dissipated, it forms for a convalescent the preliminary to a fuller diet and stronger wine; at another time it actually takes off the fever; and it is generally adopted, when it is expedient to relax the skin and draw out corrupt humor and change the bodily habit. The ancients used it rather timidly, Asclepiades more boldly. There is indeed nothing to be apprehended from its use, if it be timely; before the proper time it does harm. A patient who has become free from fever can be safely bathed,[b] as soon as there has been no paroxysm for one whole day, on the next day after the time for a paroxysm. But where the fever has a regular periodicity so that it recurs every third or fourth day, when there has not been a recurrence, the bath is safe. Even whilst fevers are persisting, if they are slow and mild, and have lasted a long while, this treatment may properly be tried, so long as the parts below the ribs are neither indurated nor swollen, the tongue not furred, there is no pain in the trunk or head, and the fever is not then on the increase. And in those fevers which have a definite periodicity, there are two opportunities for the bath, one before the shivering, the other after the fever has ended: but in the case of those who have been the subjects for a long while of slow and slight fever, the time for the bath is when the paroxysm has wholly remitted, or if that does not occur, at any rate when it has mitigated, and the

corpus tam integrum est, quam maxime esse in eo genere valetudinis solet. Inbecillus homo iturus in balneum vitare debet, ne ante frigus aliquod 6 experiatur. Ubi in balneum venit, paulisper resistere experirique, num tempora adstringantur, et an sudor aliqui oriatur: si illud incidit, hoc non secutum est, inutile est eo die balineum perunguendusque is leniter et auferendus est, vitandumque omni 7 modo frigus, et abstinentia utendum. At si temporibus integris primum ibi, deinde alibi sudor incipit, fovendum os aqua calida; tum in solio descendendum est, atque ibi quoque videndum est, num sub primo contactu aquae calidae summa cutis inhorrescat, quod vix fieri potest, si priora recte cesserunt: certum autem signum inutilis balinei est. Ante vero an postea quam in aquam calidam se demittat aliquis perungui debeat, ex ratione 8 valetudinis suae cognoscat. Fere tamen, nisi ubi nominatim ut postea fiat praecipietur, moto sudore leviter corpus perunguendum, deinde in aquam calidam demittendum est. Atque hic quoque habenda virium ratio est; neque committendum, ut per aestum anima deficiat; sed maturius is auferendus curioseque vestimentis involvendus est, ut neque ad eum frigus adspiret, et ibi quoque, 9 antequam aliquid adsumat, insudet. Fomenta quoque calida sunt milium, sal, harena, quidlibet eorum calfactum et in linteum coniectum: si minore vi opus est, etiam solum linteum, at si maiore, exstincti

---

[a] See note, p. 52.

patient's body has already become as sound as it possibly can in this sort of complaint. A weakly patient who is about to go to the bath should avoid exposing himself to cold beforehand. On arriving [a] at the bath, he should sit for a while to try whether his temples become tightened, and whether any sweat arises: if the former happens without the latter, for that day the bath is unsuitable, and he should be anointed lightly and carried home, and cold is to be avoided in every possible way and abstinence practised. But if the temples are unaffected, and sweating starts there first, and then elsewhere, his face is to be fomented with hot water; then he should go down into the hot bath, where it is to be noted whether there is shrivelling of the skin at the first touch of hot water, which can hardly happen when the indications noted above have been attended to properly: it is, however, a sure sign of the bath being injurious. Whether he should be anointed before entering, or after the hot bath, should be decided from the degree of his convalescence. Generally, however, unless it has been definitely prescribed that it is to be done afterwards, when the sweating begins the body should be slightly anointed, and then he is to get into the hot water. And whilst in it also regard should be had to his strength; he ought not to be kept in the bath until he faints from the heat, but be taken out earlier and carefully wrapped up so that no cold reaches him, and so that he may sweat there also before taking anything. There are hot foments: millet, salt, or sand, any of which is heated, and put into a linen cloth: when less heat is required the linen cloth may be used alone, but

titiones involutique panniculis et sic circumdati. Quin etiam calido oleo replentur utriculi, et in vasa fictilia a similitudine[1] quas lenticulas vocant, aqua coicitur; et sal sacco linteo excipitur, demittiturque in aquam bene calidam, tum super id membrum, quod fovendum est, conlocatur . . .[2] iuxtaque ignem ferramenta duo sunt, capitibus paulo latioribus, alterumque ex iis demittitur in eum salem, et super aqua leviter aspergitur: ubi frigere coepit, ad ignem refertur, et idem in altero fit, deinde invicem in utroque: inter quae descendit salsus et calidus sucus, qui contractis aliquo morbo nervis opitulatur. His omnibus commune est digerere id, quod vel praecordia onerat, vel fauces strangulat, vel in aliquo membro nocet. Quando autem quoque utendum sit, in ipsis morborum generibus dicetur (III, IV).

18. Cum de iis dictum sit, quae detrahendo iuvant, ad ea veniendum est, quae alunt, id est, cibum et potionem. Haec autem non omnium tantum morborum sed etiam secundae valetudinis communia praesidia sunt; pertinetque ad rem omnium proprietates nosse, primum ut sani sciant, quomodo his utantur, deinde ut exsequentibus nobis morborum curationes (III, IV) liceat species rerum, quae adsumendae erunt, subicere, neque necesse sit subinde singulas eas nominare.

2 Scire igitur oportet omnia legumina, quaeque ex frumentis panificia sunt, generis valentissimi esse

[1] *Constantine's suggestion for the MSS.* similitudine, *which Marx brackets as a gloss.*

[2] *Marx suggests:* itemque in igne pelvis sale repleta conlocatur ' *again a basin full of salt is laid on the fire.*'

---

[a] Some words seem to have fallen out here: see critical note.

[b] For a list of the foodstuffs given and the probable identification of those which are doubtful, see list of Alimenta, pp. 483 ff.

if greater heat, firebrands are extinguished, wrapped up in rags, and so put round him. Further, leathern bottles are be filled with hot oil, or hot water is poured into earthenware vessels, called from their shape "lentils"; and salt is put into a linen bag, and dipped into very hot water, then laid upon the limb to be fomented.[a] . . . and two broad-ended cautery irons are heated near the fire, and one of them is dipped into that salt, and water is lightly sprinkled upon the iron held over the part. When the iron begins to cool, it is put back into the fire, and the second iron made use of in the same way as the first, turn and turn about: during the procedure a hot brine drips down, which is beneficial for sinews contracted by disease of any kind. The common effect of all these measures is to disperse whatever is oppressing the parts over the heart, or strangling the throat, or harming some limb. It will be stated under particular maladies when use is to be made of each (III, IV).

18. After having spoken of those things which benefit by depleting, we come to those which nourish, namely food and drink. Now these are of general assistance not only in diseases of all kinds but in preserving health as well; and an acquaintance with the properties of all is of importance, in the first place that those in health may know how to make use of them, then, as we follow on to the treatment of diseases (III, IV), we can state the species of aliments to be consumed, without the necessity every time of naming them singly.[b]

So then it should be known that all pulses, and all bread-stuffs made from grain, form the strongest

(valentissimum voco, in quo plurimum alimenti est);
item omne animal quadrupes domi natum; omnem
grandem feram, quales sunt caprea, cervus, aper,
onager; omnem grandem avem, quales sunt anser
et pavo et grus; omnes beluas marinas, ex quibus
cetus est quaeque his pares sunt; item mel et
caseum. Quo minus mirum est opus pistorium
valentissimum esse, quod ex frumento, adipe, melle,
caseo constat.

3 In media vero materia numerari ex holeribus
debere ea, quorum radices vel bulbos adsumimus;
ex quadrupedibus leporem; aves omnes a minimis
ad phoenicopterum; item pisces omnes, qui salem
non patiuntur solidive saliuntur.

Inbecillissimam vero materiam esse omnem caulem
holeris et quicquid in caule nascitur, qualis est
cucurbita et cucumis et capparis, omnia poma, oleas,
cochleas, itemque conculia.

4 Sed quamvis haec ita discreta sunt, tamen etiam
quae sub eadem specie sunt, magna discrimina
recipiunt, aliaque res alia vel valentior est vel infirmior: siquidem plus alimenti est in pane quam
in ullo alio, firmius est triticum quam milium, id
ipsum quam hordeum; et ex tritico firmissima siligo,
deinde simila, deinde cui nihil demptum est, quem
αὐτόπυρον Graeci vocant: infirmior est ex polline,
infirmissimus cibarius panis.

5 Ex leguminibus vero valentior faba vel lenticula
quam pisum. Ex holeribus valentior rapa napique
et omnes bulbi, in quibus cepam quoque et alium
numero, quam pastinaca vel quae specialiter

---

[a] *i.e.* the radish (radicula).

BOOK II. 18. 2-5

kind of food (I call strongest that which has most nourishment). To the same class of food belong: all domesticated quadruped animals; all large game such as the wild she-goat, deer, wild boar, wild ass; all large birds, such as the goose and peacock and crane; all sea monsters, among which is the whale and such like; also honey and cheese. Hence it is not wonderful that pastry made of grain, lard, honey and cheese is very strong food.

Among food materials of the middle class ought to be reckoned: of pot-herbs, those of which the roots or bulbs are eaten; of quadrupeds, the hare; birds of all kinds from the smallest up to the flamingo; likewise all fish which do not bear salting or are salted whole.

The weakest of food materials are: all vegetable stalks and whatever forms on a stalk, such as the gourd and cucumber and caper, all orchard fruits, olives, snails, and likewise shellfish.

But although these are so divided, nevertheless even those of the same species admit of great differentiation, one thing being stronger or weaker than another: whilst there is more nutriment in bread than in anything else, wheat is a stronger food than millet, and that again than barley; and of wheat the strongest is siligo, next simila, then the meal from which nothing is extracted, which the Greeks call autopuros; weaker is bread made from pollen, weakest the common grey bread.

Among pulses, beans and lentils are stronger food than peas. Of vegetables the turnip and navew and all bulbs, among which I include the onion also and garlic, are stronger than the parsnip, or that which is specially called a root.[a]

CELSUS

radicula appellatur. Item firmior brassica et beta et porrum quam lactuca vel cucurbita vel asparagus. 6 At ex fructibus surculorum valentiores uvae, ficus, nuces, palmulae quam quae poma proprie nominantur; atque ex his ipsis firmiora quae sucosa quam quae fragilia sunt.

Itemque ex iis avibus, quae in media specie sunt, valentior ea, quae pedibus quam quae volatu magis nititur: et ex iis, quae volatu fidunt, firmiores quae grandiores aves quam quae minutae sunt, ut ficedula et turdus. Atque eae quoque, quae in aqua degunt, leviorem cibum praestant, quam quae natandi scientiam non habent.

7 Inter domesticas vero quadrupedes levissima suilla est, gravissima bubula. Itemque ex feris, quo maius quodque animal, eo robustior ex eo cibus est.

Pisciumque eorum, qui ex media materia sunt, quibus maxime utimur, tamen gravissimi sunt, ex quibus salsamenta quoque fieri possunt, qualis lacertus est; deinde ii qui, quamvis teneriores, tamen duri sunt, ut aurata, corvus, sparus, oculata, tunc plani, post quos etiamnum leviores lupi mullique, et post hos omnes saxatiles.

8 Neque vero in generibus rerum tantummodo discrimen est, sed etiam in ipsis; quod et aetate fit et membro et solo et caelo et habitu. Nam quadrupes omne animal, si lactens est, minus alimenti praestat, itemque quo tenerior pullus cohortalis est; in piscibus quoque media aetas, quae non summam

---

[a] 24, 2. 27.
[b] Pork and bacon, tabooed by Asiatics, were the chief animal food of the Greeks, Romans and Celts; next, lamb and kid; there is no mention of mutton and very limited use of beef. See list of Alimenta, pp. 483 ff.

Also cabbage and beet and leek are stronger than lettuce or gourd or asparagus. But of fruit growing on twigs, grapes, figs, nuts, dates are stronger than orchard fruit properly so-called; and of these last, the juicy are stronger than the mealy.[a]

Likewise of those birds, which belong to the middle class, those which rely more on their feet are stronger food than those which rely more on their wings; and of those birds which depend on flight, the larger birds yield stronger food than the smaller, such as the fig-eater and thrush. And those also which pass their time in the water yield a weaker food than those which have no knowledge of swimming.

Among food from domesticated quadrupeds pork is the weakest, beef the strongest.[b] And so also of game, the larger the animal the stronger the food it yields.

The fish most in use belong to the middle class; the strongest are, however, those from which salted preparations can be made, such as the mackerel; next come those which, although more tender, are nevertheless firm, such as the gilthead, gurnard, sea bream, eyefish, then the flat fish, and after these still softer, the bass and mullets, and after these all rock fish.

Not only is there differentiation in the classes of nutriments, but also as much in the actual species of nutriment; which is due both to age and part of the animal and to soil and to climate and to habit. For every four-footed animal yields less nutriment while it is a suckling, likewise a chicken in a coop, the more tender it is; also a half-grown fish, which has not filled out to its full

magnitudinem inplevit. Deinde ex eodem sue ungulae, rostrum, aures, cerebellum, ex agno haedove cum petiolis totum caput aliquanto quam cetera membra leviora sunt, adeo ut in media
9 materia poni possint. Ex avibus colla alaeve recte infirmis adnumerantur. Quod ad solum vero pertinet, frumentum quoque valentius est collinum quam campestre; levior piscis inter saxa editus quam in harena, levior in harena quam in limo. Quo fit, ut ex stagno vel lacu vel flumine eadem genera graviora sint, leviorque qui in alto quam qui in vado vixit. Omne etiam ferum animal domestico levius, et quodcumque umido caelo quam quod sicco natum est. Deinde etiam omnia pinguia
10 quam macra, recentia quam salsa, nova quam vetusta plus alimenti habent. Tum res eadem magis alit iurulenta quam assa, magis assa quam elixa. Ovum durum valentissimae materiae est, molle vel sorbile inbecillissimae. Cumque panificia omnia firmissima sint, elota tamen quaedam genera frumenti, ut halica, oryza, tisana, vel ex isdem facta sorbitio aut pulticula, et aqua quoque madens panis inbecillissimis adnumerari potest.
11 Ex potionibus vero quaecumque ex frumento facta est, itemque lac, mulsum, defrutum, passum, vinum aut dulce aut vehemens aut mustum aut magnae vetustatis valentissimi generis est. At acetum et id vinum, quod paucorum annorum vel austerum vel pingue est, in media materia est; ideoque infirmis numquam generis alterius dari debet.

---

[a] *e.g.* the spring wheat grown in south Italy, rich in gluten, makes macaroni, spaghetti and vermicelli.

size. Then likewise in the same pig, trotters, chaps, ears or brain, in a lamb or kid the whole head, also the pettitoes, are somewhat less nutritious than other parts, and so can be placed in the middle class. In birds, the neck and wings are rightly counted as weak nutriment. As regards soil, grain is also more nutritious grown in hilly[a] than in flat districts; fish living among rocks are less nutritious than those in sand, and these again less than those in mud. Hence it is that the same classes of fish from a pond or lake or river are heavier, and those which live in deep water are lighter food than those which live in shallows. Every wild animal is a lighter food than the same species domesticated, and the product of a damp climate is lighter than that of a dry one. Again, all kinds of fat meat have more nutriment than lean, fresh meat than salted, recently killed than stale. Then the same meat is more nourishing stewed than roasted, more so roasted than boiled. A hard-boiled egg is a very substantial material, a soft cooked or raw egg very light. And while all bread-stuffs are among the most solid, yet some kinds of grain after being soaked, such as spelt, rice, pearl barley, or the gruel or porridge made out of these, and also bread soaked in water, can be reckoned among the weakest of food.

Of drinks too the strongest class are: whatever can be made from grain, likewise milk, mead, must boiled down, raisin wine, wine either sweet or heady or still fermenting or of great age. But vinegar, and that wine which is a few years old, whether dry or rich, are intermediate in quality; and therefore to weak patients nothing of the other class

Aqua omnium inbecillissima est; firmiorque ex frumento potio est, quo firmius est ipsum frumentum; firmior ex eo vino, quod bono solo quam quod tenui, quodque temperato caelo quam quod nimis aut umido aut nimis sicco nimiumque aut frigido aut 12 calido natum est. Mulsum, quo plus mellis habet, defrutum, quo magis incoctum est, passum, ex quo sicciore uva est, eo valentius est. Aqua levissima pluvialis est, deinde fontana, tum ex flumine, tum ex puteo, post haec ex nive aut glacie: gravior his ex lacu, gravissima ex palude. Facilis etiam et necessaria cognitio est naturam eius requirentibus. Nam levis pondere apparet, et ex is, quae pondere pares sunt, eo melior quaeque est, quo celerius et calfit et frigescit, quoque celerius in ea legumina 13 percoquuntur. Fere vero sequitur, ut, quo valentior quaeque materia est, eo minus facile concoquatur, sed, si concocta est, plus alat. Itaque utendum est materiae genere pro viribus, modusque omnium pro genere sumendus. Ergo inbecillis hominibus rebus infirmissimis opus est: mediocriter firmos media materia optime sustinet, et robustis apta validissima est. Plus deinde aliquis adsumere ex levioribus potest: magis in iis, quae valentissima sunt, temperare sibi debet.

19. Neque haec sola discrimina sunt; sed etiam aliae res boni suci sunt, aliae mali, quas

should be given. Water is of all the weakest; and drink from grain is the more nutritious, according as the grain itself is nutritious; wine coming from a good soil is more nutritious than from a poor one, that from a temperate climate more nutritious than from an extreme one, whether too wet or too dry, whether excessively cold or hot. Mead, the more honey it contains, must the more it is boiled down, raisin wine the drier the grapes—are the stronger. Rain water is the lightest, then spring water, next water from a river, then from a well, after that from snow or ice; heavier still is water from a lake, the heaviest from a marsh. The recognition of water[a] is as easy as it is necessary for those who want to know its nature. For by weighing, the lightness of water becomes evident, and of water of equal weight, that is the better, which most quickly heats or cools, also in which pulse is most quickly cooked. It is generally the case too that the more substantial the material, the less readily it is digested, but once digested it nourishes the more. Thus the quality of the food administered should be in accordance with the patient's strength, and the quantity in accordance with its quality. For weak patients, therefore, there is needed the lightest food; food of the middle class best sustains those moderately strong, and for the robust the strongest is the fittest. Finally, of the lightest foods more can be taken; it is rather with the strongest food that moderation should be observed.

19. The foregoing are not the only differentiations; but as well some materials have good juices,

[a] *i.e.* the recognition of the amount of organic or inorganic contamination in it.

εὐχύλους vel κακοχύλους Graeci vocant; aliae lenes, aliae acres; aliae crassiorem pituitam in nobis faciunt, aliae tenuiorem; aliae idoneae stomacho, aliae alienae sunt; itemque aliae inflant, aliae ab hoc absunt; aliae calfaciunt, aliae refrigerant; 2 aliae facile in stomacho acescunt, aliae non facile intus corrumpuntur; aliae movent alvum, aliae supprimunt; aliae citant urinam, aliae tardant; quaedam somnum movent, quaedam sensum excitant. Quae omnia ideo noscenda sunt, quoniam aliud alii vel corpori vel valetudini convenit.

20. Boni suci sunt triticum, siligo, halica, oryza, amulum, tragum, tisana, lac, caseus mollis, omnis venatio, omnes aves, quae ex media materia sunt, ex maioribus quoque eae, quas supra (18, 2) nominavi; medii inter teneros durosque pisces, ut mullus, ut lupus; verna lactuca, urtica, malva, [cucumis,][1] cucurbita, ovum sorbile, portulaca, cocleae, 2 palmulae; ex pomis quodcumque neque acerbum neque acidum est; vinum dulce vel lene, passum, defrutum; oleae, quae ex his duobus in altero utro servatae sunt; vulvae, rostra, trunculique suum, omnis pinguis caro, omnis glutinosa, omne iecur.

21. Mali vero suci sunt milium, panicium, hordeum, legumina; caro domestica permacra omnisque caro salsa, omne salsamentum, garum, vetus caseus; siser, radicula, rapa, napi, bulbi; brassica magisque etiam cyma eius, asparagus, beta, cucumis, porrus, eruca, nasturcium, thymum, nepeta, satureia, hysopum, ruta, anetum, feniculum, cuminum, anesum, lapatium, sinapi, alium, cepe;

---

[1] *Omitted by Targa, cf.* 21.

---

[a] *i.e.* digestible and indigestible.

others bad, what the Greeks call euchylous[a] and kakochylous; some are bland, others acrid; some render our phlegm thicker, others thinner; some agree with the stomach, others are alien; also some cause flatulence, others are free from that; some warm, others cool; some readily turn sour in the stomach, others do not readily decompose inside; some move the bowels, others check motions; some excite urination, others retard it; some promote sleep, others excite the senses. All these, then, should be known because one suits one body or constitution, one another.

20. Of good juice are: wheat, siligo, spelt, rice, starch,[b] frumenty, pearl barley gruel, milk, soft cheese, all sorts of game, all birds of the middle class, also the larger birds named above; fish intermediate between the soft and hard, such as mullet and bass; spring lettuce, nettle-tops, mallow, gourd, raw egg, purslane, snails, dates; orchard fruit which is neither bitter nor sour; wine sweet or mild, raisin wine, must boiled down; olives preserved either in wine or must; sow's womb, pig's chaps and trotters, all fatty or glutinous meat, and the liver of all animals.

21. Of bad juice are: millet, panic, barley, pulse; very lean meat from domesticated animals and all salted meat; all pickled fish, fish sauce, old cheese; skirret, radish, turnip, navew, bulbs; cabbage and even more its sprouts, asparagus, beet, cucumber, leek, rocket, cress, thyme, catmint, savory, hyssop, rue, dill, fennel, cummin, anise, sorrel, mustard,

---

[b] Wheat decorticated by soaking, a fine meal so called because it was not ground in a mill in the ordinary course, cf. *Cato* 87 (L.C.L. p. 89, note 2).

lienes, renes, intestina; pomum quodcumque acidum
vel acerbum est; acetum, omnia acria, acida, acerba,
oleum; pisces quoque saxatiles, omnesque, qui ex
tenerrimo genere sunt, aut qui rursus nimium duri
virosique sunt, ut fere quos stagna, lacus limosique rivi
ferunt, quique in nimiam magnitudinem excesserunt.

22. Lenes autem sunt sorbitio, pulticula,
laganus, amylum, tisana, pinguis caro et quaecumque glutinosa est; quod fere quidem in omni domestica fit, praecipue tamen in ungulis trunculisque
suum, in petiolis capitulisque haedorum et vitulorum
et agnorum, omnibusque cerebellis; item qui
proprie bulbi nominantur, lac, defrutum, passum,
2 nuclei pinei. Acria sunt omnia nimis austera,
omnia acida, omnia salsa, et mel quidem, quo melius
est, eo magis. Item alium, cepa, eruca, ruta,
nasturcium, cucumis, beta, brassica, asparagus,
sinapi, radicula, intubus, ocimum, lactuca, maximaque holerum pars.

23. Crassiorem autem pituitam faciunt ova
sorbilia, halica, oriza, amulum, tisana, lac, bulbi,
omniaque fere glutinosa. Extenuant eandem omnia
salsa, atque acria atque acida.

24. Stomacho autem aptissima sunt, quaecumque austera sunt; etiam quae acida sunt, quaeque contacta sale modice sunt; item panis sine fermento, et elota halica, vel oriza vel tisana; omnis
avis, omnis venatio; atque utraque vel assa vel
2 elixa: ex domesticis animalibus bubula: si quid ex
ceteris sumitur, macrum potius quam pingue;

garlic, onion; spleens, kidneys, chitterlings; orchard fruit when sour or bitter; vinegar, everything acrid, sour, bitter, oily; also rock fish, and all fish of the very soft kind, or on the other hand those which are very hard and strong-flavoured, mostly such as live in ponds, lakes and muddy rivers, and which have become excessively large.

22. The following are bland materials: broth, porridge, pancake, starch, pearl barley gruel, fat and glutinous meat, generally all that belonging to domesticated animals, particularly, however, the trotters and titbits of pigs, the pettitoes and heads of kids, calves, and lambs, and the brains of all animals; likewise all bulbs properly so-called, milk, must boiled down, raisin wine and pine kernels. The following are acrid: everything especially harsh, everything sour, everything salt, and even honey, and the better it is the more it is so. Likewise garlic, onion, rocket, rue, cress, cucumber, beet, cabbage, asparagus, mustard, radish, endive, basil, lettuce and most pot-herbs.

23. Now the following make phlegm thicker: raw eggs, spelt, rice, starch, pearl barley gruel, milk, bulbs, and generally all glutinous substances. Phlegm is rendered thinner by: all salted and acrid and acid materials.

24. But best suited to the stomach are: whatever is harsh, even what is sour, and that which has been sprinkled moderately with salt; so also unleavened bread, and spelt or rice or pearl barley which has been soaked; birds and game of all kinds, and both of these whether roasted or boiled; among domesticated animals, beef; of other meat the lean rather than the fat; the trotters, chaps,

ex sue ungulae, rostra, aures, volvae sterilesque; ex holeribus intubus, lactuca, pastinaca, cucurbita elixa, siser; ex pomis cerasium, morum, sorbum, pirum fragile, quale Crustuminum vel Mevianum est; item pira, quae reponuntur, Tarentina atque Signina, malum orbiculatum aut Scandianum vel Amerinum vel Cotoneum vel Punicum, uvae ex 3 olla; molle ovum, palmulae, nuclei pinei, oleae albae ex dura muria, eaedem aceto intinctae, vel nigrae, quae in arbore bene permaturuerunt, vel quae in passo defrutove servatae sunt; vinum austerum, licet etiam asperum sit, item resinatum; duri ex media materia pisces, ostrea, pectines, murices, purpurae, cocleae, cibi potionesque vel frigidae vel ferventes, apsinthium.

25. Aliena vero stomacho sunt omnia tepida, omnia salsa, omnia iurulenta, omnia praedulcia, omnia pinguia, sorbitio, panis fermentatus, itemque vel ex milio vel ex hordeo, radices holerum, et quodcumque holus ex oleo garove estur, mel, mulsum, defrutum, passum, lac, omnis caseus, uva recens, ficus et virides et aridae, legumina omnia, quaeque 2 inflare consuerunt; item thymum, nepeta, satureia, hysopum, nasturcium, lapatium, lapsanum, iuglandes. Ex his autem intellegi potest non, quicquid boni suci est, protinus stomacho convenire, neque, quicquid stomacho convenit, protinus boni suci esse.

26. Inflant autem omnia fere legumina, omnia pinguia, omnia dulcia, omnia iurulenta, mustum,

---

a Murex brandaris and Purpura haemostoma, univalve molluscs, the chief sources of the purple dye, were extensively consumed.

ears, and the sterile womb of a pig; among pot-herbs, endive, lettuce, parsnip, cooked gourd, skirret; among orchard fruit, the cherry, mulberry, service fruit, the mealy pear from Crustumeria, or the Mevian; also keeping-pears, Tarentine or Signian, the round or Scandian apple or that of Ameria or the quince or pomegranate, raisins preserved in jars; soft egg, dates, pine kernels, white olives preserved in strong brine, or the same steeped in vinegar, or black olives which have been well ripened on the tree, or which have been preserved in raisin wine, or in boiled-down must; dry wine is allowable even although it may have become harsh, also that doctored with resin; hard-fibred fish of the intermediate class, oysters, scallops, the shellfish murex and purpura,[a] snails; food and drink either very cold or very hot; wormwood.

25. But on the other hand materials alien to the stomach are: all things tepid, all things salted, all things stewed, all things over-sweetened, all things fatty, broth, leavened bread, and likewise that made from either millet or barley, pot-herb roots, and pot-herbs eaten with oil or fish sauce, honey, mead, must boiled down, raisin wine, milk, cheese of all kinds, fresh grapes, figs both green and dry, pulse of all sorts, and whatever causes flatulence; likewise thyme, catmint, savory, hyssop, cress, sorrel, charlock, walnuts. But it can be understood from the above that what has good juice does not necessarily agree with the stomach, and that whatever agrees with the stomach has not necessarily good juice.

26. Now flatulence is produced by: almost all food which is leguminous, fatty, sweet, everything

atque etiam id vinum, cui nihil adhuc aetatis accessit;
ex holeribus alium, cepa, brassica, omnesque radices,
2 excepto sisere et pastinaca; bulbi, ficus etiam aridae
sed magis virides, uvae recentes, nuces omnes,
exceptis nucleis pineis, lac, omnisque caseus; quicquid deinde subcrudum aliquis adsumpsit. Minima
inflatio fit ex venatione, aucupio, piscibus, pomis,
oleis, conchyliisve, ovis vel mollibus vel sorbilibus,
vino vetere. Feniculum vero et anetum inflationes
etiam levant.

27. At calfaciunt piper, sal, caro omnis iurulenta, alium, cepa, ficus aridae, salsamentum,
vinum, et quo meracius est, eo magis. Refrigerant
holera, quorum crudi caules adsumuntur, ut intubus
et lactuca, et item coriandrum, cucumis, elixa cucurbita, beta, mora, cerasia, mala austera, pira
fragilia, caro elixa, praecipueque acetum, sive
cibus ex eo sive potio adsumitur.

28. Facile autem intus corrumpuntur panis fermentatus, et quisquis alius quam ex tritico est,
lac, mel; ideoque etiam lactentia atque omne
pistorium opus, teneri pisces, ostrea, holera, caseus
et recens et vetus, crassa vel tenera caro, vinum
dulce, mulsum, defrutum, passum; quicquid deinde
vel iurulentum est vel nimis dulce vel nimis tenue.
2 At minime intus vitiantur panis sine fermento, aves
et eae potius duriores, duri pisces: neque solum
aurata tuta aut sparus sed etiam lolligo, lucusta,
polypus; item bubula omnisque dura caro; eademque aptior est, si macra, si salsa est; omniaque

stewed, must, and also that wine which has not as yet matured; among pot-herbs, garlic, onion, cabbage, and all roots except skirret and parsnip; bulbs, figs even when dried but especially when green, fresh grapes, all nuts except pine kernels, milk, cheese of all kinds; lastly anything eaten half-cooked. The least flatulence comes from what is got by hunting and birding, from fish, orchard fruit, olives, or shellfish, from eggs whether cooked soft or raw, from old wine. Fennel and anise in particular even relieve flatulence.

27. Again the heating foods are: pepper, salt, all stewed meat, garlic, onion, dried figs, pickled fish, wine, and the stronger this is, the more heating it is. Cooling foods are: pot-herbs the stalks of which are eaten uncooked, such as endive and lettuce, and also coriander, cucumber, cooked gourds, beet, mulberries, cherries, sour apples, mealy pears, boiled meat, and in particular vinegar, whether taken with food or as a drink.

28. Foods that readily decompose inside are: leavened bread, and any sort other than that made of wheat, flour, milk, honey, and therefore also all things made with milk and all pastry, soft fish, oysters, vegetables, cheese both new and old, meat fat or tender, sweet wine, mead, must boiled down, raisin wine; finally everything stewed or over-sweetened or over-thin. But the following decompose the least within: unleavened bread, birds, especially those with harder flesh, hard fish, not only for instance the gilthead or the sea bream, but also the squid, lobster and octopus; likewise beef and hard meat of all kinds; and the same is better if lean or salted; all pickled fish,

salsamenta, cocleae, murices, purpurae; vinum austerum vel resinatum.

29. At alvum movent panis fermentatus, magisque si cibarius vel hordiacius est, brassica, si subcruda est, lactuca, anetum, nasturcium, ocimum, urtica, portulaca, radicula, capparis, alium, cepa, malva, lapatium, beta, asparagus, cucurbita, cerasia, mora, uva ex olla, omnia mitia, ficus etiam arida, 2 sed magis viridis, uvae recentes; pingues minutae aves, cocleae, garum, salsamentum, ostrea, pelorides, echini, musculi, et omnes fere conchulae, maximeque ius earum, saxatiles et omnes teneri pisces, sepiarum atramentum; si qua caro adsumitur, pinguis, eadem vel iurulenta vel elixa, aves quae natant, mel crudum, lac, lactentia omnia, mulsum, vinum dulce vel salsum, aqua tenera; omnia dulcia, tepida, pinguia, elixa, iurulenta, salsa vel diluta.

30. Contra astringunt panis ex siligine vel ex simila, magis si sine fermento est, magis etiam si ustus est, intenditurque vis eius etiam, si bis coquitur, pulticula vel ex halica vel ex panicio vel ex milio, itemque ex isdem sorbitio, et magis si haec ante fricta sunt; lenticula, cui vel beta vel intubus vel ambubeia vel plantago adiecta est, magisque etiam si illa ante fricta est, per se etiam intubus vel ex plantagine vel ambubeia fricta, minuta holera, 2 brassica bis decocta; dura ova, magisque si assa sunt; minutae aves, merula, palumbus, magisque in posca si decoctus est, grues, omnes aves, quae magis currunt quam volant; lepus, caprea, iecur ex eis, quae sebum habent, maximeque bubulum,

BOOK II. 28. 2–30. 2

snails, the shellfish, murex purpura; and wine which is harsh or resinated.

29. Again, the bowels are moved by: leavened bread, and especially if it is the grey wheaten or barley bread, cabbage if lightly cooked, lettuce, dill, cress, basil, nettle-tops, purslane, radish, caper, garlic, onion, mallow, sorrel, beet, asparagus, gourds, cherries, mulberries, raisins preserved in jars, all ripe fruit, a fig even dried, but especially a green one, fresh grapes; fat small birds, snails, fish sauce, pickled fish, oysters, giant mussels, sea-urchins, sea-mussels, almost all shellfish, especially the soup made from them, rock fish and all soft fish, cuttle-fish ink; any meat eaten when fat, either stewed or boiled, waterfowl, uncooked honey, milk, all things made with milk, mead, wine sweet or salted, soft water; all food sweetened, tepid, fatty, boiled, stewed, salted or watery.

30. On the contrary the bowels are confined: by bread made from siligo or simila flour, especially when unleavened, and particularly so when toasted, and this property is even increased by baking twice. porridge either from spelt or panic or millet, as well as gruel from the same, and especially if these have been parched beforehand; lentil porridge to which beet or endive or chicory or plantain has been added, and especially when these have been previously toasted, or endive by itself, or roasted with plantain, or chicory, the smaller pot-herbs, cabbage twice boiled; eggs rendered hard, especially by poaching; small birds, the blackbird and wood-pigeons especially when cooked in diluted vinegar, cranes, all birds which run rather than fly; the hare, wild she-goat, the liver of animals which yield suet,

ac sebum ipsum; caseus, qui vehementior vetustate fit vel ea mutatione, quam in eo transmarino videmus, 3 aut si recens est, ex melle mulsove decoctus; item mel coctum, pira inmatura, sorva, magisque ea, quae torminalia nominantur, mala Cotonea et Punica, oleae vel albae vel permaturae, murta, palmulae, purpurae, murices, vinum resinatum vel asperum, item meracum, acetum, mulsum quod infervuit, item defrutum, passum, aqua vel tepida vel praefrigida, dura, id est ea, quae tarde putrescit, ideoque pluvia potissimum; omnia dura, macra, austera, aspera, tosta, et in eadem carne assa potius quam elixa.

31. Urinam autem movent quaecumque in horto nascentia boni odoris sunt, ut apium, ruta, anetum, ocimum, menta, hysopum, anesum, coriandrum, nasturcium, eruca, feniculum; praeter haec asparagus, capparis, nepeta, thymum, satureia, lapsanum, pastinaca, magisque agrestis, radicula, siser, cepa; ex venatione maxime lepus; vinum tenue, piper et rotundum et longum, sinapi, absinthium, nuclei pinei.

32. Somno vero aptum est papaver, lactuca, maximeque aestiva, cuius coliculus iam lacte repletus est, morum, porrus. Sensus excitat nepeta, thymum, satureia, hysopum, praecipueque puleium, ruta et cepe.

33. Evocare vero materiam multa admodum

---

[a] 'Good against the gripes.' Pyrus sorbus torminalis, an intestinal astringent for dysentery.

[b] Then follows a list of local applications placed here rather

particularly the ox, and suet itself; cheese which has become rather strong in taste, either from age or because of that change which we note in cheese from across the sea, or, if it is new, after it has been cooked in honey or mead; also cooked honey, unripe pears, service fruit, especially those called torminalia,[a] quinces and pomegranates, olives either white or over-ripe, myrtleberries, dates, the purpura and murex, wine resinated or harsh, and that undiluted, vinegar, mead which has been heated, also must boiled down, raisin wine, water tepid or very cold, hard water (that is, which decomposes late), hence principally rain water; everything hard, lean, harsh, rough, grilled, and in the case of the same meat the flesh roasted rather than boiled.

31. The following increase the urine: garden herbs of good odour, as parsley, rue, dill, basil, mint, hyssop, anise, coriander, cress, rocket, fennel; and besides these asparagus, capers, catmint, thyme, savory, charlock, parsnip, especially growing wild, radish, skirret, onion; of game especially the hare; thin wine, pepper both round and long, mustard, wormwood, pine kernels.

32. For producing sleep the following are good: poppy, lettuce, and mostly the summer kinds in which the stalk is very milky, the mulberry, the leek. For exciting the senses: catmint, thyme, savory, hyssop, and especially pennyroyal, rue and onion.

33. For drawing out[b] the material of the disease

than in Book V, because the materials were included under the heading of Ailments.

possunt, sed ea cum ex peregrinis medicamentis maxime constent, aliisque magis, quam quibus ratione victus succurritur (V. *Proem.*, 1, 2), opitulentur, in praesentia differam: ponam vero ea, quae prompta et iis morbis, de quibus protinus (III, IV) dicturus sum, apta corpus erodunt, et sic eo quod mali est extrahunt. Habent autem hanc facultatem semen erucae, nasturcii, radiculae, praecipue tamen omnium sinapi. Salis quoque et fici eadem vis est.

2 Leniter vero simul et reprimunt et molliunt lana sucida quo cum aceto vel vino oleum adiectum est, contritae palmulae, furfures in salsa aqua vel aceto decoctae.

At simul et reprimunt et refrigerant herba muralis (παρθένιον vel περδείκιον appellant), serpullum, puleium, ocimum, herba sanguinalis, quam Graeci πολύγονον vocant, portulaca, papaveris folium, capriolique vitium, coriandrum, folia hyocimum, muscus, siser, apium, solanum (quam strychnon Graeci vocant), brassicae folia, intubus, 3 plantago, feniculi semen; contrita pira vel mala, praecipueque Cotonea, lenticula; aqua frigida, maximeque pluvialis, vinum, acetum, et horum aliquo madens vel panis vel farina vel spongia vel cinis vel lana sucida vel etiam linteolum; creta Cimolia, gypsum; melinum, murteum, rosa, acerbum oleum; verbenarum contusa cum teneris colibus 4 folia; cuius generis sunt olea, cupressus, myrtus, lentiscus, tamarix, ligustrum, rosa, rubus, laurus, hedera, Punicum malum.

Sine frigore autem reprimunt cocta mala Cotonea,

---

[a] From Cimolus in the Cyclades.

BOOK II. 33. 1-4

certainly many things can be used, but as they are mostly composed of foreign medicaments and are more useful in other affections than in those relieved by the dietetic regimen, I will defer their consideration for the present (V. Proem., 1, 2): but I will mention here those which are at hand, and are suitable to the diseases of which I am about to speak (III, IV), since they blister the body and thus extract from it the material of disease. Now those which have this faculty are the seeds of rocket, cress, radish, and most of all mustard. The same faculty exists in salt and figs.

Those which gently both repress and mollify at the same time are greasy wool to which has been added oil with vinegar or wine, crushed dates, bran boiled in salt water or vinegar.

But those which simultaneously repress and cool are pellitory, which the Greeks call parthenion or perdeikion, thyme, pennyroyal, basil, the blood-herb which the Greeks call polygonon, purslane, poppy-leaf, vine-tendril, coriander, hyocyamus-leaves, moss, skirret, parsley, solanum, which the Greeks call strychnos, cabbage-leaves, endive, plantain, fennel-seed; crushed pears and apples and especially quinces, lentils; cold water, especially rain water, wine and vinegar, and everything soaked in these, whether bread or meal or sponge or ashes, or greasy wool or even lint; Cimolian chalk,[a] gypsum; oil perfumed with quince, myrtle, rose; unripe olive oil; vervains, the leaves crushed along with their young twigs, of which sort are the olive, cypress, myrtle, mastic, tamarisk, privet, rose, bramble, laurel, ivy, and pomegranate.

Those which repress without cooling are cooked

malicorium, aqua calda, in qua verbenae coctae sunt, quas supra (§ 3, 4) posui, pulvis vel ex faece vini vel ex murti foliis, amarae nuces.

5 Calfacit vero ex qualibet farina cataplasma sive ex tritici sive farris sive hordei sive ervi sive lolii sive milii vel panicii vel lenticulae vel fabae vel lupini vel lini vel feni Graeci, ubi ea defervuit calidaque inposita est. Valentior tamen ad id omnis farina est ex mulso quam ex aqua cocta. Praeterea cyprinum, irinum, medulla, adeps ex fele, oleum, magisque si vetus est, iunctaque oleo sal, nitrum, git, piper, quinquefolium.

6 Fereque quae vehementer reprimunt et refrigerant, durant: quae calfaciunt, digerunt et emolliunt, praecipueque ad emolliendum potest cataplasma ex lini vel feni Graeci semine.

His autem omnibus, et simplicibus et permixtis, varie medici utuntur, ut magis quid quisque persuaserit sibi appareat, quam quid evidenter compererit.

BOOK II. 33. 4-6

quinces, pomegranate rind, hot water in which the vervains enumerated above have been boiled, powdered wine lees or myrtle leaves, bitter almonds.

But those which are heating are poultices made of meal, whether of wheat or spelt or barley or bitter vetches or darnel or millet or panic or lentil or bean or lupin or linseed or fenugreek, when one of these has been boiled and applied hot. All forms of meal poultices, however, are rendered more efficacious by cooking in mead instead of in water. Besides there are: cyprus or iris oil, marrow, cat's fat, olive oil, especially if it is old, and there has been added to the oil salt, soda, black cummin, pepper, cinquefoil.

Generally those which are powerful to repress inflammation, and cool, harden the tissues; those which are heating, disperse inflammation and soften, and this last property belongs especially to plasters of linseed or fenugreek seeds.

But as regards all these medicaments, whether used as simples or in mixtures, their uses by medical men vary, so that it is clear that each man follows his own ideas rather than what he has found to be true by actual fact.

BOOK III

LIBER TERTIUS

1. Provisis omnibus, quae pertinent ad universa genera morborum, ad singulorum curationes veniam. Hos autem in duas species Graeci diviserunt, aliosque ex his acutos, alios longos esse dixerunt. Idemque quoniam non semper eodem modo respondebant, eosdem alii inter acutos, alii inter longos rettulerunt; ex quo plura eorum genera esse manifestum est. 2 Quidam enim breves utique sunt, qui cito vel tollunt hominem, vel ipsi cito finiuntur; quidam longi, sub quibus neque sanitas in propinquo neque exitium est; tertiumque genus eorum est, qui modo acuti, modo longi sunt, idque non in febribus tantummodo, in quibus frequentissimum est, sed in aliis quoque 3 fit. Atque etiam praeter hos quartum est, quod neque acutum dici potest, quia non peremit, neque utique longum, quia, si occurritur, facile sanatur. Ego cum de singulis dicam, cuius quisque generis sit indicabo. Dividam autem omnes in eos, qui in totis corporibus consistere videntur, et eos, qui oriuntur in partibus. Incipiam a prioribus, pauca de omnibus praefatus.

4 In nullo quidem morbo minus fortuna sibi vindicare quam ars potest: ut pote quom repugnante natura

BOOK III

1. HAVING dealt with all that pertains to whole classes of diseases taken together, I come to the treatment of diseases one by one. Now the Greeks divided these into two species, terming some acute, others chronic. But because maladies did not always respond in the same way to treatment, some of the Greek writers have placed among the acute what others have placed among the chronic; from this it is clear that there are more than two classes. For some diseases are certainly of short duration, which carry off the patient quickly, or themselves come quickly to an end; some are chronic, in which neither recovery is near at hand nor death; and there is a third class, at one time acute, at another time chronic, and that occurs not only in fevers, where it is most frequent, but in other affections also. And besides the above there is a fourth class which cannot be said to be acute, because it is not fatal, nor really chronic, because if treated it is readily cured. When I come myself to speak of diseases singly, I will point out to which class each belongs. But I shall divide all diseases into those which appear to have their seat in the body as a whole, and into those which originate in particular parts. I shall begin with the former, after a few words of preface concerning all.

Whatever the malady luck no less than the art can claim influence for itself; seeing that with nature

nihil medicina proficiat. Magis tamen ignoscendum medico est parum proficienti in acutis morbis quam in longis: hic enim breve spatium est, intra quod, si auxilium non profuit, aeger extinguitur: ibi et deliberationi et mutationi remediorum tempus patet, adeo ut raro, si inter initia medicus accessit, obsequens 5 aeger sine illius vitio pereat. Longus tamen morbus cum penitus insedit, quod ad difficultatem pertinet, acuto par est. Et acutus quidem quo vetustior est, longus autem quo recentior, eo facilius curatur.

Alterum illud ignorari non oportet, quod non omnibus aegris eadem auxilia conveniunt. Ex quo incidit, ut alia atque alia summi auctores quasi sola 6 vindicarint, prout cuique cesserat. Oportet itaque, ubi aliquid non respondet, non tanti putare auctorem quanti aegrum, et experiri aliud atque aliud, sic tamen ut in acutis morbis cito mutetur quod nihil prodest, in longis, quos tempus ut facit sic etiam solvit, non statim condemnetur, si quid non statim profuit, minus vero removeatur, si quid paulum saltem iuvat, quia profectus tempore expletur.

2. Protinus autem inter initia scire facile est, quis acutus morbus, quis longus sit, non in iis solis, in quibus semper ita se habet, sed in iis quoque, in quibus variat. Nam ubi sine intermissionibus accessiones

---

ᵃ The metaphor is from a candle, cf. Lucr. VI, 791.

in opposition the art of medicine avails nothing. There is, however, for a practitioner who is unsuccessful, more excuse in acute than in chronic diseases: for acute diseases are of short duration, within which the patient is snuffed out,[a] if not benefited by the treatment: chronic diseases give time for deliberation, and for change of remedies, so that when the practitioner is in attendance from the commencement, it is seldom that a docile patient should perish unless by the practitioner's default. A chronic disease, nevertheless, when it has become deep-seated, is no less difficult to deal with than an acute one. And indeed the older an acute malady, the more recent a chronic one, the more easily it is treated.

There is another point which should be borne in mind, that the same remedies do not suit all patients. Hence it is that the highest authorities proclaim as if they were the only remedies, now some, now others, each in accordance with what he has found successful. It is well, then, when any one remedy fails, to look not so much to the authority as to the patient, and to make trial, now of one, now of another remedy, taking care, however, that in acute diseases what is doing no good is changed quickly; in chronic diseases which it takes time to produce as well as to remove, if a remedy does not succeed at once, it should not be condemned at once, much less should it be discontinued if it is beneficial, though only to a small extent, because the progress is completed by time.

2. Now at their commencements, it is easy to recognize at once what is an acute disease, and what a chronic one, not only as regards those which take a uniform course, but also when the course is variable. For when severe paroxysms and pains are causing

et dolores graves urgent, acutus est morbus: ubi lenti dolores lentaeve febres sunt et spatia inter accessiones porrigunt, acceduntque ea signa, quae priore volumine (II. 5, 2, 3) exposita sunt, longum 2 hunc futurum esse manifestum est. Videndum etiam est, morbus an increscat, an consistat, an minuatur, quia quaedam remedia increscentibus morbis, plura inclinatis conveniunt; eaque, quae decrescentibus apta sunt, ubi acutus increscens urget, in remissionibus potius experienda sunt. Increscit autem morbus, dum graviores dolores accessionesque veniunt, eaeque et ante, quam proximae, revertuntur 3 et postea desinunt. Atque in longis quoque morbis etiam tales notas non habentibus scire licet increscere, si somnus incertus est, si deterior concoctio, si foediores deiectiones, si tardior sensus, si pigrior mens, si percurrit corpus frigus aut calor, si id magis pallet. Ea vero, quae contraria his sunt, decedentis eius notae sunt.[1] . . . Praeter haec in acutis morbis serius aeger alendus est, nec nisi iam iis inclinatis, ut primo dempta materia impetum frangat, in longis 4 maturius, ut sustinere spatium adfecturi mali possit. Ac si quando is non in toto corpore sed in parte est, magis tamen ad rem pertinet vim totius corporis moliri quam proprie partis aegrae sanitatem. Multum

[1] *Marx suggests that something is lost here.*

---

[a] Book II. 5, 2, 3.
[b] Periodic fevers. See below, 3. 2, desinens, desistit.

BOOK III. 2. 1-4

distress without intermission, the disease is acute: it becomes evident that the future course will be prolonged when there are but slight pains and fever, and when there are long intervals between the paroxysms, and there are in addition the signs which have been described in the preceding book.[a] It is also to be noted whether the disease is increasing, or stationary, or lessening, because some remedies are suitable for increasing, more for declining maladies; and when an acute fever is increasing in urgency, remedies which are suitable in decreasing affections are to be tried rather during the remissions. A disease is increasing in urgency when pains and paroxysms occur with more severity, and when they both recur at shorter intervals, and desist[b] later than before. And, in chronic diseases too, even if they do not present such characteristic signs, it may be recognized that the affection is increasing: if sleep is irregular; if digestion deteriorates; if the stools become more foul; the senses duller; the mind more sluggish; if a feeling of cold, or of heat, runs through the body, if the body becomes more pale. But opposite signs mark a decline in the disease. . . . In acute diseases, moreover, the patient is to be given food after more delay, and not until the paroxysm is already declining so that its force may be broken primarily by the withholding of nutriment; in chronic diseases, earlier, so that it may support the patient for the duration of his coming illness. But if sometimes, not the whole body, but a part only, is affected, still the support of the strength of the whole body rather than the curing by itself of the part diseased is of more importance. It makes a great difference also

etiam interest, ab initio quis recte curatus sit an perperam, quia curatio minus iis prodest, in quibus adsidue frustra fuit. Si qui temere habitus adhuc integris viribus vivit, admota curatione momento restituitur.

5 Sed cum ab iis coeperim (*lib.* II. 2), quae notas quasdam futurae adversae valetudinis exhibent, curationum quoque principium ab animadversione eiusdem temporis faciam. Igitur si quid ex is, quae proposita sunt (*lib.* II. 2), incidit, omnium optima sunt quies et abstinentia: si quid bibendum est, aqua, idque interdum uno die fieri satis est, interdum, si terrentia manent, biduo; proximeque abstinentiam sumendus est cibus exiguus, bibenda aqua, postero die etiam vinum, deinde invicem alternis diebus modo aqua modo vinum, donec omnis [causae] 6 metus finiatur. Per haec enim saepe instans gravis morbus discutitur. Plurimique falluntur, dum se primo die protinus sublaturos languorem aut exercitatione aut balneo aut coacta deiectione aut vomitu aut sudationibus aut vino sperant; non quo non interdum incidat aut non deceperit sed quo saepius fallat, solaque abstinentia sine ullo periculo medeatur: cum praesertim etiam pro modo terroris moderari liceat, et, si leviora indicia fuerunt, satis sit a vino tantum abstinere, quod subtractum plus, quam si 7 cibo quid dematur, adiuvat; si paulo graviora, facile sit non aquam tantum bibere sed etiam cibo

whether from the commencement the patient has been treated correctly or incorrectly, because treatment has less advantage in those cases in which a course has been persisted in without effect. If a patient lives through indiscreet treatment with his strength unimpaired, an appropriate treatment may restore him forthwith.

But as I commenced (II. 2) with those symptoms which show some signs of impending illness, I shall make a beginning as to treatment by noticing the same period. If, therefore, any of the signs then referred to occur, the best treatment is rest and abstinence; if anything at all is to be drunk, let it be water, and it is sufficient for this to be continued sometimes for one day, sometimes, when alarming signs persist, for two days; on the day following the fast, food should be taken sparingly, and water drunk; the next day even wine, and then in turn, on alternate days, water and wine, until all anxiety is at an end. For often in this way a severe disease is dispersed while it is impending. And many deceive themselves with the hope of getting rid of the languor straightaway on the first day, either by exercise, or by a bath, or by a purge, or by an emetic, or by sweating, or by drinking wine: not but that such a procedure may succeed or not disappoint, but more often it fails, and abstinence by itself is a remedy without any risk; especially since it also admits of being modified in accordance with the degree of apprehension, and if the indications are of the slighter kind, it is enough to abstain from wine alone, its withdrawal being more advantageous than if something were subtracted from the food; if they are of somewhat greater severity, it is easy to limit the drink to water, and at

carnem subtrahere, interdum panis quoque minus quam pro consuetudine adsumere, umidoque cibo esse contentos et holere potissimum, satisque sit tum ex toto a cibo, a vino, ab omni motu corporis abstinere, cum vehementes notae terruerunt. Neque dubium est, quin vix quisquam, qui non dissimulavit sed per haec mature morbo occurrit, aegrotet.

3. Atque haec quidem sanis facienda sunt tantum cum causa metuentibus. Sequitur vero curatio febrium quod et in toto corpore et vulgare maxime morbi genus est. Ex his una cotidiana, altera tertiana, altera quartana est. Interdum etiam longiore circuitu quaedam redeunt, sed id raro fit. In prioribus et morbi . . .[1] sunt et medicina.

Et quartanae quidem simpliciores sunt. Incipiunt fere ab horrore, deinde calor erumpit, finitaque febre biduum integrum est: ita quarto die revertitur.

2 Tertianarum vero duo genera sunt. Alterum eodem modo, quo quartana, et incipiens et desinens, illo tantum interposito discrimine, quod unum diem praestat integrum, tertio redit. Alterum longe perniciosius, quod tertio quidem die revertitur, ex quadraginta autem et octo horis fere triginta et sex per accessionem occupat (interdum etiam vel minus vel plus), neque ex toto in remissione desistit, sed tantum levius est. Id genus plerique medici ἡμιτριταῖον appellant.

3 Cottidianae vero variae sunt et multiplices.

[1] *Marx notes that some word has fallen out, e.g., varii; this suggestion is translated.*

the same time to withdraw meat from the diet, sometimes also to use less bread than usual, and to be content with moist food, especially pot-herbs; and it is sufficient to abstain entirely from food and wine, and from all bodily movement, only when serious symptoms have given rise to alarm. Nor is there a doubt that scarcely anyone falls ill who has hidden nothing but has countered disease in good time by these measures.

3. These then are the things to be done by those who, being in health, have cause merely to be apprehensive. Now there follows the treatment of fevers, a class of disease which both affects the body as a whole, and is exceedingly common. Of fevers, one is quotidian, another tertian, a third quartan. At times certain fevers recur in even longer cycles, but that is seldom. In the former varieties both the diseases and their medicines are of various kinds.

Now quartan fevers have the simpler characteristics. Nearly always they begin with shivering, then heat breaks out, and the fever having ended, there are two days free; thus on the fourth day it recurs.

But of tertian fevers there are two classes. The one, beginning and desisting in the same way as a quartan, has merely this distinction, that it affords one day free, and recurs on the third day. The other is far more pernicious; and it does indeed recur on the third day, yet out of forty-eight hours, about thirty-six, sometimes less, sometimes more, are in fact occupied by the paroxysm, nor does the fever entirely cease in the remission, but it only becomes less violent. This class most practitioners term *hemitritaion*.

Quotidian fevers, however, vary and have many

CELSUS

Aliae enim protinus a calore incipiunt, aliae a frigore, aliae ab horrore. Frigus voco, ubi extremae partes membrorum inalgescunt, horrorem, ubi corpus totum intremit. Rursus aliae sic desinunt, ut ex toto sequatur integritas; aliae sic, ut aliquantum quidem minuatur ex febre, nihilo minus tamen quaedam reliquiae remaneant, donec altera accessio accedat; ac saepe aliae ...[1] vix quicquam aut nihil
4 remittant sed continuent. Deinde aliae fervorem ingentem habent, aliae tolerabilem: aliae cotidie pares sunt, aliae inpares, atque invicem altero die lenior, altero vehementior ...:[2] aliae tempore eodem postridie revertuntur, aliae vel serius vel celerius: aliae diem noctemque accessione et decessione implent, aliae minus, aliae plus: aliae cum decedunt, sudorem movent, aliae non movent; atque alias per sudorem ad integritatem venitur, alias tantum
5 corpus inbecillius redditur. At accessiones etiam modo singulae singulis diebus fiunt, modo binae pluresve concurrunt. Ex quo saepe evenit, ut cotidie plures accessiones remissionesque sint, sic tamen, ut unaquaeque alicui priori respondeat. Interdum vero accessiones quoque confunduntur, sic ut notari neque tempora earum neque spatia possint. Neque verum est, quod dicitur a quibusdam, nullam febrem inordinatam esse, nisi aut ex vomica aut ex inflammatione aut ex ulcere: facilior enim semper curatio foret, si hoc verum esset: sed quod evidentes causae
6 faciunt, facere etiam abditae possunt. Neque de re sed de verbo controversiam movent, qui, cum

---

[1] *Marx supplies* 'iunguntur ut' *after* aliae *which is translated*.

[2] *Marx suggests* 'accessio' *after* vehementior *which is translated:* cf. 5, 3.

forms. For some begin straightaway with a feeling of heat, others of chill, others with shivering. I call it a chill when the extremities become cold, shivering when the whole body shakes. Again, some desist so that complete freedom follows, others so that there is some diminution of the fever, yet none the less some remnants persist until the onset of the next paroxysm; and others often run together so that there is little or no remission, but the attacks are continuous. Again, some have a vehement hot stage, others a bearable one; some are every day equal, others unequal, and the paroxysm in turn slighter one day, more severe another: some recur at the same time the day following, some either earlier or later; some take up a day and a night with the paroxysm and the remission, some less, others more; some set up sweating as they remit, others do not; and in some, freedom is arrived at through sweating, in others the body is only made the weaker. But the paroxysms also occur sometimes once on any one day, sometimes twice or more often.[a] Hence it often comes about that daily there are several paroxysms and remissions, yet so that each corresponds to one which has preceded it. But at times the paroxysms also become so confused together, that neither their durations nor intermissions can be observed. It is not true, as some say, that no fever is irregular unless as the outcome either of an abscess or of inflammation or of ulceration; for if this were true, the treatment always would be the easier, but what evident causes bring about, hidden ones can bring about also. And men are not arguing about facts but about words

[a] Cf. III. 5, 6.

aliter aliterque in eodem morbo febres accedunt, non easdem inordinate redire, sed alias aliasque subinde oriri dicunt; quod tamen ad curandi rationem nihil pertineret, etiamsi vere diceretur. Tempora quoque remissionum modo liberalia, modo vix ulla sunt.

4. Et febrium quidem ratio maxime talis est: curationum vero diversa genera sunt, prout auctores aliquos habent. Asclepiades officium esse medici dicit, ut tuto, ut celeriter, ut iucunde curet. Id votum est, sed fere periculosa esse nimia et festinatio et voluptas solet. Qua vero moderatione utendum sit, ut, quantum fieri potest, omnia ista contingant prima semper habita salute, in ipsis partibus curationum considerandum erit.

2 Et ante omnia quaeritur, primis diebus aeger qua ratione continendus sit. Antiqui medicamentis quibusdam datis concoctionem moliebantur, eo quod cruditatem maxime horrebant: deinde eam materiem, quae laedere videbatur, ducendo saepius alvum subtrahebant. Asclepiades medicamenta sustulit; alvum non totiens sed fere tamen in omni morbo subduxit; febre vero ipsa praecipue se ad remedium eius uti professus est: convellendas enim vires aegri putavit luce, vigilia, siti ingenti, sic ut ne os
3 quidem primis diebus elui sineret. Quo magis

if, when during the same illness fevers come on in different ways, they say that these are not irregular returns of the same fever, but other different ones arising in succession; even though it were true, it would have nothing to do with the mode of treatment. The duration of remissions also is at times considerable, at other times scarcely of any length.

4. Such for the most part is the account of fevers; but there are different sorts of treatment in accordance with what is held by the several authorities. Asclepiades said that it is the office of the practitioner to treat safely, speedily, and pleasantly. That is our aspiration, but there is generally danger both in too much haste and too much pleasure. But what moderation must be shown, in order that as far as possible all those blessings may be attained, the patient's safety being always kept first, will be considered among the actual details of the treatment.

Before everything is the question as to what regimen the patient should keep to during the first days. The ancients tried to ensure assimilation by administering certain medicaments, because they dreaded indigestion most of all; next by the repetition of clysters they extracted the matter which appeared to be doing harm. Asclepiades did away with medicaments; he did not clyster the bowel with such frequency but still he generally did this in every disease; but the actual fever, he professed to use as a remedy against itself: for he deemed that the patient's forces ought to be reduced by daylight, by keeping awake, by extreme thirst, so that during the first days he would not allow even the mouth to be swilled out. Therefore those are quite wrong who

falluntur, qui per omnia iucundam eius disciplinam esse concipiunt: is enim ulterioribus quidem diebus cubantis etiam luxuriae subscripsit, primis vero tortoris vicem exhibuit. Ego autem medicamentorum dari potiones et alvum duci non nisi raro debere concedo: non ideo tamen id agendum, ut aegri vires convellantur, existimo, quoniam ex inbecillitate
4 summum periculum est. Minui ergo tantum materiam superantem oportet, quae naturaliter digeritur, ubi nihil novi accedit. Itaque abstinendus a cibo primis diebus est; in luce habendus aeger, nisi infirmus, interdiu est, quoniam corpus ista quoque digerit, isque cubare quam maxime...[1] conclavi debet. Quod ad sitim vero somnumque pertinet, moderandum est, ut vigilet interdiu. Noctu, si fieri potest, conquiescat: ac neque potet,
5 neque nimium siti crucietur; os etiam eius elui potest, ubi et siccum est, et ipsi faetet, quamvis id tempus potioni aptum non est. Commodeque Erasistratus dixit saepe interiore parte umorem non requirente os et fauces requirere, neque ad rem male haberi aegrum pertinere. Ac primo quidem sic tenendus est.
6 Optimum vero medicamentum eius est oportune cibus datus; qui quando primum dari debeat, quaeritur. Plerique ex antiquis tarde dabant, saepe quinto die, saepe sexto; et id fortasse vel in Asia vel in Aegypto caeli ratio patitur. Asclepiades ubi aegrum triduo per omnia fatigarat, quartum diem cibo destinabat. At Themison nuper, non

[1] *Marx refers to* I. 2, 7, conclavi quam maxime et alto et lucido et spatioso. *Probably the same epithets have fallen out here.*

---

[a] See critical note. "In a room as lofty, light and airy as possible."

believe that his regimen was a pleasant one in all respects; for in the later days he allowed even luxuries to his patient, but in the first days of the fever he played the part of torturer. Now in my opinion medicinal draughts and clysters should only be administered occasionally; and I consider that they should not be used as to pull to pieces the patient's strength, since the greatest danger is from weakness. There ought to be, therefore, only such a diminution of superfluous matter as is dispersed by natural processes when nothing is being added afresh. Hence for the first days there is to be abstinence from food; the patient is to keep in the light during the day unless weak, for this also clears the body; and so he ought to lie up in a room as ... as possible.[a] As regards indeed thirst and sleep, it should be so managed that he keeps awake during the day; at night as far as possible he should be at rest; and he should neither drink much nor be too much distressed by thirst; his mouth also can be swilled out when dry, if he has a bad taste in it, even though that is not the time suitable for a drink. And Erasistratus said appropriately that often whilst the inside does not require fluid, the mouth and throat require it, and it does not help to keep the patient in suffering. And for the first days, such ought to be the regimen.

But his best medicament is food opportunely given; the question is when it should first be given. Most of the ancients gave it late, often on the fifth, often on the sixth day of illness, which the climate of Asia or of Egypt may perchance permit. Asclepiades, after he had for three days harassed the patient in every way, destined the fourth day for food. But Themison, recently, took into account

CELSUS

quando coepisset febris, sed quando desisset, aut certe levata esset, considerabat; et ab illo tempore expectato die tertio, si non accesserit febris, statim; si accesserat, ubi ea vel desierat, vel si adsidue inhaerebat, certe si se inclinaverat, cibum dabat.
7 Nihil autem horum utique perpetuum est. Nam potest primo die primus cibus dandus esse, potest secundo, potest tertio, potest non nisi quarto aut quinto, potest post unam accessionem, potest post duas, potest post plures. Refert enim qualis morbus sit, quale corpus, quale caelum, quae aetas, quod tempus anni; minimeque in rebus inter se multum differentibus perpetuum esse praeceptum temporis
8 potest. Ex morbo, qui plus virium aufert, celerius cibus dandus est, itemque eo caelo, quo magis digerit. Ob quam causam in Africa nulla die aeger abstineri recte videtur. Maturius etiam puero quam iuveni, aestate quam hieme dari debet. Unum illud est, quod semper, quod ubique servandum est, ut aegri vires subinde adsidens medicus inspiciat; et quamdiu supererunt, abstinentia pugnet; si inbecillitatem vereri coeperit, cibo subveniat. Id enim eius officium est, ut aegrum neque supervacua materia oneret, neque inbecillitatem fame
9 prodat. Idque apud Erasistratum quoque invenio; qui quamvis parum docuit, quando venter, quando corpus ipsum exinaniretur, dicendo tamen haec esse visenda et tum cibum dandum, cum corpori

not when the fever began, but when it ceased, or at any rate was alleviated; and awaiting the third day from that time, if there was no return of the fever, gave food at once; if fever recurred, he gave food when it ceased, or if it obstinately persisted, he certainly gave it if the fever abated. But on none of these matters is there actually an invariable precept. For it may be that the first food should be given on the first day, it may be on the second, it may be on the third, it may be not until the fourth or fifth day; it may be after one paroxysm, it may be after two, it may be after several. For it all depends upon the kind of disease, the patient's body, the climate, his age, and the time of year; where circumstances differ so greatly, there cannot be an invariable rule of time by any means. In the case of a disease which takes away more of the patient's strength, food is to be given earlier, and the same in a climate in which he uses up more. Hence in Africa it seems right that a patient should never fast over a day. Food should also be given sooner to a child than to an adolescent, sooner in summer than in winter. There is one thing that should be observed, always, and everywhere, that the patient's strength should be continually under the eye of the attending practitioner; and so long as there is a superfluity, he should counter it by abstinence; if he begins to fear weakness, he should assist with food. For it is his business to see that the patient is neither burdened by superfluous material nor rendered weak by hunger. And this I find also in the writings of Erasistratus; who although he did not direct when the bowels should be emptied, or when the body in general, nevertheless, by saying that such things should be seen to, and food given

CELSUS

deberetur, satis ostendit, dum vires superessent, dari non oportere: ne deficerent, consulendum esse. Ex his autem intellegi potest ab uno medico multos non posse curari, eumque, si artifex sit, idoneum
10 esse, qui non multum ab aegro recedit. Sed qui quaestui serviunt, quoniam is maior ex populo est, libenter amplectuntur ea praecepta, quae sedulitatem non exigunt, ut in hac ipsa re. Facile est enim dies vel accessiones numerare is quoque, qui aegrum raro vident: ille adsideat necesse est, qui quod solum opus est visurus est, quando nimis inbecillus futurus sit, nisi cibum acceperit. In pluribus tamen ad initium cibi dies quartus aptissimus esse consuevit.

11 Est autem alia etiam de diebus ipsis dubitatio, quoniam antiqui potissimum impares sequebantur, eosque, tamquam tum de aegris iudicaretur, κρίσιμους nominabant. Hi erant dies tertius, quintus, septimus, nonus, undecimus, quartus decimus, unus et vicesimus, ita ut summa potentia septimo, deinde quarto decimo, deinde uni et
12 vicensimo daretur. Igitur sic aegros nutriebant, ut dierum inparium accessiones expectarent, deinde postea cibum quasi levioribus accessionibus instantibus darent, adeo ut Hippocrates, si alio die febris desisset recidivam timere sit solitus. Id Asclepiades iure ut vanum repudiavit, atque in nullo die, qua par inparve esset, iis vel maius vel minus periculum

---

[a] This doctrine of numbers was derived from Pythagoras who may have learnt of the importance attached by Eastern

when it was needed by the body, showed sufficiently that food should not be given while the strength was in excess, but that care should be taken not to let it become deficient. Hence it can be understood that it is not possible for many patients to be cared for by one practitioner, and provided that he is skilled in the art, he is a suitable one who does not much absent himself from the patient. But they who are slaves to gain, since more is to be got out of a crowd, are glad to adopt those precepts which do not exact a sedulous attendance, as in this very instance. For even those who see the patient but seldom find it easy to count days or paroxysms; a physician must be always at hand, if he is to see the one thing that matters, the point when the patient is about to become too weak unless he gets food. The fourth day, however, is generally the most suitable date for beginning to give food.

But there is another uncertainty which concerns even the days themselves, since the ancients chiefly preferred the odd days and termed them critical,[a] as though then the fate of the sick man was decided. These were the third, fifth, seventh, ninth, eleventh, fourteenth and twenty-first days, the most importance being attached to the seventh, next to the fourteenth and then to the twenty-first. Therefore they administered food to their patients as follows: they awaited paroxysms on odd days, and after that they gave food, as though slighter paroxysms were impending, insomuch that Hippocrates, when the fever desisted on any other than an odd day, was accustomed to fear a recurrence. Asclepiades has justly repudiated this as false, and he said that no day was more or less dangerous

esse dixit. Interdum enim peiores dies pares fiunt, et oportunius post eorum accessiones cibus
13 datur. Nonnumquam etiam in ipso morbo dierum ratio mutatur, fitque gravior, qui remissior esse consuerat; atque ipse quartus decimus par est, in quo magnam vim esse antiqui fatebantur. Qui cum octavum primi die naturam habere contenderent, ut ab eo secundus septenarius numerus inciperet, ipsi sibi repugnabant non octavum, neque decimum, neque duodecimum diem sumendo quasi potentiorem:
14 plus enim tribuebant nono et undecimo. Quod cum fecissent sine ulla probabili ratione, ab undecimo non ad tertium decimum sed ad quartum decimum transibant. Est etiam apud Hippocraten eis, quos septimus dies liberaturus sit, quartum esse gravissimum. Ita illo quoque auctore in die pari et gravior febris esse potest et certa futuri nota. Atque idem alio loco quartum quemque diem ut in utrumque efficacissimum adprehendit, id est quartum,

---

Sages to numbers, both in relation to religion and to philosophy—the predominance of numbers in the rhythm of life and of Nature and of the heavens, as exemplified in the doctrine of the Sevens.

The doctrine of Crisis and of Critical days constituted an early attempt to coordinate medical opinion on the course of periodic fevers in malarial countries. We now explain that the origin of the doctrine was related to the cycles in the life-history of the malarial parasites. Benign tertian fever is set up by the Plasmodium vivax, which has a cycle of 48 hours and which occasions a recurrence of the fever every third day; quartan fever is due to the Plasmodium malariae, with a cycle of 72 hours, the fever recurring every fourth day; subtertian fever in hot countries is an infection by Plasmodium falciforme with a cycle of 48 hours; besides there are a variety of permutations, whether owing to a repetition of inoculation by the same parasite, or to a mixed infection,

to patients for being even or odd. For sometimes even days are the worse, and it is more suitable to give food after paroxysms on these days. Sometimes even in the course of the same fever the daily order changes, and that day becomes graver which had wont to have more of a remission; and besides, the fourteenth day itself, which the ancients confessed to be of great importance, is an even day. Since they held that the eighth day had the character of the first day, because from it began the second numbering of seven, they contradicted themselves in not giving more importance to the eighth, tenth and twelfth days, for they gave more to the ninth and eleventh. After doing this without any rational probability, they went on from the eleventh, not to the thirteenth, but to the fourteenth day. There is even in Hippocrates this statement, that the fourth day is the gravest in the case of those whom the seventh day is to liberate. So according to that very authority, there may be on an even day both a graver fever and a certain sign of what will happen. In another passage the same authority regarded each fourth day, namely, the fourth, seventh, eleventh,

---

etc. Thus two generations of the tertian parasite, maturing on succeeding days, produce a quotidian fever; two generations of the quartan parasite cause a fever on two successive days, with one day of freedom from fever.

Celsus argues against the theory of critical days by quoting irreconcilable statements in the *Aphorisms* (cf. Hipp. IV. 112, 144, 150, 154, *Aphorisms* II. 23, IV. 36, IV. 61, and IV. 71. See also II. 42, *Prog.* 20).

Hippocrates had used an ancient system of counting a continued series which repeated the last figure at the beginning of the next (1, 2, 3, 4—4, 5, 6, 7—7, 8, 9, 10); but this resulted in some series beginning with even numbers; the

CELSUS

septimum, undecimum, quartum decimum, septimum
15 decimum. In quo et ab inparis ad paris numeri rationem transit et ne hoc quidem propositum conservavit, cum a septimo die undecimus non quartus sed quintus sit. Adeo apparet, quacumque ratione ad numerum respeximus, nihil rationis sub illo quidem auctore reperiri. Verum in his quidem antiquos tum celebres admodum Pythagorici numeri fefellerunt, cum hic quoque medicus non numerare dies debeat, sed ipsas accessiones intueri, et ex his coniectare, quando dandus cibus sit.

16 Illud autem magis ad rem pertinet scire, tum oporteat dari, cum iam bene venae conquieverunt, an etiamnum manentibus reliquiis febris. Antiqui enim quam integerrimis corporibus alimentum offerebant: Asclepiades inclinata quidem febre sed etiamnum tamen inhaerente. In quo vanam rationem secutus est, non quo non sit interdum maturius cibus dandus, si mature timetur altera accessio, sed quo scilicet quam sanissimo dari debeat: minus enim conrumpitur quod integro
17 corpori infertur. Neque tamen verum est, quod Themisoni videbatur, si duabus horis integer futurus esset aeger, satius esse tum dare, ut ab integro potissimum corpore diduceretur. Nam si diduci tam celeriter posset, id esset optimum: sed cum id breve tempus non praestet, satius est principia

---

fourth series would begin with 10 instead of the important 11. In order to avoid this he started a new series beginning with 11 (11, 12, 13, 14). However, Celsus points out that even so no rational system can be arrived at, and that the attempt to connect the Pythagorean numbers, especially 7 and 11, with the course of fevers is absurd.

fourteenth, and seventeenth, as the most effective in both respects. Thus he passed from an odd system of reckoning to an even one, yet did not, even then, keep to his proposition; for the eleventh is not the fourth day after the seventh, but the fifth. It is clear enough that by whatever reasoning we view this numbering, there is to be found nothing rational in that authority at least. But in these matters indeed the Pythagorean numbers, then quite famous, deceived the ancients, since here also the practitioner ought not to count days, but observe the actual paroxysms, and from these infer when food should be given.

But it is much more pertinent to this subject to know whether food should be given when the pulse has well quieted down, or while remnants of the fever still persist. For the ancients proffered food when the bodies were free as far as possible from fever: Asclepiades did so when the fever was beginning to abate although present. In this he followed false reasoning; not that food may not be given earlier sometimes, if another paroxysm is feared soon, but it certainly ought to be given when the patient is at his soundest: for food is less corrupted when introduced into a body free from fever. Nor however, is that true, which Themison held, that if the patient was likely to be free from fever for a couple of hours, it was better to give food then, in order that the food might be distributed when the body was as far as possible fever-free. For if it were possible for it to be distributed so quickly, that would be the best plan; but since that short time does not allow of it, it is better that the first food should be received by a declining fever, rather than

cibi a decedente febre quam reliquias ab incipiente excipi. Ita si longius tempus secundum est, quam integerrimo dandum est; si breve, etiam antequam ex toto integer fiat. Quo loco vero integritas est, eodem est remissio, quae maxime in febre continua potest esse.

18 Atque hoc quoque quaeritur, utrum tot horae expectandae sint, quot febrem habuerunt, an satis sit primam partem earum praeteriri, quo[d] aegris iucundius insidat [si interdum non vacat].[1] Tutissimum est autem ante totius accessionis tempus praeterire, quamvis, ubi longa febris fuit, potest indulgeri aegro maturius, dum tamen ante minime pars dimidia praetereatur. Idque non in ea sola febre, de qua proxime dictum est, sed in omnibus ita servandum est.

5. Haec magis per omnia febrium genera perpetua sunt: nunc ad singulas earum species descendam. Igitur si semel tantum accessit, deinde desiit, eaque vel ex inguine vel ex lassitudine vel ex aestu aliave re simili fuit, sic ut interior nulla causa metum fecerit, postero die, cum tempus accessionis ita transiit,

2 ut nihil moverit, cibus dari potest. At si ex alto calor venit et gravitas vel capitis vel praecordiorum secuta est neque apparet quid corpus confuderit, quamvis unam accessionem secuta integritas est, tamen quia tertiana timeri potest, expectandus est

---

[1] *The text here appears to be corrupt*; si interdum non vacat *does not seem relevant. One MS. reads* quibus *for* si.

---

[a] Hipp. IV. 150 (*Aph.* IV. 55).

that remnants of food should be received by a recommencing fever. In this case, if the favourable time is longer, it should be given when the body is as free as possible; if short, even before it becomes quite free. But what also holds good for a full freedom does so also for a remission, which can occur, especially in the course of a continuous fever.

And there is the further question, whether it is necessary to wait for the same number of hours as the fever lasted, or if it is sufficient to suffer the first part of them to elapse so that the food may settle down more comfortably for the patient, [if sometimes there is no intermission]. It is safest, however, first to let pass the period of the whole preceding paroxysm, although in the case of a prolonged fever the patient may be indulged earlier, provided that half at least of that time has first passed. And this is to be observed not only in the fever just mentioned but in all.

5. The foregoing rules are rather of general application to fevers of all sorts: now I pass to their particular kinds. If, therefore, there has been only one paroxysm, then an intermission, and the fever arises either from the groin,[a] or from fatigue, or from hot weather, or some other similar thing, and so that it gives no apprehension of a more internal cause, then on the day following, when the time for the recurrence of a paroxysm has elapsed without any disturbance, food can be given. But when there supervenes a deeply seated heat and a sense of weight, whether in the head or in the parts below the ribs, and it is not evident what is disturbing the system, even although freedom follows upon a single paroxysm, nevertheless the third day is to be awaited because a

dies tertius; et ubi accessionis tempus praeteriit, cibus dandus est, sed exiguus, quia quartana quoque timeri potest; et die quarto demum, si corpus integrum est, eo cum fiducia utendum. Si vero postero tertiove aut quarto die secuta febris est, scire licet morbum esse. Sed tertianarum vel quartanarum, quarum et certus circumitus est et finis in integritate et liberaliter quieta tempora sunt, expeditior ratio est; de quibus suo loco dicam
3 (III. **14, 15**). Nunc vero eas explicabo, quae cotidie urgent. Igitur tertio quoque die cibus aegro commodissime datur, ut alter febrem minuat, alter viribus subveniat. Sed is dari debet, si cotidiana febris est, quae ex toto desinat, simul atque corpus integrum factum est: si quamvis non accessiones, febres tamen iunguntur et cotidie quidem increscunt sed sine integritate tamen remittunt, cum corpus ita se habet, ut maior remissio non expectetur; si altero die gravior, altero levior accessio est, post
4 graviorem. Fere vero graviorem accessionem levior nox sequitur; quo fit, ut graviorem accessionem nox quoque tristior antecedat. At si continuatur febris neque levior umquam fit et dari cibum necesse est, quando dari debeat, magna dissensio est. Quidam, quia fere remissius matutinum tempus aegris est, tum putant dandum. Quod si respondet, non quia

---

[a] Hipp. IV. 110 (*Aph.* II. 13).

tertian is to be apprehended; and when the time for such a paroxysm has passed, food is to be given, but in small amount, because a quartan may yet be apprehended; and not until the fourth day, if the body is still free, may it be used with confidence. But if on the second, or third, or fourth day fever has recurred, the disease can be recognized. But tertian and quartan fever in which there is both a definite cycle ending in freedom from fever, and ample periods of quiet, are more quickly dealt with, and of these I will speak in their proper place (III. **14, 15**). Now, however, I will explain the treatment of those fevers which cause trouble every day. Food, therefore, is more suitably given to the patient upon alternate days, in order one day to diminish the fever, the other to recruit his strength. But if it be that sort of quotidian fever in which there is a complete intermission, food should be given immediately upon the body becoming fever-free: if, although there are no paroxysms, the fever is nevertheless continuous and daily increasing, but with remissions that are not complete, food should be given when the system is in that state that no major remission is expected; if the paroxysm on one day is more severe, on the next day milder, food is to be given after the more severe paroxysm. For generally [a] an easier night follows upon the more severe paroxysm, and consequently a worse night precedes the more severe paroxysm. But if the fever continues without ever becoming milder, and it is necessary to give food, there is a great controversy as to the time when it should be given. Some, because patients generally have more of a remission early in the morning, think that food should be given then. But if this answers, the

mane est, sed quia remissior aeger est, dari debet.
5 Si vero ne tum quidem ulla requies aegris est, hoc ipso peius id tempus est, quod, cum sua natura melius esse debeat, morbi vitio non est; simulque insequitur tempus meridianum, a quo cum omnis aeger fere peior fiat, timeri potest, ne ille magis etiam quam ex consuetudine urgeatur. Igitur alii vespere tali aegro cibum dant: sed cum eo tempore fere pessimi sint qui aegrotant, verendum est, ne, si quid tunc moverimus, fiat aliquid asperius.
6 Ob haec ad mediam noctem decurro, id est, finito iam gravissimo tempore eodemque longissime distante, secuturis vero antelucanis horis, quibus omnes fere maxime dormiunt, deinde matutino tempore, quod natura sua levissimum est. Si vero febres vagae sunt, quia verendum est, ne cibum statim subsequantur, quandocumque quis ex accessione levatus
7 est, tunc debet adsumere. At si plures accessiones eodem die veniunt, considerare oportet, paresne per omnia sint, quod vix fieri potest, an inpares. Si per omnia pares sunt, post eam potius accessionem cibus dari debet, quae non inter meridiem et vesperum desinit. Si inpares sunt, considerandum est, quo distent: nam si gravior altera, altera levior est, post graviorem dari debet; si altera longior, altera brevior, post longiorem: si altera gravior, altera

reason for giving food is not the fact that it is morning, but the fact that the patient has more of a remission. But if the patient has no relief even in the morning, it becomes all the worse time for food, just because, although by itself that time should be better, owing to the fault of the disease it is not so; and at the same time, it is followed by midday, after which generally patients become worse, and so it may be feared that the patient may become more distressed than usual. To such a patient, therefore, others give food in the evening: but since at that time those who are ill are generally at their worst, there is fear that any action we may then take may exasperate the fever somewhat. For these reasons I delay until midnight, that is, when one critical time is over, and the next the furthest off, whilst the hours which follow before dawn are those during which all patients generally sleep the most; after that comes early morning, naturally a period of greatest relief. If, however, fevers are erratic, since there is apprehension that paroxysms may immediately follow food, whenever a patient begins to have relief after a paroxysm, then food ought to be taken. But if several paroxysms occur on the same day, it should be noted whether they are equal in all respects, which can scarcely ever be the case, or unequal. If they are equal in all respects, food should be given rather after any paroxysm which does not desist between midday and evening. If they are unequal, it is to be considered in what way they differ; for if one is more severe and another slighter, food should be given after the more severe; if one lasts longer, another a shorter time, after the longer; if one is more severe, another more prolonged, it is to be observed which of the

longior est, considerandum est, utra magis adfligat, illa vi, an haec tempore, et post eam dandum. 8 Sed paene plurimum interest, quantae qualesque inter eas remissiones sint: nam si post alteram febrem motio manet, post alteram integrum corpus est, integro corpore cibo tempus aptius est. Si semper febricula manet, sed alterum tamen longius tempus remissionis est, id potius eligendum est, adeo ut, ubi accessiones continuantur, protinus inclinata priore dandus cibus sit. Etenim perpetuum est, ad quod omne consilium derigi potest, cibum quam maxime semper ab accessione futura reducere, et hoc salvo, dare quam integerrimo 9 corpore. Quod non inter duas tantum sed etiam inter plures accessiones servabitur. Sed cum sit aptissimum tertio quoque die cibum dare, tamen si corpus infirmum est, cotidie dandus est; multoque magis, si continentes febres sine remissione sunt, quanto magis corpus adfligunt; aut si duae pluresve accessiones eodem die veniunt. Quae res efficit, ut et a primo die protinus cibus dari cotidie debeat, si protinus venae conciderunt; et saepius eodem die, si inter plures accessiones subinde vis corpori 10 deest. Illud tamen in his servandum est, ut post eas febres minus cibi detur, post quas, si per corpus liceret, omnino non daretur. Cum vero febris instet, incipiat, augeatur, consistat, decedat, deinde

---

[a] Hipp. IV. 102, *Aph.* I. 11.

two causes most distress, the former by its severity, or the latter by its length, and food must be given after the one which causes most distress. But what matters almost more than anything is, how long and of what kind are the remissions between them: for if after one paroxysm shivering persists, after another the body is free from this, the more suitable time for food is when the body is free. If a slight feverishness persists all the time, but a longer period of remission occurs at one time than at another, that is the time to be selected; so that, when paroxysms are continuous, straightway, when the first one has begun to pass off, food may be given. For it is the general rule[a] to which every plan of treatment should be directed, to give food always as long as possible before the next ensuing paroxysm, and while keeping this rule, to give the food when the body is most free from fever. This should be observed not merely with two paroxysms but also with several. But although it is most proper to give food on alternate days, yet if the system is weak, it should be given every day; and far more so if the fevers continue without remissions, inasmuch as they distress the patient more; or when two or more paroxysms occur on the same day. This occurrence renders it necessary that immediately from the first day, food must be administered daily if the pulse has immediately become weak, and several times on the same day, if in the course of several paroxysms there is progressive diminution of the bodily strength. However, in these cases we must keep to this rule, that less food is to be given after paroxysms of such a kind that no food at all would be given after them if the bodily condition allowed it. When, however, a fever threatens, begins,

in decessione consistat aut finiatur, scire licet optimum cibo tempus esse febre finita; deinde, cum decessio eius consistit; tertium, si necesse est, quandocumque decedit: cetera omnia periculosa esse. Si tamen propter infirmitatem necessitas urget, satius esse consistente iam incremento febris aliquid offerre quam increscente, satius esse instante quam incipiente, cum eo tamen, ut nullo tempore is, qui 11 deficit, non sit sustinendus. Neque Hercules satis est ipsas tantum febres medicum intueri, sed etiam totius corporis habitum et ad eum derigere curationem, sive supersunt vires seu desunt seu quidam alii affectus interveniunt. Cum vero semper aegros securos agere conveniat, ut corpore tantum, non etiam animo laborent, tum praecipue, ubi cibum sumpserunt. Itaque si qua sunt, quae exasperatura eorum animos sunt, optimum est ea, dum aegrotant, eorum notitiae subtrahere: si id fieri non potest, sustinere tamen post cibum usque somni tempus, et cum experrecti sunt, tum exponere.

6. Sed de cibo quidem facilior cum aegris ratio est, quorum saepe stomachus hunc respuit, etiamsi mens concupiscit: de potione vero ingens pugna est, eoque magis, quo maior febris est. Haec enim sitim accendit, et tum maxime aquam exigit, cum illa periculosissima est. Sed docendus aeger est,

increases, continues stationary, declines, then persists at a low level, or terminates, it should be recognized that the best time for food is after the fever has terminated; next, when it is continuing at a diminished level; and thirdly, if need be, whenever there is a decline; all other times are dangerous. If, however, there is urgent necessity on account of weakness, it is better to give some food when the increase in the fever has become stationary, rather than whilst it is increasing, better whilst the paroxysm is as yet imminent, rather than after it has commenced, nevertheless with this proviso, that there is no time at which a patient who is failing should not be supported. Most emphatically, it is not enough for the practitioner to pay attention merely to the actual fevers, but also he must look to the habit of the body as a whole, and direct treatment to that, whether patients have superabundance or deficiency of strength, or whether there are other intervening affections. While, however, it is always of advantage for patients to be free from care, so that they may suffer in body alone, and not also in spirit, it is so especially after food has been taken. Therefore if there are any things which might exasperate their emotions, it is best to withhold these from notice whilst they are ill: if this cannot be done, nevertheless to keep all back after food, until the time of sleep, and to tell them when they wake up.

6. But the rationing of patients' food is the easier because often the stomach spues it back, although the appetite is eager for it; over drink, however, there is a mighty battle, the more so the greater the fever. For fever inflames thirst, and then most demands water when it is most dangerous. But

CELSUS

ubi febris quierit, protinus sitim quoque quieturam, longioremque accessionem fore, si quod ei datum fuerit alimentum: ita celerius eum desinere sitire, 2 qui non bibit. Necesse est tamen, quanto facilius etiam sani famem quam sitim sustinent, tanto magis 3 aegris in potione quam in cibo indulgere. Sed primo quidem die nullus umor dari debet, nisi subito sic venae ceciderunt, ut cibus quoque dari debeat, secundo vero ceterisque etiam, quibus cibus non dabitur, tamen si magna sitis urgebit, potio dari 4 debet. Ac ne illud quidem ab Heraclida Tarentino dictum ratione caret: ubi aut bilis aegrum aut cruditas male habet, expedire quoque per modicas potiones misceri novam materiam corruptae. Illud videndum est, ut qualia tempora cibo leguntur, talia potioni quoque, ubi sine illo datur, deligantur . . .,[1] aut cum aegrum dormire cupiemus, quod fere sitis prohibet. Satis autem convenit, cum omnibus febricitantibus nimius umor alienus sit, tum praecipue esse feminis, quae ex partu in febres inciderunt.

5 Sed cum tempora cibo potionique febris et remissionis ratio det, non est expeditissimum scire, quando aeger febricitet, quando melior sit, quando deficiat; sine quibus dispensari illa non possunt. Venis enim maxime credimus, fallacissimae rei, quia saepe istae leniores celerioresve sunt et aetate et sexu et corporum natura. Et plerumque satis sano corpore, si stomachus infirmus est, nonnumquam

[1] *Ritter and other editors have noted that there is a lacuna here.*

the patient is to be taught that when the fever quiets down, thirst also will become quiet at once, and that the paroxysm will be prolonged if any sustenance is given to it: thus he who does not drink will the sooner cease to be thirsty. It is necessary, however, seeing that even in health hunger is more easily borne than thirst, to indulge patients more as to drink than food. But on the first day, at any rate, no fluid at all should be given, unless the pulse sinks so suddenly that food as well ought to be given: on the second day too and even on later days upon which food is not given, yet if great thirst oppresses, drink should be given. And indeed that dictum of Heraclides of Tarentum was not wanting in reason: whenever either bile or indigestion disorders the patient, it is also expedient by draughts in moderation to mingle fresh material with the decomposing. We must see that, just as times are appointed for food, so they are appointed also for drink when given apart from food, . . . or when we want the patient to get the sleep which thirst usually prevents. But there is sufficient agreement that for all who are feverish an excess of fluid is unsuitable, and especially for women who have lapsed into fever after childbirth.

But although the character of the fever, and of its remission, fixes the time for giving food and drink, yet it is not very easy to know when the patient has fever, when he is better, when he is becoming worse: without which food and drink cannot be administered. For the pulse upon which we mostly rely (III. **4, 16**) is a very deceptive thing, because often it is rendered slower or faster by age and by sex and by constitution. And very frequently when the body is fairly healthy, if the stomach is weak,

CELSUS

etiam incipiente febre, subeunt et quiescunt, ut inbecillus is videri possit, cui facile laturo gravis 6 instat accessio. Contra saepe eas concitare solet balneum et exercitatio et metus et ira et quilibet alius animi adfectus, adeo ut, cum primum medicus venit, sollicitudo aegri dubitantis, quomodo illi se habere videatur, eas moveat. Ob quam causam periti medici est non protinus ut venit adprehendere manu brachium, sed primum desidere hilari vultu percontarique, quemadmodum se habeat, et si quis eius metus est, eum probabili sermone lenire, tum deinde eius corpori manum admovere. Quas venas autem conspectus medici movet, quam facile 7 mille res turbant. Altera res est, cui credimus, calor, aeque fallax: nam hic quoque excitatur aestu, labore, somno, metu, sollicitudine. Intueri quidem etiam ista oportet, sed eis non omnia credere. Ac protinus quidem scire est, non febricitare eum, cuius venae naturaliter ordinatae sunt, teporque talis est, qualis esse sani solet: non protinus autem sub calore motuque febrem esse concipere, sed ita: si summa quoque arida inaequaliter cutis est; si calor et in fronte est et ex imis praecordiis oritur; si spiritus ex naribus cum fervore prorumpit; si color aut rubore aut pallore novo mutatus est; si oculi graves et aut persicci aut subumidi sunt; si sudor, cum sit, inaequalis est; si venae non

BOOK III. 6. 5–7

also at times when a fever is beginning, the pulse is low and quiescent, so that possibly a patient may seem weak who will yet easily support the impending severe paroxysm. On the contrary, the bath and exercise and fear and anger and any other feeling of the mind is often apt to excite the pulse; so that when the practitioner makes his first visit, the solicitude of the patient who is in doubt as to what the practitioner may think of his state, may disturb his pulse. On this account a practitioner of experience does not seize the patient's forearm with his hand, as soon as he comes, but first sits down and with a cheerful countenance asks how the patient finds himself; and if the patient has any fear, he calms him with entertaining talk, and only after that moves his hand to touch the patient. If now the sight of the practitioner makes the pulse beat, how easily may a thousand things disturb it! Another thing which we put faith in, a sensation of heat, is equally fallacious: for it may be excited by hot weather, by work, by sleep, by fear, by anxiety. Such things also should be noted indeed, but not altogether relied on. And we know at once that he is not feverish, whose pulse is of natural regularity, and his warmth such as is customary in health: we must not, however, at once assume fever if there is heat and high pulse, but under the following conditions: if also the surface of the skin is dry in patches; if both the forehead feels hot, and it feels hot deep under the heart; if the breath streams out of the nostrils with burning heat; if there is a change of colour whether to unusual redness or to pallor; if the eyes are heavy and either very dry or somewhat moist; if sweat, when there is any, comes in patches; if the

8 aequalibus intervallis moventur. Ob quam causam medicus neque in tenebris neque a capite aegri debet residere, sed inlustri loco adversus, ut omnes notas ex voltu quoque cubantis percipiat. Ubi vero febris fuit ac decrevit, expectare oportet, num tempora partesve corporis aliae paulum madescant, quae sudorem venturum esse testentur; ac si qua nota est, tum demum dare potui aquam calidam, cuius salubris effectus est, si sudorem per omnia membra 9 diffundit. Huius autem rei causa continere aeger sub veste satis multa manus debet, eademque crura pedesque contegere; qua male plerique aegros in ipso febris impetu, pessimeque, ubi ardens ea est, male habent. Si sudare corpus coepit, linteum tepefacere oportet paulatimque singula membra detergere. At ubi sudor omnis finitus est, aut si is non venit, ubi quam maxime potuit idoneus esse cibo aeger videtur, leviter sub veste ungendus est, 10 tum detergendus, deinde ei cibus dandus. Cibus autem febricitantibus umidus est aptissimus aut umori certe quam proximus, utique ex materia quam levissima maximeque sorbitio; eaque, si magnae febres fuerint, quam tenuissima esse debet. Mel quoque despumatum huic recte adicitur, quo corpus magis nutriatur: sed id si stomachum offendit, supervacuum est, sicut ipsa quoque sorbitio. Dari vero in vicem eius potest vel intrita ex aqua calida vel halica elota; si firmus est stomachus et

---

[a] Hipp. IV. 154 (*Aph.* IV. 54).
[b] intrita sc. farina, crumbled bread, caudle, sop, cf. Pliny, *N.H.* IX. 8, and Varro, *R.R.* III. ix. 21.

BOOK III. 6. 7–10

pulse is irregular. On this account the practitioner should not take his seat in a dark part of the room, nor at the patient's head, but he should face the patient in a good light, so that he may note all the signs from his face as he lies in bed. Now when there has been fever and it has decreased, one should observe whether the temples or other parts of the body are becoming a little moist, which is evidence that sweating is about to set in; and if there is any sign of it, then and not before hot water should be given to drink, of which the effect is salutary if it causes a general sweating all over the body. Now to promote this the patient should keep his hands well covered under the bedclothes, and do the same with his legs and feet. But it is a mistake to torment patients with bed-clothes, as many do, at the very paroxysm of the fever, worst of all when it is an ardent fever. If the body begins to sweat, a linen towel should be warmed, and each part gradually wiped over. But when the sweating has quite ended, or if none has come, when the patient seems in the most fit state for food, he should be anointed lightly under the bedclothes, next wiped over, and then given food. For patients in fever, liquid [a] food is best, or whatever approximates to fluid, and that of the lightest possible kind, barley gruel in particular; and if there have been high fevers, that should be of the thinnest. Honey also which has been freed from the comb may be correctly added to give the body more nutriment; but if it upsets the stomach this is unnecessary, as also is the gruel itself. But in its place can be given either crumbled bread [b] or washed spelt groats in hot water; in hydromel if the stomach is firm and the bowels tight,

compressa alvus, ex aqua mulsa; si vel ille languet
vel haec profluit, ex posca. Et primo quidem cibo
id satis est: secundo vero aliquid adici potest, ex
eodem tamen genere materiae, vel holus vel conchylium vel pomum. Et dum febres quidem increscunt,
hic solus cibus idoneus est: ubi vero aut desinunt aut
levantur, semper quidem incipiendum est ab aliquo
ex materia levissima, adiciendum vero aliquid ex
media, ratione habita subinde et virium hominis
et morbi. Ponendi vero aegro varii cibi, sicut
Asclepiades praecepit, tum demum sunt, ubi fastidio
urgetur neque satis vires sufficiunt, ut paulum ex
singulis degustando famem vitet. At si neque
vis neque cupiditas deest, nulla varietate sollicitandus
aeger est, ne plus adsumat quam concoquat. Neque
verum est, quod ab eo dicitur, facilius concoqui
cibos varios: eduntur enim facilius, ad concoctionem
autem materiae genus et modus pertinent. Neque
inter magnos dolores neque increscente morbo
tutum est aegrum cibo impleri, sed ubi inclinata
iam in melius valetudo est.

Sunt aliae quoque in febribus observationes
necessariae. Atque id quoque videndum est, quod
quidam (*prohoem.* 55) solum praecipiunt, adstrictum
corpus sit an profluat; quorum alterum strangulat,
alterum digerit.

Nam si adstrictum est, ducenda alvus est, movenda
urina, eliciendus omni modo sudor. In hoc genere

---

^a Hipp. IV. 102 (*Aph.* I. 7.)

or in vinegar and water if the former is weak and the latter loose. And indeed this will suffice for food on the first day; then on the next day some addition can be made, yet from the same class of food, either pot-herbs or shell-fish or orchard fruit. And whilst fevers are on the actual increase, this is the only suitable food; but when the fevers have subsided or abated, a beginning indeed is to be made always with something of the lightest kind, then something to be added of the middle class, regard being had throughout both to the patient's strength and to his disease. A variety of food may be placed before the patient as Asclepiades prescribed, only when he is troubled by loss of appetite, and insufficiency of strength, in order that by tasting a little of each he may avoid starvation. But if there is no lack of strength nor loss of appetite, the patient should not be tempted by a variety of food, lest he take more than he can digest. And there is no truth in what Asclepiades said, that a variety of food is more easily digested; for it is eaten more readily, but digestion depends upon what the food is, and how much.[a] Nor is it safe for the patient to be filled up with food whilst there are great pains, nor during an increase of the malady, but only after his illness has turned towards improvement.

In fevers there are also other things that have to be observed. And this also must be noted, which some give as their sole precept, whether the body is constricted or relaxed; the first condition chokes it, the second wastes it away.

For if there is constriction, the bowels are to be moved by a clyster, urination promoted, and sweating elicited in every way. In this class of maladies it is

CELSUS

morborum emisisse sanguinem, concussisse vehementibus gestationibus corpus, in lumine habuisse, 14 imperasse famem, sitim, vigiliam prodest. Utile est etiam ducere in balneum, prius demittere in solium, tum ungere, iterum ad solium redire multaque aqua fovere inguina; interdum etiam oleum in solio cum aqua calida miscere; uti cibo serius et rarius, tenui, simplici, molli, calido, exiguo, maximeque holeribus, qualia sunt lapatium, urtica, malva, vel iure etiam concharum musculorumve aut lucustarum: neque danda caro nisi elixa est. At potio esse debet magis liberalis, et ante cibum et post hunc et cum hoc ultra quam sitis coget. Poteritque a balineo etiam pinguius aut dulcius dari vinum; poterit semel aut bis interponi Graecum salsum.

15 Contra vero si corpus profluit, sudor coercendus, requies habenda erit, tenebris somnoque, quandoque volet, utendum, non nisi leni gestatione corpus agitandum, et pro genere mali subveniendum. Nam si venter fluit, aut si stomachus non continet, ubi febris decrevit, liberaliter oportet aquam tepidam potui dare, et vomere cogere, nisi aut fauces aut praecordia aut latus dolet, aut vetus morbus est.

16 Si vero sudor exercet, duranda cutis est nitro vel sale, quae cum oleo miscentur; ac si levius id vitium est, oleo corpus ungendum; si vehementius, rosa vel melino vel murteo, cui vinum austerum sit adiectum. Quisquis autem fluore aeger est, cum venit in balineum, prius ungendus, deinde in solium

---

a The hot bath, II. 17, 7.; cf p. 52 note.

beneficial to let blood, to shake up the body by vigorous rocking, to keep the patient in the light, to impose hunger and thirst and wakefulness. It is also useful to take the patient to the bath, putting him first into the solium,[a] next to anoint him, then to return him to the solium again and foment his groins with plenty of water; at times also oil may be mixed with the water in the solium; food is to be used later and not too often: it is to be thin, plain, soft, hot, scanty, consisting mainly of pot-herbs, such as sorrel, nettle-tops, mallow, and also of soup made from shell-fish, mussels or spiny lobsters. No meat should be given unless boiled. But as to drink, there should be more freedom, both before and after and along with food, beyond what thirst demands. Again, after the bath wine of fuller body and sweeter can also be given; once or twice Greek salted wine can be used.

On the contrary, however, if the system is relaxed, sweating is to be suppressed, rest in a dark room resorted to, and sleep allowed at will; the body is to be rocked only in the lightest fashion, and helped as may suit the illness. For if the patient has loose motions, or if the stomach does not retain its contents, when the fever has subsided he should be given a large drink of tepid water, and be induced to vomit, unless the throat or chest or the side is painful, or the disease is of long standing. But if sweating is troublesome, the skin should be hardened by nitre or salt, mixed with oil; and if the sweating is rather slight, the body is to be anointed with olive oil: if more profuse, with rose, quince, or myrtle oil, to which a dry wine should be added. But any patient with loose motions, when he reaches the bath, should

17 demittendus est. Si in cute vitium est, frigida quoque quam calida aqua melius utetur. Ubi ad cibum ventum est, dari debet is valens, frigidus, siccus, simplex, qui quam minime corrumpi possit, panis tostus, caro assa, vinum austerum vel certe subausterum; si venter profluit, calidum, si sudores nocent vomitusve sunt, frigidum.

7. 1. Desiderat quoque propriam animadvorsionem in febribus pestilentiae casus. In hac utile minime est aut fame aut medicamentis uti, aut ducere alvum. Si vires sinunt, sanguinem mittere optimum est, praecipueque si cum dolore febris est: si id parum tutum est, ubi febris aut tenuata est aut levata est, vomitu pectus purgare. Sed in hoc maturius quam in aliis morbis ducere in balineum opus est, vinum calidum et meracius dare, et omnia glutinosa; inter quae carnem quoque generis eiusdem. Nam quo celerius eiusmodi tempestates corripiunt, eo maturius auxilia etiam cum quadam temeritate rapienda sunt. Quod si puer est qui laborat, neque tantum robur eius est, ut ei sanguis mitti possit, siti[1] ei utendum est, ducenda alvus vel aqua vel tisanae cremore, tum denique is levibus cibis nutriendus. Et ex toto non sic pueri ut viri curari debent. Ergo, ut in alio quoque genere morborum, parcius in his agendum est: non facile sanguinem mittere, non facile ducere alvum, non cruciare vigilia fameque aut nimia siti, non vino curare satis convenit. Vomitus post febrem eli-

---

[1] *v. d. Linden suggested* cucurbitulis, 'cupping,' *for* siti.

a See I. 10.

be first anointed, then put into the solium. When there is anything wrong with the skin, it is better to use cold rather than hot water. Coming to the food, this should be nutritious, cold, dry, plain, with the least possible tendency to decomposition, bread toasted, meat roasted, wine dry or at any rate somewhat dry; if the bowels are loose, the wine should be hot, but cold when there is trouble from sweating or vomiting.

7. Among fevers the case of pestilence[a] demands special consideration. In this it is practically useless to prescribe fasting or medicine or clysters. If strength permits of it, blood-letting is best, and especially if there is fever with pain: but if that is hardly safe, after the fever has either declined or remitted, the chest is cleared by an emetic. But in such cases the patient requires to be taken to the bath earlier than in other affections, to be given hot and undiluted wine, and all food glutinous, including that sort of meat. For the more quickly such violent disorders seize hold, the earlier are remedies to be taken in hand, even with some temerity. But if a child is the sufferer, and not robust enough for blood-letting to be possible, thirst is to be used in his case, the bowels are to be moved by a clyster whether of water or of pearl-barley gruel; then and not before he is to be sustained by light food. Indeed in general children ought not to be treated like adults. Therefore, as in any other sort of disease, we must set to work with more caution in these cases; not let blood readily, not readily clyster, not torment by wakefulness and by hunger or excess of thirst, nor is a wine treatment very suitable. After the remission of the fever a vomit is to be elicited, then

ciendus est, deinde dandus cibus ex levissimis, tum is dormiat; posteroque die, si febris manet, abstineatur; tertio ad similem cibum redeat. Dandaque opera est, quantum fieri potest, ut inter oportunam abstinentiam cibosque oportunos, omissis ceteris, nutriatur.

2. Si vero febris ardens extorret, nulla medicamenti danda potio est, sed in ipsis accessionibus oleo et aqua refrigerandus est; quae miscenda manu sunt, donec albescant. Eo conclavi tenendus, quo multum et purum aerem trahere possit; neque multis vestimentis strangulandus, sed admodum levibus tantum velandus est. Possunt etiam super stomachum inponi folia vitis in aqua frigida tincta. Ac ne siti quidem nimia vexandus est. Alendus maturius est, id est a tertio die, et ante cibum idem perungendus. Si pituita in stomachum coit, inclinata iam accessione vomere cogendus est; tum dandum frigidum holus, aut pomum ex iis, quae stomacho conveniunt. Si siccus manet stomachus, protinus vel tisanae vel halicae vel orizae cremor dandus est, cum quo recens adeps cocta sit. Cum vero in summo incremento morbus est, utique non ante quartum diem, magna siti antecedente, frigida aqua copiose praestanda est, ut bibat etiam ultra satietatem. Cum iam venter et praecordia ultra modum repleta satisque refrigerata sunt, vomere debet. Quidam ne vomitum quidem exigunt, sed ipsa aqua frigida tantum ad satietatem data pro medicamento utuntur. Ubi utrumlibet factum est, multa veste

BOOK III. 7. 1. c—7. 2. c

food of the lightest nature is given, after which let the child sleep; next day, if the fever persists, let the child be kept without food, and on the third day return to food as above. Our aim should be, as far as possible to sustain the child, by food when suitable, with abstinence in between when suitable, omitting all else.

But if an ardent fever is parching up the patient, no medicinal draught is to be given, but during the paroxysms he is to be cooled by oil and water, mixed by the hand until they turn white. He should be kept in a room where he can inhale plenty of pure air; he is not to be stifled by a quantity of bed-clothes, but merely covered by light ones. Vine leaves also which have been dipped in cold water can be laid over the stomach. He is not even to be distressed by too much thirst; he should get food fairly soon, namely from the third day, and after being anointed beforehand. If phlegm collects in the stomach, when the paroxysm has already declined he is to be made to vomit; then to be given cold salads, or orchard fruit agreeable to the stomach. If the stomach remains dry, there should be given to begin with either pearl barley or spelt or rice gruel with which fresh lard has been boiled. Whilst the fever is at its height, certainly not before the fourth day, and if there is already great thirst, cold water is to be administered copiously so that the patient may drink even beyond satiety. As soon as the stomach and chest have become replete beyond measure and sufficiently cooled, he should vomit. Some do not even insist on the vomit, but use the cold water by itself, given up to satiety, as the medicament. When either of the above has been done, the patient is to

CELSUS

operiendus est, et collocandus ut dormiat; fereque post longam sitim et vigiliam, post multam satietatem, post infractum calorem plenus somnus venit; D per quem ingens sudor effunditur, idque praesentissimum auxilium est, sed in iis tamen, in quibus praeter ardorem nulli dolores, nullus praecordiorum tumor, nihil prohibens vel in thorace vel in pulmone vel in faucibus, non ulcera, non deiectio,[1] non profluvium alvi fuit. Si quis autem in huiusmodi febre leviter tussit, is neque vehementi siti conflictatur, neque bibere aquam frigidam debet, sed eo modo curandus est, quo in ceteris febribus praecipitur.

8. At ubi id genus tertianae est, quod emitritaeon medici appellant, magna cura opus est, ne id fallat: habet enim plerumque frequentes accessiones decessionesque, ut aliud genus morbi videri possit porrigique febris inter horas XXIIII et XXXVI, 2 ut quod idem est, non idem esse videatur. Et magnopere necessarium est neque dari cibum nisi in ea remissione, quae vera est, et ubi ea venit, protinus dari. Plurimique sub alterutro curantis errore subito moriuntur. Ac nisi magnopere res aliqua prohibet, inter initia sanguis mitti debet, tum dari cibus, qui neque incitet febrem, et tamen longum eius spatium sustineat.

9. Nonnumquam etiam lentae febres sine ulla remissione corpus tenent, ac neque cibo neque ulli remedio locus est. In hoc casu medici cura esse

---

[1] *Some MSS. read* non defectio non deiectio *and some editors keep* defectio (*fainting*) *deleting* deiectio *as a gloss on* profluvium alvi.

---

[a] Hipp. IV. 154 (*Aph.* IV. 54).
[b] Cf. p. 227. ff.

be well wrapped up and put to bed so that he may sleep; and generally, after prolonged thirst and wakefulness, after full sating with water, after making a break in the heat, there comes abundant sleep: which brings on a profuse sweat, and this is an immediate relief, but only to those who have no pains accompanying the ardent fever, no swelling of the parts below the ribs, nothing prohibitory either in the chest or in the lung or in the throat, no ulcerations, no diarrhoea, no flux from the bowel. But if in fever [a] of this sort the patient coughs readily, he is not to be distressed by severe thirst, nor ought he to drink water cold, but he is to be treated in the way prescribed for other fevers.

8. But when the fever is that kind of tertian which the physicians call hemitritaion,[b] great care is required to avoid a mistake, for it has a number of frequently recurring paroxysms and remissions, so that it can appear to be some other class of disease and the fever may last from twenty-four to thirty-six hours, so that what is really the same paroxysm may not seem to be the same. And it is then exceedingly important not to give food except in that remission which is a real one, and when that does come, to give it at once. Many die suddenly from error one way or the other on the part of the practitioner. And unless something strongly prohibits, blood should be let at the onset, then food is to be given, which, without exciting the fever, should yet sustain a long course of it.

9. Sometimes also slow fevers hold the body without any remission, and give no place for either food or any medicament. In that case it should be the aim of the practitioner to change the disease, for

debet, ut morbum mutet; fortasse enim curationi oportunior fiet. Saepe igitur ex aqua frigida, cui oleum sit adiectum, corpus eius pertractandum est, quoniam interdum sic evenit, ut horror oriatur et fiat initium quoddam novi motus exque eo, cum magis corpus incaluit, sequatur etiam remissio. In his 2 frictio quoque ex oleo et sale salubris videtur. At si diu frigus est et torpor et iactatio corporis, non alienum est in ipsa febre mulsi dare tres aut quattuor cyathos, vel cum cibo vinum bene dilutum. Intenditur enim saepe ex eo febris, et maior ortus calor simul et priora mala tollit et spem remissionis inque ea curationis ostendit. Neque Hercules ista curatio nova est, qua nunc quidem traditos sibi aegros, qui sub cautioribus medicis trahebantur, interdum contrariis remediis sanant. Siquidem apud antiquos quoque ante Herophilum et Erasistratum maximeque post Hippocratem fuit ... Petro quidam, qui febricitantem hominem ubi acceperat, multis vestimentis operiebat, ut simul calorem ingentem 3 sitimque excitaret. Deinde ubi paulum remitti coeperat febris, aquam frigidam potui dabat, ac si moverat sudorem, explicuisse se aegrum iudicabat; si non moverat, plus etiam aquae frigidae ingerebat et tum vomere cogebat. Si alterutro modo febre liberaverat, protinus suillam assam et vinum homini dabat; si non liberaverat, decoquebat aquam sale

¹ *Marx suggests that the words* 'celebris. Fuit' *have fallen out.*

---

[a] If Marx's suggestion (cf. crit. note) is right the words, 'This treatment was well known,' must be inserted.
[b] Galen (XV. 436) also copied from Erasistratus this prescription of meat and wine by Petron of Aegina, also known for his work on pharmacy (Galen, XIII. 642).

perhaps that will make it more amenable to treatment. For this object cold water, to which oil has been added, should be sprinkled at frequent intervals over the patient's body, for it thus comes about now and again that a shivering follows, and some beginning of a fresh pulse motion, and after this when, after the body has become hotter, there may even follow a remission. Rubbing with oil and salt appears also to benefit such cases. But if for a long while there is a chill, and a numbness and a tossing of the body, it is not unfitting to administer three or four cups of honeyed wine even while the fever is present, or food along with wine well diluted. For often in this way the fever is augmented, and the increased heat which arises simultaneously both relieves the pre-existing disorders and offers hope of a remission, and through that of treatment. Assuredly that treatment is no novelty by which some nowadays at times cure by contrary remedies patients who have been handed over to them, after dragging on under more cautious practitioners. Even among the ancients, before Herophilus and Erasistratus, but especially after Hippocrates. . . .[a] There was a certain Petron, who on taking over a patient with fever, covered him with a quantity of clothes in order simultaneously to excite great heat and thirst. Then when the fever began to remit somewhat, he gave cold water to drink; and if this raised a sweat, he declared that the patient was recovering; if it did not, he administered even more cold water and then forced him to vomit. If by either of the above ways he had rendered the patient free from fever, he at once gave him roast pork and wine;[b] if he had not so freed him, he boiled water

adiecto eamque bibere cogebat, ut movendo ventrem purgaret. Et intra haec omnis eius medicina erat: eaque non minus grata fuit is, quos Hippocratis successores non refecerant, quam nunc est is, quos Herophili vel Erasistrati aemuli diu tractos non adiuverunt. Neque ideo tamen non est temeraria ista medicina, quia, si plures protinus a principiis excepit, interemit. Sed, cum eadem omnibus convenire non possint, fere quos ratio non restituit, temeritas adiuvat; ideoque eiusmodi medici melius alienos aegros quam suos nutriunt. Sed est circumspecti quoque hominis et novare interdum et augere morbum et febres accendere, quia curationem ubi id quod est non recipit, potest recipere id quod futurum est.

10. Considerandum est etiam, febresne solae sint, an alia quoque his mala accedant, id est, num caput doleat, num lingua aspera, num praecordia intenta sint.

Si capitis dolores sint, rosam cum aceto miscere oportet et in id ingerere; deinde habere duo pittacia, quae frontis latitudinem longitudinemque aequent, ex his invicem alterum in aceto et rosa habere, alterum in fronte; aut intinctam iisdem lanam sucidam inponere. Si acetum offendit, pura rosa utendum est; si rosa ipsa laedit, oleo acerbo. Si ista parum iuvant, teri potest vel iris arida vel nuces amarae vel quaelibet herba ex refrigerantibus;

• II. 33, 2, 6.

BOOK III. 9. 3–10. 2

with salt, and obliged the patient to drink it, in order that by moving the bowels he might cleanse the stomach. And the above formed the whole of this man's practice; and it pleased those whom successors of Hippocrates had failed to cure, no less than in our time it pleases those, who, after they have dragged on for a long while under disciples of Herophilus and Erasistratus, have not been benefited. Yet it is rash treatment none the less, for if it is adopted forthwith at the commencement, it kills many patients. Since, however, it is impossible for the same remedies to suit everybody, rashness helps those whom the usual regimen has not made well; hence it is that practitioners of this class manage other people's patients better than their own. Yet it is the part also of a circumspect man at times to renew and increase a disease and to inflame fevers, for when the existing condition does not answer to a treatment, that which is to come may do so.

10. We must also take into consideration whether fevers exist alone, or whether there are additional troubles, namely whether the head aches, whether the tongue is roughened, whether the chest is tight.

If there is headache, rose oil should be mixed with vinegar and poured over the head; next two strips of linen are taken, each corresponding in length and breadth to the forehead, of which in turn one is placed in the rose oil and vinegar, the other on the forehead; or unscoured wool is soaked in the same and applied. If the vinegar hurts, the rose oil alone is to be used; if rose oil itself irritates, then bitter olive oil. If there is little relief from the above, we may pound up either dried orris root, or bitter almonds, or some other from among refrigerant [a]

quorum quidlibet ex aceto inpositum dolorem minuit, sed magis aliud in alio. Iuvat etiam panis cum papavere iniectus, vel cum rosa, cerussa spumave argenti. Olfacere quoque vel serpullum vel anethum non alienum est.

At si in praecordiis inflammatio et dolor est, primo superimponenda sunt cataplasmata reprimentia, ne, si calidiora fuerint, plus eo materiae concurrat; deinde si prima inflammatio se remisit, tum demum ad calida et umida veniendum est, ut ea, quae
3 remanserunt, discutiant. Notae vero inflammationis sunt quattuor: rubor et tumor cum calore et dolore. Quo magis erravit Erasistratus, cum febrem nullam esse sine hac dixit. Ergo si sine inflammatione dolor est, nihil imponendum est: hunc enim statim ipsa febris solvet. At si neque inflammatio neque febris sed tantum praecordiorum dolor est, protinus calidis et siccis fomentis uti licet.

4 Si vero lingua sicca et scabra est, detergenda primum penicillo est ex aqua calida, deinde unguenda mixtis inter se rosa et melle. Mel purgat, rosa reprimit simulque siccescere non sinit. At si scabra non est sed arida, ubi penicillo detersa est, ungi rosa debet, cui cerae paulum sit adiectum.

11. Solet etiam ante febres esse frigus idque vel molestissimum morbi genus est. Ubi id expectatur,

herbs; any one of these applied with vinegar lessens pain, one more in one case, another in another. There is benefit from the application of bread soaked in poppy head decoction, or in rose oil containing cerussa or litharge. Also it is not unsuitable to snuff up thyme or dill.

But if there is inflammation and pain in the chest, the first thing is to apply to it repressing plasters, lest more diseased matter should gather there, if hotter ones were applied; next, when the primary inflammation has subsided, and not before, we must go on to hot and moist plasters, in order to disperse what remains of the matter. Now the signs of an inflammation are four: redness and swelling with heat and pain. Over this Erasistratus [a] greatly erred, when he said that no fever occurred apart from inflammation. Therefore if there is pain without inflammation, nothing is to be put on: for the actual fever at once will dissolve the pain. But if there is neither inflammation nor fever, but just pain in the chest, it is allowable to use hot and dry foments from the first.

Again if the tongue is dry and scabrous, it is to be wiped over first with a pledget of wool dipped in hot water, then to be smeared with a mixture of rose oil and honey. Honey cleans, rose oil represses and at the same time does not allow the tongue to dry. But if the tongue is not scabrous, only dry, after being wiped over with the pledget of wool, it should be smeared with rose oil to which a little wax has been added.

11. Generally also preceding fevers there is a chill, and that is a most troublesome class of malady.

[a] Galen, XV. 159.

omni potione prohibendus est aeger: haec enim paulo ante data multum malo adicit. Item maturius veste multa tegendus est; admovenda partibus iis, pro quibus metuimus, sicca et calida fomenta sic, ne statim vehementissimi calores incipiant, sed 2 paulatim increscant. Perfricandae quoque eae partes manibus unctis ex vetere oleo sunt eique adiciendum aliquid ex calefacientibus. Contentique medici quidam una frictione etiam ex quolibet oleo sunt. In harum febrium remissionibus nonnulli tres aut quattuor cyathos sorbitionis etiamnum manente febre dant, deinde ea bene finita reficiunt stomachum cibo frigido et levi. Tum hoc ego puto temptandum, quom parum cibus semel et post febrem datus prodest. 3 Sed curiose prospiciendum, ne tempus remissionis decipiat: saepe enim in hoc quoque genere valetudinis iam minui febris videtur, et rursus intenditur. Aliqua ei remissioni credendum est, quae etiam moratur et iactationem foetoremque quendam oris, quem ozenam Graeci vocant, minuit. Illud satis convenit, si cottidie pares accessiones sunt, cotidie parvum cibum dandum: si inpares, post graviorem, cibum; post leviorem, aquam mulsam.

12. Horror autem eas fere febres antecedit, quae certum habent circuitum et ex toto remittuntur; ideoque tutissimae sunt maximeque curationes admittunt. Nam ubi incerta tempora sunt, neque

---

[a] II. 33, 5.
[b] See Introduction, p. xi.

When it is expected, the patient is to be prohibited from drinking anything: for this, given a little while beforehand, adds much to the illness. Likewise he is to be covered up quite soon with a quantity of bed-clothes; to the parts about which we feel concern there are to be applied such dry and hot foments as will not immediately set up a very vehement heat, but gradually increase it. The said parts are also to be rubbed by hands anointed with old olive oil, to which has been added one of the heating agents.[a] And some practitioners are satisfied with one rubbing of any kind of oil. During remissions of these fevers, some give three or four cupfuls of barley water even although some fever still persists; then, the fever having definitely ended, they reinvigorate the stomach with cold and light food. This, I think,[b] should be tried only when there has been little benefit from food given once and at the end of the paroxysm. It must be carefully looked to, however, that the time of the remission is not deceptive; for often in this class of illness the fever seems to diminish, and then again becomes intense. Some degree of trust must be placed in that remission which is prolonged, and diminishes restlessness and the foulness of the mouth which the Greeks term ozaena. This is pretty generally agreed, that if the daily paroxysms are equal, a little food should be given every day: if the paroxysms are unequal, food should be given after the more severe, after the slighter ones hydromel.

12. Now shivering usually precedes those fevers which have a fixed cycle and a complete remission; hence they are the most safe, and specially admit of treatment. For when periodicity is uncertain,

CELSUS

alvi ductio neque balineum neque vinum neque medicamentum aliud recte datur: incertum est enim, quando febris ventura sit: ita fieri potest, ut, si subito venerit, summa in eo pernicies sit, quod
2 auxilii causa sit inventum. Nihilque aliud fieri potest, quam ut primis diebus bene abstineatur aeger, deinde sub decessu febris eius, quae gravissima est, cibum sumat. At ubi certus circumitus est, facilius illa omnia temptantur, quia magis proponere nobis accessionum et decessionum vices possumus. In his autem, cum inveteraverunt, utilis fames non est: primis tantummodo diebus ea pugnandum est; deinde dividenda curatio est, et ante
3 horror, tum febris discutienda. Igitur cum primum aliquis inhorruit, et ex horrore incaluit, dare oportet ei potui tepidam aquam subsalsam et vomere eum cogere: nam fere talis horror ab iis oritur, quae biliosa in stomacho resederunt. Idem faciendum est, si proximo quoque circuitu aeque accessit: saepe enim sic discutitur, iamque, quod genus febris sit, scire licet. Itaque sub expectatione proximae accessionis, quae instare tertia potest, deducendus in balineum est, dandaque opera, ut per tempus
4 horroris in solio sit. Si ibi quoque . . .[1] senserit, nihilo minus idem sub expectatione quartae accessionis faciat: siquidem eo quoque modo saepe id discutitur. Si ne balneum quidem profuit, ante accessionem alium edat, aut bibat aquam calidam cum pipere:

[1] *Marx supplies* frigus *which is translated.*

neither clyster nor bath, nor wine nor other medicament, is administered at the right moment: for it is uncertain when the fever will supervene, so that if it comes on suddenly, it may happen that there is the greatest harm in what is intended to serve as an aid. And there is nothing else that can be done, except for the patient to abstain strictly for the first days, then, upon the decline of that paroxysm which is the severest, to take food. When, however, there is an assured cycle, all those remedies are more easily tried, because we are more able to inform ourselves of the alternations between paroxysms and remissions. In those fevers, however, which have become inveterate, starving is not of service; it is only in the first days that the fever is to be thus countered; later the treatment is to be divided, first to disperse the shivering, then the fever. Therefore, as soon as the patient shivers, and after the shivering grows hot, he should be given to drink tepid water with a little salt in it, and so made to vomit: for generally such shivering arises from a bilious sediment in the stomach. Likewise if shivering recurs at the next cycle, the same should be done; for often the fever is thus shaken off, and now we may learn to what class it belongs. And so in view of the possibility of the next paroxysm, the third which may be threatening, the patient should be conducted to the bath, and it should be so arranged that he is already in the solium at the moment for the shivering. If there also he feels chilled, yet none the less he should do the same again in view of a fourth paroxysm, for often in that way the shivering is shaken off. If there is no benefit even from the bath, before the paroxysm let him eat garlic, or drink hot water containing pepper;

siquidem ea quoque adsumpta calorem movent, qui horrorem non admittit. Deinde eodem modo, quo in frigore praeceptum est (11, 1), antequam inhorrescere possit, operiatur, fomentisque, sed protinus validioribus, totum corpus circumdet maximeque involutis extinctis testis et titionibus. Si nihilo minus horror perruperit, multo oleo calefacto inter ipsa vestimenta perfundatur, cui aeque ex calfacientibus aliquid sit adiectum; adhibeaturque frictio, quantam is sustinere poterit, maximeque in manibus et cruribus; et spiritum ipse contineat. Neque desistendum est, etiamsi horret: saepe enim pertinacia iuvantis malum corporis vincit. Si quid evomuit, danda aqua tepida, iterumque vomere cogendus est; utendumque eisdem est, donec horror finiatur. Sed praeter haec ducenda alvus est, si tardius horror quiescit: siquidem id quoque exonerato corpori prodest. Ultimaque post haec auxilia sunt gestatio et fricatio. Cibus autem in eiusmodi morbis maxime dandus est, qui mollem alvum praestet, caro glutinosa: vinum, cum dabitur, austerum.

13. Haec ad omnes circuitus febrium pertinent: discernendae tamen singulae sunt, sicut rationem habent dissimilem. Si cottidiana est, triduo primo magnopere abstineri oportet, tum cibis altero quoque die uti: si res inveteraverit, post febrem experiri balneum et vinum, magisque si horrore sublato haec superest.

14. Si vero tertiana, quae ex toto intermittit,

---

[a] Cf. III. 11, 1.   [b] II. 17, 9.   [c] Cf. III. 3, 2 ff.

to see if these when taken excite heat which prevents the shivering. Further in the same way as prescribed for a chill before shivering can come on, the patient should be covered up, and the whole body surrounded with foments—but the stronger ones are to be used at once [a]—and thoroughly encompassed by wraps which enclose hot tiles [b] and cinders. If, notwithstanding, shivering breaks out, let the patient be anointed freely under the wraps with hot oil, to which add one of the heating elements: let rubbing be applied, so far as he can bear it, especially of the arms and legs, whilst he holds his breath. Nor should it be stopped even if he shivers; for often the pertinacity of the rubber overcomes the body's malady. If he vomits somewhat, tepid water is to be given him, and he is to be forced to vomit again; the same measures must be used until shivering comes to an end. But if the shivering is too slow in subsiding, in addition to the above, a clyster should be given; for that also is of good effect by unloading the body. The last remedies after these are rocking and rubbing. Now in such illness the food to be given is such chiefly as will secure a soft motion, meat glutinous, wine, when any is given, dry.

13. The foregoing remarks apply to all periodic [c] fevers: but they are to be distinguished, according to the dissimilar characters of each. If it is a daily fever, it is particularly important to abstain for the first three days, then to make use of food upon alternate days: if this fever has become inveterate, the bath and wine are to be tried at the end of the paroxysm, and especially so when the fever persists after the shivering has been removed.

14. But if it be a tertian, when there are complete

aut quartana est, mediis diebus et ambulationibus uti oportet aliisque exercitationibus et unctionibus. Quidam ex antiquis medicis Cleophantus in hoc genere morborum multo ante accessionem per caput aegrum multa calida aqua perfundebat, deinde vinum dabat. Quod, quamvis pleraque praecepta eius viri secutus est Asclepiades, recte tamen
2 praeteriit: est enim anceps. Ipse, si tertiana febris est, tertio die post accessionem dicit alvum duci oportere, quinto post horrorem vomitum elicere, deinde post febrem, sicut illi mos erat, adhuc calidis dari cibum et vinum, sexto die in lectulo detineri: sic enim fore, ne septimo die febris accedat. Id saepe posse fieri verisimile est. Tutius tamen est, ut hoc ipso ordine utamur, tria remedia, vomitus, alvi ductionis, vini per triduum, id est tertio die et quinto et septimo temptare, ne vinum nisi post
3 accessionem die septimi bibat. Si vero primis diebus discussus morbus non est, inciditque in vetustatem, quo die febris expectabitur, in lectulo se contineat, post febrem confricetur, tum cibo adsumpto bibat aquam: postero die, qui vacat, ab exercitatione unctioneque aqua tantum contentus conquiescat. Et id quidem optimum est: si vero inbecillitas urgebit, et post febrem vinum et medio die paulum cibi debebit adsumere.

intermissions, or a quartan, on the intermediate days the patient should make use both of walking and of other exercises and of anointings. In this kind of malady, well before the paroxysm, a certain Cleophantus, one of the ancient physicians, poured over the patient's head quantities of hot water; and then gave wine. Asclepiades, although he followed many of this man's precepts, rejected this one, and rightly, for it is of doubtful effect. In the case of a tertian fever, Asclepiades said that on the third day following the paroxysm, the bowels should be moved by a clyster; on the fifth day after the shivering a vomit should be elicited; then, after the paroxysm, according to the custom of Cleophantus, patients whilst still heated were to be given food and wine, on the sixth day to be kept in bed; for so he hoped to prevent a paroxysm on the seventh day. It is likely that this may often happen. It is safer, however, so that we may use the exact order laid down, to try the three remedies, vomiting, clystering, and wine-drinking, on three several days, that is, on the third, fifth and seventh days, with this proviso that on the seventh day wine is not to be drunk until after the time for the paroxysm. But if a tertian fever is not dispersed within the first days, but is becoming chronic on the day that the paroxysm is expected, the patient should keep his bed; after the paroxysm he should be rubbed, then, having taken food, drink water; on the day following, which is free from fever, the patient should keep quiet, avoid exercise and anointing, and be content with water only. And that indeed is the best procedure; but if there is urgent weakness, he may both take wine after the paroxysm and a little food on the intermediate day.

CELSUS

15. Eadem in quartana facienda sunt. Sed cum haec tarde admodum finiatur nisi primis diebus discussa est, diligentius ab initio praecipiendum est, quid in ea fieri debeat. Igitur si cui cum horrore febris accessit eaque desiit, eodem die et postero tertioque continere se debebit, et aquam tantummodo calidam primo die post febrem sumere; biduo proximo, quantum fieri potest, ne hanc quidem: si quarto die cum horrore febris revertitur, vomere, sicut ante (**12**, 3) praeceptum est; deinde post febrem modicum cibum sumere, vini quadrantem. 2 Postero tertioque die abstineri, aqua tantummodo calida, si sitiet, adsumpta. Septimo die balineo frigus praevenire; si febris redierit, ducere alvum; ubi ex eo corpus conquieverit, in unctione vehementer perfricari; eodem modo sumere cibum et vinum; biduo proximo se abstinere, frictione servata. Decimo die rursus balneum experiri; et si postea febris accessit, aeque perfricari, vinum copiosius bibere. Ac proximum est, tot dierum ut abstinentia[1] 3 cum ceteris, quae praecipiuntur, febrem tollant. Si vero nihilo minus remanet, aliud ex toto sequendum est curationis genus, idque agendum, ut id, quod diu sustinendum est, corpus facile sustineat. Quo minus etiam probari curatio Heraclidis Tarentini debet, qui primis diebus ducendam alvum, deinde abstinendum in septimum diem dixit. Quod ut

[1] *The text is uncertain. One MS. reads* proximum est quies tot dierum et abstinentia, *etc.*

---

[a] $\frac{1}{4}$ sextarius, i.e. 125 c.cm.

BOOK III. 15. 1-3

15. In a quartan fever the same should be done. But seeing that unless it has been shaken off within the first days, it is a long while in terminating, we must be more careful from the very first to lay down what should be done in it. Therefore if a paroxysm has set in with shivering and has remitted, the patient ought to observe a regimen on the same day and on the following and on the third day; on the first day after the paroxysm he should take only hot water; on the next two days abstain if possible even from that; on the fourth day, if the fever recurs with shivering, he should vomit, as was prescribed before; then after the paroxysm he should take a limited quantity of food and of wine four ounces.[a] On the next two days he should fast, taking only hot water if thirsty. On the seventh day the cold stage should be anticipated by the bath; if a paroxysm recurs, the bowels should be moved by a clyster; having settled down after the clyster, the patient should be anointed and rubbed vigorously; then take food and wine as above; on the next two days abstain, and undergo rubbing. On the tenth day trial is again made of the bath; and if after that a paroxysm follows, he should in the same way be rubbed, and drink wine more freely. And it is likely that so many days of fasting, along with the other measures prescribed, will get rid of the fever. But if the quartan fever persists notwithstanding, a totally different line of treatment is to be pursued, the aim being that the body may easily bear what has to be borne for a long while. Therefore we cannot approve the practice of Heraclides of Tarentum, who said that in the first days the bowel was to be clystered and then there was to be abstinence until the seventh day.

sustinere aliquis possit, tamen etiam febre liberatus vix refectioni valebit: adeo, si febris saepius acces-
4 serit, concidet. Igitur, si tertio decimo die morbus manebit, balineum neque ante febrem neque postea temptandum erit, nisi interdum iam horrore discusso. Horror ipse per ea, quae supra (12) scripta sunt expugnandus; deinde post febrem oportebit ungui et vehementer perfricari, cibum et validum et fortiter adsumere, vino uti quantum libebit; postero die, cum satis quieverit, ambulare, exerceri, ungui, perfricari fortiter, cibum capere sine vino, tertio die
5 abstinere. Quo die vero febrem expectabit, ante surgere, exerceri dareque operam, ut in ipsa exercitatione febris tempus incurrat: sic enim saepe illa discutitur. At si in opere occupavit, tum demum se recipere. In eiusmodi valetudine medicamenta sunt
6 oleum, frictio, exercitatio, cibus, vinum. Si venter adstrictus est, solvendus est. Sed haec facile validiores faciunt: si inbecillitas occupavit, pro exercitatione gestatio est; si ne hanc quidem sustinet, adhibenda[1] tamen frictio est. Si haec quoque vehemens[1] onerat, intra quietem et unctionem et cibum sistendum est; dandaque opera est, ne qua

[1] *Bracketed by Marx as a gloss.*

BOOK III. 15. 3–6

Even supposing a man could endure this, yet if he does become freed from fever, he will have scarcely strength enough to recover; therefore if there be more frequent recurrences of the fever he will sink. If, therefore, the disease shall remain on the thirteenth day, the bath should not be tried, either before or after the paroxysm, except occasionally when the shivering has been thrown off. The actual shivering is to be driven off by the measures above prescribed; then after the paroxysm it will be proper that the patient be anointed, and rubbed vigorously, and take food both nourishing and abundant, with as much wine as he likes; on the day following, when sufficiently rested, he is to walk, to take exercise, to be anointed and vigorously rubbed, then to take food without wine, and on the third day to abstain. On the day that a recurrence of the paroxysm is expected, he should get up beforehand, and so arrange the performance of the exercises that the time for the onset of the fever concurs with that of the exercise; for often in this way the paroxysm is thrown off. But if attacked during the exercise, he should thereupon return home to bed. In this kind of sickness the remedies are: anointing, rubbing, exercise, food, wine. If constipated, the bowels are to be clystered. But whilst the stronger patients can easily carry out the above, if weakness has supervened, rocking should replace exercise; if even that cannot be borne, nevertheless rubbing should be applied. If this also, when vigorous, is trying to the patient, treatment should be restricted to rest and anointing and food; care being taken that indigestion does not convert the quartan into a

cruditas in cotidianam id malum vertat. Nam quartana neminem iugulat: sed si ex ea cotidiana facta est, in malis aeger est; quod tamen nisi culpa vel aegri vel curantis numquam fit.

16. At si duae quartanae sunt, neque eae, quas proposui (**15,** 4–6), exercitationes adhiberi possunt, aut ex toto quiescere opus est, aut, si id difficile est, leviter ambulare, considere diligenter involutis pedibus et capite; quotiens febris accessit et desiit, cibum modicum sumere et vinum, reliquo tempore, nisi inbecillitas urguet, abstinere. At si duae febres paene iunguntur, post utramque cibum sumere, deinde vacuo tempore et moveri aliquid et post unctionem cibo uti. Cum vero vetus quartana raro nisi vere solvatur, utique eo tempore attendendum 2 est, ne quid fiat, quod valetudinem impediat. Prodestque in vetere quartana subinde mutare victus genus, a vino ad aquam, ab aqua ad vinum, a lenibus cibis ad acres, ab acribus ad lenes transire; esse radicem, deinde vomere; iure vel . . . pulli gallinacei ventrem resolvere; oleo ad frictiones adicere calfacientia; ante accessionem sorbere vel aceti cyathos duos, vel unum sinapis cum tribus Graeci vini salsi,[1] vel mixta paribus portionibus et in aqua diluta piper, castoreum, laser, murram. Per haec enim similiaque corpus agitandum est, ut moveatur ex eo statu, quo detinetur. Si febris quievit, diu meminisse eius diei convenit, eoque vitare frigus, calorem, cruditatem,

---

[1] *Marx supplies* 'concharum vel,' *and this has been translated. See* II. *29*; III. **6,** *14.*

---

[a] Cf. III. **5,** 6 and note on p. 236.

BOOK III. 15. 6–16. 2

quotidian fever. For a quartan kills no one, but when a quotidian is made out of it, the patient is in a bad way; this, however, does not happen unless through the fault either of the patient or of the practitioner.

16. But if there is a double quartan fever, and those exercises which I have mentioned cannot be adopted, either the patient should rest entirely, or if that is difficult walk quietly, then sit with his feet and head carefully wrapped up; as often as a paroxysm has recurred and has remitted, he should take food in moderation and wine; for the remainder of the remission, unless there is urgent weakness, he should fast. But if two paroxysms are almost continuous,[a] he should take food after both are over; then in the intermission he should move about a little, and after being anointed take food. Now since an inveterate quartan is seldom got rid of except in the spring, it is at that season especially that attention is to be given, lest something occur to hinder recovery. And it is of advantage in an old quartan to alter now and then the class of diet, and change from wine to water, from water to wine, from bland food to acrid, from acrid to bland; to eat radish, then to vomit; to move the bowels by shell-fish or chicken broth; to add heating agents to the oil for rubbing; before the paroxysm to sip two cups of vinegar, or one cup of mustard in three of Greek salted wine, or pepper, castoreum, laser and myrrh mixed in equal proportions in water. For by these and such-like remedies the system is to be stirred up in order that it may be moved from the state in which it is being held. When the fever has quieted down, for a long while it is well to keep in mind the day on which it occurred, and on that day to avoid cold, heat,

lassitudinem: facile enim revertitur, nisi a sano quoque aliquamdiu timetur.

17. At si ex quartana facta cotidiana est, cum id vitio inciderit, per biduum abstinere oportet, et frictione uti, vespere tantummodo aquam potui dare: tertio die saepe fit, ne febris accedat. Sed sive fuit sive non fuit, cibus post accessionis tempus est dandus. At si manet, per biduum abstinentia, quanta maxime imperari corpori potest, frictione cotidie utendum est.

18. Et febrium quidem ratio exposita est. Supersunt vero alii corporis adfectus, qui huic superveniunt, ex quibus eos, qui certis partibus adsignari non possunt, protinus iungam.

Incipiam ab insania, primamque huius ipsius partem adgrediar, quae et acuta et in febre est: φρένησιν
2 Graeci appellant. Illud ante omnia scire oportet, interdum in accessione aegros desipere et loqui aliena. Quod non quidem leve est neque incidere potest nisi in febre vehementi; non tamen aeque pestiferum est: nam plerumque breve esse consuevit levatoque accessionis impetu protinus mens redit. Neque id genus morbi remedium aliud desiderat, quam quod in curanda febre praeceptum est.
3 Phrenesis vero tum demum est, cum continua dementia esse incipit, cum aeger, quamvis adhuc sapiat, tamen quasdam vanas imagines accipit:

---

[a] Insania is used in its widest sense unsoundness (of mind).
[b] II. 1, 15, Hipp. IV. 194 (*Aph.* VII. 12.)

indigestion, fatigue: for fever readily recurs unless even a convalescent patient fears that day for some time to come.

17. But if a quotidian has been made out of a quartan fever, since this may have happened from mismanagement, the patient ought to fast for two days, make use of rubbing, and be given only a drink of water in the evening: it often happens that on the third day there is no paroxysm. But whether or not, food should be given after the time for the paroxysm. But if this fever persists, a two days' fast should be enjoined so far as the system can bear it, and rubbing used every day.

18. The regimen of fevers has now been expounded; there are, however, other affections of the body which follow upon this, among which I subjoin in the first place those which cannot be assigned to any definite part.

I shall begin with insanity,[a] and first that form of it which is both acute and found in fever. The Greeks call it phrenesis.[b] Before all things it should be recognized, that at times, during the paroxysm of a fever, patients are delirious and talk nonsense. This is indeed no light matter, and it cannot occur unless in the case of a severe fever; it is not, however, always equally dangerous; for commonly it is of short duration, and when the onslaught of the paroxysm is relieved, at once the mind comes back. This form of the malady does not require other remedy than that prescribed for the curing of the fever. But insanity is really there when a continuous dementia begins, when the patient, although up till then in his senses, yet entertains certain vain imaginings; the insanity becomes established when the mind becomes

perfecta est, ubi mens illis imaginibus addicta est. Eius autem plura genera sunt: siquidem ex phreneticis alii tristes sunt; alii hilares; alii facilius continentur et intra verba desipiunt; alii consurgunt et violenter quaedam manu faciunt; atque ex his ipsis alii nihil nisi impetu peccant, alii etiam artes adhibent summamque speciem sanitatis in captandis malorum operum occasionibus praebent, sed exitu
4 deprenduntur.—Ex his autem eos, qui intra verba desipiunt, aut leviter etiam manu peccant, onerare asperioribus coercitionibus supervacuum est: eos, qui violentius se gerunt, vincire convenit, ne vel sibi vel alteri noceant. Neque credendum est, si vinctus aliqui, dum levari vinculis cupit, quamvis prudenter et miserabiliter loquitur, quoniam is dolus
5 insanientis est. Fere vero antiqui tales aegros in tenebris habebant, eo quod iis contrarium esset exterreri, et ad quietem animi tenebras ipsas conferre aliquid iudicabant. At Asclepiades, tamquam tenebris ipsis terrentibus, in lumine habendos eos dixit. Neutrum autem perpetuum est: alium enim lux, alium tenebrae magis turbant; reperiunturque, in quibus nullum discrimen deprehendi vel hoc vel illo modo possit. Optimum itaque est utrumque experiri, et habere eum, qui tenebras horret, in luce, eum qui lucem, in tenebris. At ubi nullum tale discrimen est, aeger, si vires habet, loco lucido;
6 si non habet, obscuro continendus est. Remedia vero adhibere, ubi maxime furor urget, supervacuum

BOOK III. 18. 3-6

at the mercy of such imaginings. But there are several sorts of insanity; for some among insane persons are sad, others hilarious; some are more readily controlled and rave in words only, others are rebellious and act with violence; and of these latter, some only do harm by impulse, others are artful too, and show the most complete appearance of sanity whilst seizing occasion for mischief, but they are detected by the result of their acts. Now that those who merely rave in their talk, or who make but trifling misuse of their hands, should be coerced with the severer forms of constraint is superfluous; but those who conduct themselves more violently it is expedient to fetter, lest they should do harm either to themselves or to others. Anyone so fettered, although he talks rationally and pitifully when he wants his fetters removed, is not to be trusted, for that is a madman's trick. The ancients generally kept such patients in darkness, for they held that it was against their good to be frightened, and that the very darkness confers something towards the quieting of the spirit. But Asclepiades said that they should be kept in the light, since the very darkness was terrifying. Yet neither rule is invariable: for light disturbs one more, darkness another; and some are met with in whom no difference can be observed, either one way or the other. It is best, therefore, to make trial of both, and to keep that patient in the light who is frightened by darkness, and him in darkness who is frightened by light. And when there is no such difference, the patient if strong should be kept in a light room, if not strong he should be kept in a dim one. Now it is useless to adopt remedies when

est: simul enim febris quoque increscit. Itaque tum nihil nisi continendus aeger est: ubi vero res patitur, festinanter subveniendum est. Asclepiades perinde esse dixit his sanguinem mitti ac si trucidentur: rationem hanc secutus, quod neque insania esset, nisi febre intenta, neque sanguis nisi in 7 remissione eius recte mitteretur. Sed ipse in his somnum multa frictione quaesivit, cum et intentio febris somnum impediat et frictio non nisi in remissione eius utilis sit. Itaque hoc quoque auxilium debuit praeterire. Quid igitur est? Multa in praecipiti periculo recte fiunt alias omittenda. Et continuata quoque febris habet tempora, quibus, etsi non remittit, non tamen crescit, estque hoc ut non optimum, sic tamen secundum remediis tempus, quo, si vires aegri patiuntur, sanguis quoque mitti debet. 8 Minus deliberari potest, an alvus ducenda sit. Tum interposito die convenit caput ad cutem tondere, aqua deinde fovere, in qua verbenae aliquae decoctae sint [vel ex reprimentibus]; aut prius fovere, deinde radere, et iterum fovere; ac novissime rosa caput naresque implere, offerre etiam naribus rutam ex aceto contritam, movere sternumenta medica9 mentis in id efficacibus. Quae tamen facienda sunt in iis, quibus vires non desunt: si vero inbecillitas est, rosa tantum caput adiecto serpullo similique

the delirium is at its height; for simultaneously fever is also increasing. So then there is nothing else to do than to restrain the patient, but when circumstances permit, relief must be given with haste. Asclepiades said that in such cases to let blood is just to commit murder; following the line of reasoning, that there was no insanity unless with high fever, and that properly blood was let only during the stage of remission. But he himself in these cases sought to bring on sleep by prolonged rubbing, though it is the intensity of the fever which hinders sleep, and it is only during the remission that rubbing is of service. Hence he ought to have passed over this remedy also. What then is there to do? Many things may be rightly done in imminent danger, which otherwise ought to be omitted. And fever also, when continuous, has times during which, although it does not remit, yet it does not increase, and this time, although not the best, yet is the second best time for remedies; and at this time blood ought to be let, if the patient's strength allow it. There can be less question as to whether a motion should be induced. Next, after a day's interval, the head should be shaved bare and then fomented with water in which vervains or other repressive herbs have been boiled; alternatively it is proper first to foment, then to shave, and again to foment; and lastly to pour rose oil over the head and into the nostrils; also to hold to the nose rue pounded up in vinegar, and to excite sneezing by drugs efficacious for the purpose. Such things, however, should be done only in the case of those who are not lacking in strength; but if there is weakness, the head is merely moistened by rose oil to which thyme or something similar has been added.

CELSUS

aliquo madefaciendum est. Utiles etiam in quibuscumque viribus duae herbae sunt, solanum et muralis, si simul ex utraque suco expresso caput impletur. Cum febris remisit, frictione utendum est, parcius tamen in iis, qui nimis hilares quam in iis, qui nimis
10 tristes sunt. Adversus autem omnium sic insanientium animos gerere se pro cuiusque natura necessarium est. Quorundam enim vani metus levandi sunt, sicut in homine praedivite famem timente incidit, cui subinde falsae hereditates nuntiabantur. Quorundam audacia coercenda est, sicut in iis fit, in quibus continendis plagae quoque adhibentur. Quorundam etiam intempestivus risus obiurgatione et minis finiendus: quorundam discutiendae tristes cogitationes; ad quod symphoniae et cymbala
11 strepitusque proficiunt. Saepius tamen adsentiendum quam repugnandum et paulatim et non evidenter ab iis, quae stulte dicentur, ad meliora mens eius adducenda. Interdum etiam elicienda ipsius intentio; ut fit in hominibus studiosis litterarum, quibus liber legitur aut recte, si delectantur, aut perperam, si id ipsum eos offendit: emendando enim convertere animum incipiunt. Quin etiam recitare, si qua meminerunt, cogendi sunt. Ad cibum quoque quosdam non desiderantes reduxerunt
12 ii, qui inter epulantes eos conlocarunt. Omnibus vero sic adfectis somnus et difficilis et praecipue necessarius est: sub hoc enim plerique sanescunt. Prodest ad id atque etiam ad mentem ipsam con-

BOOK III. 18. 9–12

Whatever the patient's strength, the two herbs, bitter-sweet and pellitory, are beneficial, if the head is wetted with the juice expressed from both simultaneously. When the fever has remitted, recourse should be had to rubbing, more sparingly, however, in those who are over-cheerful, than in those who are too gloomy. But in dealing with the spirits of all patients suffering from this type of insanity, it is necessary to proceed according to the nature of each case. Some need to have empty fears relieved, as was done for a wealthy man in dread of starvation, to whom pretended legacies were from time to time announced. Others need to have their violence restrained as is done in the case of those who are controlled even by flogging. In some also untimely laughter has to be put a stop to by reproof and threats; in others, melancholy thoughts are to be dissipated, for which purpose music, cymbals, and noises are of use. More often, however, the patient is to be agreed with rather than opposed, and his mind slowly and imperceptibly is to be turned from the irrational talk to something better. At times also his interest should be awakened; as may be done in the case of men fond of literature, to whom a book may be read, correctly when they are pleased by it, or incorrectly if that very thing annoys them; for by making corrections they begin to divert their mind. Moreover, they should be pressed to recite anything they can remember. Some who did not want to eat were induced to do so, by being placed on couches between other diners. But certainly for all so affected sleep is both difficult and especially necessary; for under it many get well. Beneficial for this, as also for composing the mind itself, is

CELSUS

ponendam crocinum unguentum cum irino in caput additum. Si nihilo minus vigilant, quidam somnum moliuntur potui dando aquam, in qua papaver aut hyoscyamos decocta sint, alii mandragorae mala pulvino subiciunt, alii vel amomum vel sycamini lacrimam fronti inducunt. Hoc nomen apud medicos
13 reperio. Sed cum Graeci morum sycaminon appellant, mori nulla lacrima est. Sic vero significatur lacrima arboris in Aegypto nascentis, quam ibi sycomoron appellant. Plurimi, decoctis papaveris corticibus, ex ea aqua spongia os et caput subinde
14 fovent. Asclepiades ea supervacua esse dixit, quoniam in lethargum saepe converterent: praecepit autem, ut primo die a cibo, potione, somno abstineretur, vespere ei daretur potui aqua, tum frictio admoveretur lenis, ut ne manum quidem qui perfricaret vehementer inprimeret; postero deinde die isdem omnibus factis vespere ei daretur sorbitio et aqua, rursusque frictio adhiberetur: per hanc enim
15 nos consecuturos, ut somnus accedat. Id interdum fit, et quidem adeo, ut illo confitente nimia frictio etiam lethargi periculum adferat. Sed si sic somnus non accessit, tum demum illis medicamentis arcessendus est, habita scilicet eadem moderatione, quae hic quoque necessaria est, ne, quem obdormire volumus, excitare postea non possimus. Confert etiam aliquid ad somnum silanus iuxta cadens, vel gestatio post cibum et noctu, maximeque suspensi
16 lecti motus. Neque alienum est, si neque sanguis

saffron ointment with orris applied to the head. If in spite of this patients are wakeful, some endeavour to induce sleep by draughts of decoction of poppy or hyoscyamus; others put mandrake apples under the pillow; others smear the forehead with cardamomum balsam or sycamine tears. This name I find used by practitioners, but there are no tears on the mulberry, although the Greeks call the mulberry sycaminon. What in fact is meant are the tears of a tree growing in Egypt, which they call in that country sycamoros. Many foment the face and head at intervals with a sponge dipped in a decoction of poppy heads. Asclepiades said that these things were of no benefit, because they often produced a change into lethargy (III. **20**); but he prescribed for the patient that during the first day he should keep from food, drink and sleep, in the evening water should be given him to drink, after which he should be rubbed with gentleness, but the rubber must not press hard even with the hand (II. **14**); during the day following the same was to be done, then in the evening gruel and water should be given and rubbing again applied: for by this he said we should succeed in bringing on sleep. This does happen sometimes, and to such a degree that Asclepiades allowed that excess of rubbing may even cause danger of lethargy. But if sleep does not thus occur, then at length it is to be procured by the above medicaments, having regard, of course, to the same moderation, which is necessary here also, for fear we may afterwards not be able to wake up the patient whom we wish to put to sleep. Sleep is also assisted by the sound of falling water near by, also rocking after food and at night, and especially the motion of a slung hammock (II. **15**). If blood

CELSUS

ante missus est, neque mens constat, neque somnus accedit, occipitio inciso cucurbitulam admovere; quae quia levat morbum, potest etiam somnum facere. Moderatio autem in cibo quoque adhibenda est: nam neque aeger implendus est, ne insaniat, neque ieiunio utique vexandus, ne inbecillitate in cardiacum incidat. Opus est cibo infirmo maximeque sorbitione, potione aquae mulsae, cuius ternos cyathos bis hieme, quater aestate dedisse satis est.

17 Alterum insaniae genus est, quod spatium longius recipit, quia fere sine febre incipit, leves deinde febriculas excitat. Consistit in tristitia, quam videtur bilis atra contrahere.—In hac utilis detractio sanguinis est: si quid hanc poterit[1] prohibere, prima est abstinentia, secunda per album veratrum vomitumque purgatio, post utrumlibet adhibenda bis die frictio est; si magis valet, frequens etiam exercitatio: in ieiuno vomitus. Cibus sine vino dandus ex media materia est; quam quotiens posuero, scire licebit etiam ex infirmissima dari posse, dum ne illa sola quis utatur: valentissima tantummodo esse remo-
18 venda. Praeter haec servanda alvus est quam tenerrima, removendi terrores, et potius bona spes

---

[1] poterit *added by Marx*.

---

[a] In the region of the skull occupied by the cerebellum, the ancients imagined an empty space, in which the *materies morbi* or *pituita* collected, and could be extracted by the cupping accompanied by scarification. Compare Juvenal XIV. 57: *Cum facias peiora senex, vacuumque cerebro iam pridem caput hoc ventosa cucurbita quaerat?*

[b] For cardiacus morbus, see III. 19, note p. 302.

[c] A cyathus was a twelfth part of a sextarius; the dose prescribed = 125 c.cm.

BOOK III. 18. 16–18

has not been let before, and the patient's mind is unstable and sleep does not occur, it is not unfitting to apply a cup over an incision into the occiput,[a] which can produce sleep because it relieves the disease. Now moderation in food is also to be observed: for the patient ought not to be surfeited lest it madden him, and he should certainly not be tormented by fasting lest he collapse[b] through debility. The food should be light, in particular gruel, and hydromel for drink, of which three cups[c] are enough, given twice a day in winter, and four times in summer.

There is another sort of insanity, of longer duration because it generally begins without fever, but later excites a slight feverishness. It consists in depression which seems caused by black bile.[d] Bloodletting is here of service; but if anything prohibit this, then comes firstly abstinence, secondly, a clearance by white hellebore and a vomit. After either, rubbing twice a day is to be adopted; if the patient is strong, frequent exercise as well: vomiting on an empty stomach. Food of the middle class should be given without wine; but as often as I indicate this class of food, it should be understood that some of the weakest class of food also may be given, provided that this is not used alone; and that it is only the strongest class of food which is excluded. In addition to the above: the motions are to be kept very soft, causes of fright excluded, good hope

[d] Black bile: μελαγχολία. II. **1**, 6. Black-bile residues from the formation of blood in the liver were supposed to pass back by the splenic vein to the spleen which then passed them on by short veins into the stomach whence they were evacuated. Disturbance of the regular procedure induced the diseases included under the term Melancholia.

offerenda; quaerenda delectatio ex fabulis ludisque, maxime quibus capi sanus adsuerat; laudanda, si qua sunt, ipsius opera et ante oculos eius ponenda; leviter obiurganda vana tristitia; subinde admonendus, in iis ipsis rebus, quae sollicitant, cur potius laetitiae quam sollicitudinis causa sit. Si febris quoque accessit, sicut aliae febres curanda est.

19 Tertium genus insaniae est ex his longissimum, adeo ut vitam ipsam non impediat; quod robusti corporis esse consuevit. Huius autem ipsius species duae sunt: nam quidam imaginibus, non mente falluntur, quales insanientem Aiacem vel Orestem percepisse poetae ferunt: quidam animo desipiunt.

20 Si imagines fallunt, ante omnia videndum est, tristes an hilares sint. In tristitia nigrum veratrum deiectionis causa, in hilaritate album ad vomitum excitandum dari debet; idque si in potione non accepit, in pane adiciendum est, quo facilius fallat: nam si bene se purgaverit, ex magna parte morbum levabit. Ergo etiamsi semel datum veratrum parum profecerit, interposito tempore iterum dari debet. Neque ignorare oportet leviorem esse morbum cum risu quam cum serio insanientium. Illud quoque perpetuum est in omnibus morbis, ubi ab inferiore parte purgandus aliquis est, ventrem eius ante solvendum esse; ubi a superiore, comprimendum.

21 Si vero consilium insanientem fallit, tormentis quibusdam optime curatur. Ubi perperam aliquid

---

[a] Sophocles *Ajax*.
[b] Euripides *Orestes*.   [c] II. 12. 1.

rather put forward; entertainment sought by storytelling, and by games, especially by those with which the patient was wont to be attracted when sane; work of his, if there is any, should be praised, and set out before his eyes; his depression should be gently reproved as being without cause; he should have it pointed out to him now and again how in the very things which trouble him there may be cause of rejoicing rather than of solicitude. When there is fever besides, it is to be treated like other fevers.

The third kind of insanity is of all the most prolonged whilst it does not shorten life, for usually the patient is robust. Now of this sort there are two species: some are duped not by their mind, but by phantoms, such as the poets say Ajax[a] saw when mad or Orestes;[b] some become foolish in spirit.

If phantoms mislead, we must note in the first place whether the patients are depressed or hilarious. For depression black hellebore[c] should be given as a purge, for hilarity white hellebore as an emetic; and if the patient will not take the hellebore in a draught, it should be put into his bread to deceive him the more easily; for if he has well purged himself, he will in great measure relieve himself of his malady. Therefore even if one dose of the hellebore has little effect, after an interval another should be given. It should be known that a madman's illness is less serious when accompanied by laughter than by gravity. This also is an invariable precept in all disease, that when a patient is to be purged downwards, his belly is to be loosened beforehand, but confined when he is to be purged upwards.

If, however, it is the mind that deceives the madman, he is best treated by certain tortures. When he

CELSUS

dixit aut fecit, fame, vinculis, plagis coercendus est. Cogendus est et attendere et ediscere aliquid et meminisse: sic enim fiet, ut paulatim metu cogatur considerare quid faciat. Subito etiam terreri et expavescere in hoc morbo prodest, et fere quicquid
22 animum vehementer turbat. Potest enim quaedam fieri mutatio, cum ab eo statu mens, in quo fuerat, abducta est. Interest etiam, ipse sine causa subinde rideat, an maestus demissusque sit: nam demens hilaritas terroribus iis, de quibus supra (§ 21) dixi, melius curatur. Si nimia tristitia, prodest lenis sed multa bis die frictio, item per caput aqua frigida infusa, demissumque corpus in aquam et oleum.
23 Illa communia sunt, insanientes vehementer exerceri debere, multa frictione uti, neque pinguem carnem neque vinum adsumere; cibis uti post purgationem ex media materia quam levissimis; non oportere esse vel solos vel inter ignotos, vel inter eos, quos aut contemnant aut neglegant; mutare debere regiones et, si mens redit, annua peregrinatione esse iactandos.
24 Raro sed aliquando tamen ex metu delirium nascitur. Quod genus insanientium specie[1] . . . similique victus genere curandum est, praeterquam quod in hoc insaniae genere solo recte vinum datur.

19. His morbis praecipue contrarium est id genus, quod cardiacum a Graecis nominatur, quamvis

[1] *There is a lacuna here. Marx suggests* species similes habet *and this has been translated.*

---

[a] καρδία usually meant the heart, but it was also used by Hippocrates for the stomach. By Galen and others the καρδιακόν πάθος was placed at the junction of the stomach and oesophagus. *Cardiacus morbus* was regarded by Celsus not as a disease of heart or stomach, but a collapse or syncope.

BOOK III. 18. 21–19. 1

says or does anything wrong, he is to be coerced by starvation, fetters and flogging. He is to be forced both to fix his attention and to learn something and to memorize it; for thus it will be brought about that little by little he will be forced by fear to consider what he is doing. To be terrified suddenly and to be thoroughly frightened is beneficial in this illness and so, in general, is anything which strongly agitates the spirit. For it is possible that some change may be effected when the mind has been withdrawn from its previous state. It also makes a difference, whether from time to time without cause the patient laughs, or is sad and dejected: for the hilarity of madness is better treated by those terrors I have mentioned above. If there is excessive depression light and prolonged rubbing twice a day is beneficial, as well as cold water poured over the head, and immersion of the body in water and oil. The following are general rules: the insane should be put to fatiguing exercise, and submitted to prolonged rubbing, and given neither fat meat nor wine: after the clearance the lightest food of the middle class is to be used; they should not be left alone or among those they do not know, or among those whom either they despise or disregard; they ought to have a change of scene, and if the mind returns, they should undergo the tossing incident to travel (II. **15**), once a year.

Rarely, yet now and then, however, delirium is the product of fright; this class of insanity, has similar sub-divisions, and is to be treated by the same species of dietetic regimen, except that, in this form of insaneness alone, wine is properly given.

19. That kind of affection which the Greeks call cardiac[a] is a complete contrast to the foregoing

saepe ad eum phrenetici transeunt: siquidem mens in illis labat, in hoc constat. Id autem nihil aliud est quam nimia inbecillitas corporis, quod stomacho languente inmodico sudore digeritur. Licetque protinus scire id esse, ubi venarum exigui inbecillique pulsus sunt, sudor autem et supra consuetudinem et modo . . .[1] et tempore ex toto thorace et cervicibus atque etiam capite prorumpit, pedibus tantummodo et cruribus siccioribus atque frigentibus; acutique id morbi genus est.

2 Curatio prima est supra praecordia imponere quae reprimant cataplasmata, secunda sudorem prohibere. Id praestat acerbum oleum vel rosa vel melinum aut myrteum, quorum aliquo corpus leviter perunguendum, ceratumque ex aliquo horum tum inponendum est. Si sudor vincit, delinendus homo est vel gypso vel spuma argentea vel Cimolia creta, vel etiam subinde horum pulvere respergendus. Idem praestat pulvis ex contritis aridi myrti vel rubi foliis, aut ex austeri et boni vini arida faece; pluraque similia sunt, quae si desunt, satis utilis 3 est quilibet ex via pulvis iniectus. Super haec vero, quo minus corpus insudet, levi veste debet esse contectus, loco non calido, fenestris patentibus, sic ut perflatus quoque aliquis accedat.

Tertium auxilium est inbecillitati iacentis cibo vinoque succurrere. Cibus non multus quidem, sed saepe tamen nocte ac die dandus est, ut nutriat,

[1] *To fill the lacuna here Marx suggests* alieno, *and this has been translated.*

a II. 33. 2–4.

diseases, although insane persons often pass over into it; in those the mind gives way, in this it holds firm. Indeed the illness is nothing other than excessive weakness of the body, which, while the stomach is languid, wastes away through immoderate sweating. And it may be recognized at once by the exiguous and weak pulsation of the blood vessels, while sweat, at once unaccustomed and excessive and untimely, breaks out all over the chest and neck, and even over the head, the feet and legs remaining more dry and cold; and it is a form of acute disease.

The primary treatment is the application over the chest of repressant [a] plasters; the secondary, to stop sweating. The latter is accomplished by bitter olive oil, or rose or quince or myrtle oil, with any of which the body is to be lightly anointed, then a salve made up of any of them is to be applied. If the sweating wins, the patient is to be smeared over with gypsum or litharge or cimolian chalk, or even powdered over with the same at intervals. A powder consisting of the pounded leaves of dried myrtle or of blackberry, or of the dried lees of dry and good wine, attains the same end; there are many similar materials, and if these are not at hand, it is useful enough to scatter on any dust from the road. In addition to this, moreover, in order that he may sweat less, the patient should be lightly covered and lie in a cool room, with the windows open, so that some breeze reaches him.

A third aid is to help his weakness whilst in bed by food and wine. The food, whilst not much in quantity, should be given often, as well by night as by day, so as to nourish without becoming onerous.

neque oneret. Is esse debet ex infirmissima materia et stomacho aptus. Nisi si necesse est, ad vinum 4 festinare non oportet. Si verendum est, ne deficiat, tum et intrita ex hoc, et hoc ipsum austerum quidem, sed tamen tenue, meraculum, egelidum subinde et liberaliter dandum est, adiecta polenta, si modo is aeger parum cibi adsumit; idque vinum esse debet neque nullarum virium, neque ingentium: recteque toto die ac nocte vel tres heminas aeger bibet; si vastius corpus est, plus etiam: si cibum non accipit, perunctum perfundere aqua frigida ante conveniet, 5 et tunc dare. Quod si stomachus resolutus parum continet, et ante cibum et post eum sponte vomere oportet, rursusque post vomitum cibum sumere. Si ne id quidem manserit, sorbere vini cyathum, interpositaque hora sumere alterum. Si id quoque stomachus reddiderit, totum corpus bulbis contritis superinlinendum est; qui ubi inaruerunt, efficiunt, ut vinum in stomacho contineatur, exque eo toti corpori calor, venisque vis redeat. Ultimum auxilium est in alvum tisanae vel halicae cremorem ex inferioribus partibus indere, siquidem id quoque vires 6 tuetur. Neque alienum est naribus quoque aestuantis admovere quod reficiat [id est rosa et vinum]: si qua in extremis partibus frigent, unctis et calidis manibus fovere. Per quae si consequi potuimus, ut et sudoris impetus minuatur, et vita prorogetur, incipit iam tempus ipsum esse praesidio. Ubi esse

---

a 'The sole hope' in this disease, according to Pliny, *N.H.* XXIII. 25. Compare Juvenal, V. 32: Cardiaco cyathum numquam missurus amico.

b Pearl barley porridge.

c This is the first classical reference to nutrient enemeta (alvus trophicus) attributed by Pliny, *N.H.* XII. end, XX. 20. 8 to Lycus of Neapolis, a commentator on Hippocrates.

It should consist of the weakest class of materials and should be suitable to the stomach. Unless there is necessity, it is not well to hurry on to wine.[a] But when fainting is apprehended, then there is given both bread crumbled into the wine and wine by itself dry indeed yet thin, undiluted, lukewarm, at intervals and liberally, and if the patient is taking but little food, polenta[b] may be scattered into the wine; and that wine should not be lacking in strength, yet not over-strong: the patient may properly drink three quarters of a litre, and even more if of large build, in the course of a day and night: if he does not take his food, the patient should be anointed, should have cold water poured over him, and after that be given food. But if the stomach has become relaxed and retains but little, let the patient vomit as he will, whether before or after food, and after vomiting take food again. If even after that food is not retained, he should sip a cupful of wine, and another after the interval of an hour. If the stomach returns this wine also, he should be rubbed all over with pounded onions, which as they dry cause the stomach to retain the wine, and as a result, cause heat to return throughout the body, and a forceful pulsation to the blood vessels. The last resource is the introduction into the bowel[c] from below of barley or spelt gruel, since that too supports the patient's strength. Should he feel hot, it is not inappropriate to hold to his nostrils a restorative such as rose oil in wine; if he has cold extremities, they should be rubbed by hands anointed and warmed. If we can by these measures obtain a diminution in the severity of the sweating, and a prolongation of life, time itself now begins to come to our aid. When the patient appears to have reached

CELSUS

in tuto videtur, verendum tamen, ne in eandem infirmitatem cito recidat; itaque vino tantummodo remoto cotidie validiorem cibum debet adsumere, donec satis virium corpori redeat.

20. Alter quoque morbus est aliter phrenetico contrarius. In eo difficilior somnus, prompta ad omnem audaciam mens est: in hoc marcor et inexpugnabilis paene dormiendi necessitas. Lethargum Graeci nominarunt. Atque id quoque genus acutum est, et nisi succurritur, celeriter iugulat. — Hos aegros quidam subinde excitare nituntur admotis iis, per quae sternutamenta evocentur, et iis, quae odore foedo movent, qualis est pix cruda, lana sucida, piper, 2 veratrum, castoreum, acetum, alium, cepa. Iuxta etiam galbanum incendunt, aut pilos aut cornu cervinum; si id non est, quodlibet aliud: haec enim cum conburuntur, odorem foedum movent. Tharrias vero quidam accessionis id malum esse dixit, levarique, cum ea decessit; itaque eos, qui subinde excitant, sine usu male habere. Interest autem, in decessione expergiscatur aeger, an aut febris non levetur aut levata quoque ea somnus urgueat. Nam si expergiscitur, adhibere . . .[1] eum sopito supervacuum est: neque enim vigilando melior fit, sed per se, si melior 3 est, vigilat. Si vero continens ei somnus est, utique excitandus est, sed iis temporibus, quibus febris

[1] *The text here is corrupt. Marx conjectures* remedium *for* eum, *and this is translated. Other MSS. readings are* ei *and* ei ut *for* eum.

---

[a] Celsus describes an acute affection which corresponds with that of pneumonia complicating remittent fever. The term *lethargia* was applied very variously. Cf. Horace, *Sat.* II. 3. 145.

safety, rapid relapse into the same state of weakness is still to be feared; hence along with a gradual withdrawal of wine, the patient ought each day to take stronger food, until a sufficiency of bodily strength is gained.

20. There is also another disease, a contrast in a different way to the phrenetic. In the latter sleep is got with great difficulty, and the mind is disposed to any foolhardiness; in this disease there is a pining away, and an almost insurmountable need of sleep. The Greeks name it lethargy.[a] And it also is an acute sort, and unless remedied, quickly kills. Some strive to excite these patients by applying at intervals medicaments to promote sneezing, and those which stimulate by their offensive odour, such as burning pitch, unscoured wool, pepper, hellebore, castoreum, vinegar, garlic, onion. Moreover, they burn near them galbanum, hair or hartshorn, or when that is not at hand, some other kind of horn, for these when burnt give out an offensive odour. One Tharrias said, indeed, that this affection is a sort of feverish paroxysm, and that the patient is relieved when that remits, hence those who keep on irritating such patients do harm uselessly. But the important point is whether the patient wakes up with the remission; or whether the fever is either not relieved, or else it is relieved and yet sleep still oppresses him. For if the patient wakes up, it is needless to treat him as if in a stupor; for he is not made better by keeping him awake, but if he is better he keeps awake of himself. If the sleepiness is uninterrupted the patient must certainly be aroused, but only at those times when the fever is of the slightest, in order that he may both make a

levissima est, ut et excernat aliquid et sumat. Excitat autem validissime repente aqua frigida infusa. Post remissionem itaque perunctum multo oleo corpus tribus aut quattuor amphoris totum per caput perfundendum est. Sed hoc utemur, si aequalis aegro spiritus erit, si mollia praecordia: sin aliter haec erunt, ea potiora, quae supra (2, 3) comprehensa sunt. Et quod ad somnum quidem pertinet, commodissima haec ratio est.

4 Medendi autem causa caput radendum, deinde posca fovendum est, in qua laurus aut ruta decocta sit. Altero die inponendum castoreum, aut ruta ex aceto contrita aut lauri bacae aut hedera cum rosa et aceto; praecipueque proficit et ad excitandum hominem naribus admotum et ad morbum ipsum 5 depellendum capiti frontive inpositum sinapi. Gestatio etiam in hoc morbo prodest, maximeque oportune cibus datus, id est in remissione, quanta maxime inveniri poterit. Aptissima autem sorbitio est, donec morbus decrescere incipiat, sic ut, si cotidie gravis accessio est, haec cotidie detur; si alternis, post graviorem sorbitio, post leviorem mulsa aqua. Vinum quoque cum tempestivo cibo datum non 6 mediocriter adiuvat. Quod si post longas febres eiusmodi torpor corpori accessit, cetera eadem servanda sunt: ante accessionem autem tribus quattuorve horis castoreum, si venter adstrictus est, mixtum cum scamonea; si non est, per se ipsum cum aqua dandum est. Si praecordia mollia sunt, cibis utendum plenioribus; si dura, in isdem sorbi-

---

a An amphora = about 30 litres.
b Convolvulus scammonia yields resin, a drastic purge; castoreum was added as a stimulant to prevent depression.

natural evacuation and take food. Now a most powerful excitant is cold water poured suddenly over him; therefore when the fever has remitted, and he has been anointed freely, he should have three or four jarfuls [a] poured over his head. But this measure should be employed only when the patient's breathing is regular, and the parts below the ribs soft: otherwise those are to be preferred which have been mentioned above. Such is the most suitable procedure, so far as concerns sleepiness.

But in order to cure, the head is to be shaved, and then fomented with vinegar and water in which laurel or rue leaves have been boiled. On the following day castoreum may be applied, or rue pounded up in vinegar, or laurel berries or ivy with rose oil and vinegar: mustard put to the nostrils is particularly efficacious both for arousing the patient, and when put on the head or forehead for driving out the disease itself. Rocking is also advantageous in this malady; and most of all food given opportunely, that is in the greatest degree of remission that can be found. Now gruel is most fitting until the disease begins to decrease; so if there is a severe paroxysm every day, it is given daily; if every other day, after a more severe paroxysm, gruel, and after a slighter paroxysm, hydromel. Wine also is of no mean service, when given at the proper time along with suitable food. But if this kind of torpor attacks the body after prolonged fevers, all the other measures are to be carried out, in the same way, and in addition three or four hours before the paroxysm castoreum is administered, mixed with scammony [b] if the bowels are costive, if not, then by itself in water. If the parts below the ribs are soft, food should be given more freely; if hard,

tionibus subsistendum, inponendumque praecordiis quod simul et reprimat et molliat.

21. Sed hic quidem acutus est morbus. Longus vero fieri potest eorum, quos aqua inter cutem male habet, nisi primis diebus discussus est: hydropa Graeci vocant. Atque eius tres species sunt. Nam modo ventre vehementer intento creber intus ex motu spiritus sonus est; modo corpus inaequale est tumoribus aliter aliterque per totum id orientibus; modo intus in unum aqua contrahitur et moto corpore 2 ita movetur, ut impetus eius conspici possit. Primum τυμπανείτην, secundum λευκοφλεγματίαν vel ὑπὸ σάρκα, tertium ἀσκείτην Graeci nominarunt. Communis tamen omnium est umoris nimia abundantia, ob quam ne ulcera quidem in his aegris facile sanescunt. Saepe vero hoc malum per se incipit, saepe alteri vetusto morbo maximeque quartanae supervenit. Facilius in servis quam in liberis tollitur, quia, cum desideret famem, sitim, mille alia taedia longamque patientiam, promptius iis succurritur, qui facile coguntur, quam 3 quibus inutilis libertas est. Sed ne ii quidem, qui sub alio sunt, si ex toto sibi temperare non possunt, ad salutem perducuntur. Ideoque . . .[1] non ignobilis medicus, Chrysippi discipulus, apud Antigonum regem, amicum quendam eius, notae intemperantiae, mediocriter eo morbo inplicitum, negavit posse

[1] *A name has probably fallen out here. Marx suggests* Zeno (*who wrote a treatise on medicaments, cf. Celsus V. Proem.* 4) *or* Meno (*a pupil of Aristotle and author of a work Iatrika*).

[a] See p. 118, note *a*.

the patient must subsist on the gruel mentioned above, whilst something is to be applied to the parts below the ribs to repress and mollify at the same time.

21. Now the foregoing is indeed an acute disease. But a chronic malady may develop in those patients who suffer from a collection of water under the skin, unless this is dispersed within the first days. The Greeks call this hydrops. And of this there are three species:[a] sometimes the belly being very tense, there is within a frequent noise from the movement of wind; sometimes the body is rendered uneven by swellings rising up here and there all over; sometimes the water is drawn all together within, and is moved with the movement of the body, so that its movement can be observed. The Greeks call the first tympanites, the second leukophlegmasia or hyposarka, the third ascites. The characteristic common to all three species is an excessive abundance of humour, owing to which in these patients ulcerations even do not readily heal. This is a malady which often begins of itself, often it supervenes upon a disease of long standing, upon quartan fever especially (III. **15, 16**). It is relieved more easily in slaves than in freemen, for since it demands hunger, thirst, and a thousand other troublesome treatments and prolonged endurance, it is easier to help those who are easily constrained than those who have an unserviceable freedom. But even those who are in subjection, if they cannot exercise complete self-control, are not brought back to health. Hence a not undistinguished physician, a pupil of Chrysippus, at the court of King Antigonus, held that a certain friend of the king, noted for intemperance, could not be cured, although but moderately affected by that

sanari; cumque alter medicus Epirotes Philippus se sanaturum polliceretur, respondit illum ad morbum aegri respicere, se ad animum. Neque eum res fefellit. Ille enim cum summa diligentia non medici tantummodo, sed etiam regis custodiretur, tamen malagmata sua devorando bibendoque suam urinam 4 in exitium se praecipitavit. Inter initia tamen non difficilis curatio est, si inperata sunt corpori sitis, requies, inedia: at si malum inveteravit, non nisi magna mole discutitur. Metrodorum tamen Epicuri discipulum ferunt, cum hoc morbo temptaretur, nec aequo animo necessariam sitim sustineret, ubi diu abstinuerat, bibere solitum, deinde evomere. Quod si redditur quicquid receptum est, multum taedio demit; si a stomacho retentum est, morbum auget; 5 ideoque in quolibet temptandum non est. Sed si febris quoque est, haec in primis summovenda est per eas rationes, per quas huic succurri posse propositum est (*capp.* IV–XVII). Si sine febre aeger est, tum demum ad ea veniendum est, quae ipsi morbo mederi solent. Atque hic quoque, quaecumque species est, si nondum nimis occupavit, iisdem auxiliis opus est. Multum ambulandum, currendum aliquid est, superiores maxime partes sic 6 perfricandae, ut spiritum ipse contineat. Evocandus est sudor non per exercitationem tantum, sed etiam in harena calida vel Laconico vel clibano similibusque aliis; maximeque utiles naturales et siccae sudationes

---

a Notes pp. 184, 185.

malady; and when another physician, Philip of Epirus, promised that he would cure him, the pupil of Chrysippus replied that Philip was regarding the disease, he the patient's spirit. Nor was he mistaken. For although the patient was watched with the greatest diligence, not only by his physician but by the king as well, by devouring his poultices and by drinking his own urine, he hurried himself headlong to his end. At the beginning, however, cure is not difficult, if there is imposed upon the body thirst, rest, and abstinence; but if the malady has become of long standing, it is not dispersed except with great trouble. They say, however, that Metrodorus, a pupil of Epicurus, when afflicted with this disease, and unable to bear with equanimity the necessary thirst, after abstaining for a long while, was accustomed to drink and then to vomit. Now if what has been drunk is then returned, distress is much reduced; but if retained in the stomach, it increases the disorder; and so it must not be tried in every case. But if there is also fever, this is first of all to be overcome by the methods which have been prescribed concerning possible relief in such cases (III. 4-17). If the patient has become free from fever, then at length we must go on to those measures by which the disorder itself is usually treated. And here, whatever the species, so long as the disease has not taken too firm a hold, the same remedies are required. The patient should walk much, run a little, and his upper parts in particular are to be rubbed while he holds his breath. Sweating is also to be procured, not only by exercise, but also by heated sand, or in the Laconicum,[a] or with a clibanus and such-like; especially serviceable are the natural and dry sweating places,

sunt, quales super Baiias in murtetis habemus. Balineum atque omnis umor alienus est. Ieiuno recte catapotia dantur, facta ex apsinthi duabus, murrae tertia parte. Cibus esse debet ex media quidem materia, sed tamen generis durioris: potio non ultra danda est quam ut vitam sustineat, opti-
7 maque est, quae urinam movet. Sed id ipsum tamen moliri cibo quam medicamento melius est. Si tamen res coget, ex iis aliquid, quae id praestant, erit decoquendum, eaque aqua potui danda. Videntur autem hanc facultatem habere iris, nardum, crocum, cinnamomum, casia, murra, balsamum, galbanum, ladanum, oenanthe, panaces, cardamomum, hebenus, cupressi semen, uva taminia quam σταφίδα ἀγρίαν Graeci nominant, habrotonum, rosae folia, acorum, amarae nuces, tragoriganum, styrax, costum, iunci quadrati et rotundi semen: illud cyperon, hoc σχοίνον Graeci vocant; quae quotiens posuero, non quae hic nascuntur, sed quae inter
8 aromata adferuntur, significabo. Primo tamen quae levissima ex his sunt id est rosae folia vel nardi spica temptanda est. Vinum quoque utile est austerum, sed quam tenuissimum. Commodum est etiam lino cotidie ventrem metiri et, qua conprehendit alvum, notam imponere, posteroque die videre, plenius corpus sit an extenuetur: id enim, quod extenuatur, medicinam sentit. Neque alienum est metiri et potionem eius et urinam: nam si plus umoris excernitur quam insumitur, ita demum secundae valetudinis spes est. Asclepiades in eo, qui ex quartana in hydropa deciderat, se abstinentia

---

[a] The list of diuretics which follows is often subsequently referred to.

such as we have in the myrtle groves above Baiae. The bath and moisture of every kind is wrong. Pills composed of wormwood two parts, myrrh one part, are given on an empty stomach. Food should be of the middle class indeed, but, of the harder kind; no more of drink is to be given than to sustain life, and the best is that which stimulates urine. But that, however, is better brought about by diet than by medicament. If, nevertheless, the matter is urgent, one of those drugs which are efficacious is to be made into a decoction, and that given as a draught. Now this faculty [a] seems to belong to iris root, spikenard, saffron, cinnamon, cassia, myrrh, balsam, galbanum, ladanum, oenanthe, opopanax, cardamomum, ebony, cypress seeds, the taminian berry which the Greeks call staphis agria, southernwood, rose leaves, sweet flag root, bitter almonds, goat's marjoram, styrax, costmary, seeds of rush, square and round (the Greeks call the former cyperon, the latter schoinon): whenever I use these terms I refer, not to native plants, but to such as are imported among spices. The mildest of these, however, are to be tried first, such as rose leaves or spikenard. A dry wine is beneficial, but it must be very thin. It is good besides to measure every day with string the circumference of the abdomen, and to put a mark where it surrounds the belly, then the day following to see whether the body is fuller or thinner, for the thinning shows a yielding to the treatment. Nor is it unserviceable to take the measure of his drink, and of his urine; for if more humour is evacuated than taken in, then at last there is hope of recovery. Asclepiades has put it on record that for a patient who had lapsed from a quartan into dropsy, he employed for two days abstinence and

CELSUS

bidui et frictione usum, tertio die iam et febre et aqua liberato cibum et vinum dedisse memoriae prodidit.

9 Hactenus communiter de omni specie praecipi[1] potest: si vehementius malum est, diducenda ratio curandi est. Ergo si inflatio et ex ea dolor creber est, utilis cotidianus aut altero quoque die post cibum vomitus est; fomentis siccis calidisque utendum est. Si per haec dolor non finitur, necessariae sunt sine ferro cucurbitulae: si ne per has quidem tormentum tollitur, incidenda cutis est; tum his utendum est. Ultimum auxilium est, si cucurbitulae nihil profuerunt, per alvum infundere copiosam aquam calidam 10 eamque recipere. Quin etiam cotidie ter quaterque opus est uti frictione vehementi, cum oleo et quibusdam calefacientibus. Sed in hac frictione a ventre abstinendum est: inponendum vero in eum crebrius sinapi, donec cutem erodat; ferramentisque candentibus pluribus locis venter exulcerandus est; servanda ulcera diutius. Utiliter etiam scilla cocta delinguitur; sed diu post has inflationes abstinendum est ab omnibus inflantibus.

11 At si id vitium est, cui λευκοφλεγματία nomen est, eas partes, quae tument, subicere soli oportet, sed non nimium, ne febricula incidat. Si is vehementior est, caput velandum est, utendumque frictione, madefactis tantum manibus aqua, cui sal et nitrum et olei paululum sit adiectum, sic ut aut pueriles aut muliebres manus adhibeantur, quo mollior earum tactus sit; idque, si vires patiuntur,

[1] *So v. d. Linden, followed by Daremberg and Marx for the* decipi, quae decipi *or* dici *of the MSS.*

rubbing, on the third day, the patient having already become freed from both the fever and the water, he gave food and wine.

So far the prescription can be common to all the species; if the disease is more severe, the method of treatment must be different. For instance, if there is flatulence and owing to that pain is frequent, a vomit is beneficial, either daily or on alternate days, after food; hot and dry foments are to be applied. If the pain is not ended by these, dry cuppings are needed, but if the torment is not relieved even by these, skin incisions are made and then the cups applied. If cupping does no good, the last resource is to infuse hot water copiously into the rectum and draw it out again. Nay even vigorous rubbing with oil and any one of the heating agents should be carried out three or four times a day, but in this rubbing the abdomen is to be left out; but to this mustard should be applied repeatedly until the skin is excoriated; and ulcerations are to be set up in many places upon the abdominal wall by means of the red hot cautery, and the ulcers to be kept open for some time. It is useful also to suck a boiled squill bulb; but for a long while after such attacks of flatulence the patient should abstain from everything that causes it.

But if the affection is that named leukophlegmasia, the swollen parts should be exposed to the sun, but not too much lest feverishness ensue. If the sun is over-strong, the head is to be covered, and rubbing is to be used with hands just moistened with water to which salt and soda and a little oil is added, taking care that the hands of children or of women are employed, for theirs is a softer touch; and this ought to be done, should the patient's strength

ante meridiem tota hora, post meridiem semihora
12 fieri oportet. Utilia etiam sunt cataplasmata, quae reprimunt, maximeque si corpora teneriora sunt. Incidendum quoque super talum quattuor fere digitis ex parte interiore, qua per aliquot dies frequens umor feratur, atque ipsos tumores incidere altis plagis oportet; concutiendumque multa gestatione corpus est; atque ubi inductae vulneribus cicatrices sunt, adiciendum et exercitationibus et cibis, donec
13 corpus ad pristinum statum revertatur. Cibus valens esse debet et glutinosus, maximeque caro. Vinum, si per stomachum licet, dulcius, sed ita ut invicem biduo triduove modo aqua, modo id bibatur. Prodest etiam lactucae marinae, quae grandis iuxta maria nascitur, semen cum aqua potui datum. Si valens est qui id accipit, et scilla cocta, sicut supra (§ 10) dixi, delinguitur. Auctoresque multi sunt inflatis vesicis pulsandos tumores esse.

14 Si vero id genus morbi est, quo in uterum multa aqua contrahitur, ambulare, sed magis modice, oportet, malagma quod digerat inpositum habere, idque ipsum superimposito triplici panno fascia, non nimium tamen vehementer, adstringere; quod a Tharria profectum esse servatum a pluribus video. Si iecur aut lienem adfectum esse manifestum est, ficum pinguem contusam adiecto melle superponere: si per talia auxilia venter non siccatur, sed umor nihilo minus abundat, celeriore via succurrere, ut is
15 per ventrem ipsum emittatur. Neque ignoro dis-

---

a As a purge. Note II. **12.** 1.
b Also used for neuralgia, III. **27.** 2B.
c Here adopted as a placebo because of its popular repute for boils. 2 Kings, XX. 7, Isaiah XXXVIII. 21.

BOOK III. 21. 11-15

allow of it, for a whole hour before noon, or for half an hour after noon. Repressing poultices are of benefit, especially for the more delicate. Also an incision should be made four fingers' breadth above the ankle on the inner side through which humour may discharge freely for some days, and the actual swellings ought to be incised by deep cuts; the body also is to be shaken up by much rocking (II. **15**); then after the incisions have formed a scab, both exercises and food must be increased, until the patient is restored to his former state of health. The food should be nutritious and glutinous, mostly meat. The wine, when the stomach permits of it, should be rather sweet, but should be drunk alternately with water every two or three days. The seeds of the wolf's-milk plant, which grows to a large size on the seacoast, may be advantageously administered [a] in a draught with water. If the patient is strong enough, he should suck a boiled squill bulb as noted above. There are many authorities who would have the swelling beaten with inflated ox-bladders.[b]

But if the form of the affection is that in which much water is drawn into the belly, the patient should take walks, but with much more moderation, and have applied a dispersive poultice, covered with three folds of linen, bandaged on not too tightly; a practice begun by Tharrias, which I see many have followed. If the liver or spleen is plainly affected, a fatty fig [c] bruised with honey should be put on over it: if the belly is not dried up by such remedies, and in spite of them the humour is in large amount, aid must be given in a quicker way, by giving issue to it through the belly itself (VII. **15**). I am quite aware

321

plicuisse Erasistrato hanc curandi viam: morbum enim hunc iocineris putavit: ita illud esse sanandum, frustraque aquam emitti, quae vitiato illo subinde nascatur. Sed primum non huius visceris unius hoc vitium est: nam et liene adfecto et in totius corporis malo habitu fit; deinde ut inde coeperit, tamen aqua, nisi emittitur, quae contra naturam ibi substitit, et iocineri et ceteris partibus interioribus nocet. Convenitque corpus nihilo minus esse curandum: neque enim sanat emissus umor, sed medicinae
16 locum facit, quam intus inclusus impedit. Ac ne illud quidem in controversiam venit, quin non omnes in hoc morbo sic curari possint, sed iuvenes robusti, qui vel ex toto carent febre, vel certe satis liberales intermissiones habent. Nam quorum stomachus corruptus est, quive ex atra bile in hoc deciderunt, quive malum corporis habitum habent, idonei huic curationi non sunt. Cibus autem, quo die primum umor emissus est, supervacuus est, nisi si vires
17 desunt. Insequentibus diebus et is et vinum meracius quidem, sed non ita multum dari debet; paulatimque evocandus aeger est ad exercitationes, frictiones, solem, sudationes, navigationes et idoneos cibos, donec ex toto convalescat. Balneum rarum res amat, frequentiorem in ieiunio vomitum. Si aestas est, in mari natare commodum est. Ubi convaluit aliquis, diu tamen alienus ei veneris usus est.

---

[a] Especially in malarious countries.
[b] Note III. 18. 17.

that such a way of treatment was disapproved of by Erasistratus, for he deemed the disease to be one of the liver, that therefore it was the liver which had to be rendered sound, and that it was of no use to let out water which, if that organ is diseased, will continually be reproduced. But firstly, the disease is not primarily one of that organ alone; for it occurs when the spleen[a] is affected, and there is a general diseased condition of the body; further, granted that it begins from the liver, the water unnaturally collected there, unless evacuated, injures both the liver and all the rest of the internal organs. And it is agreed, nevertheless, that the body generally has to be treated; for the mere evacuation of the humour does not restore health, but it affords an opportunity for medicaments which the accumulation within impedes. And indeed it is not in dispute that not everybody with this affection can be so treated, but only the young and robust, in whom fever is wholly absent, or who have sufficiently long intermissions. For those are unfitted for this treatment whose stomach is corrupted, or who have lapsed into the malady owing to black bile,[b] or who have a diseased condition of body. But food on the day the humour is first let out is not needed unless strength fails. On the days following, both food and wine, undiluted indeed, but not overmuch in quantity, should be given; little by little the patient should be submitted to exercise, rubbing, sun-heat, sweating, a sea voyage, along with a suitable diet, until he has completely recovered. Such a case requires a bath seldom, more often an emetic on an empty stomach; in summer a swim in the sea is beneficial. For a long while after his recovery, however, the practice of venery is unsuitable.

CELSUS

22. Diutius saepe et periculosius tabes eos male habet, quos invasit. Atque huius quoque plures species sunt. Una est, qua corpus non alitur, et naturaliter semper aliquis decedentibus, nullis vero in eorum locum subeuntibus, summa macies oritur, et nisi occurritur, tollit: ἀτροφίαν Graeci vocant. Ea duabus fere de causis incidere consuevit: aut enim nimio timore aliquis minus, aut aviditate nimia plus quam debet adsumit: ita quod vel deest infirmat, vel quod superat corrumpitur.
2 Altera species est quam Graeci κακεξίαν appellant, ubi malus corporis habitus est, ideoque omnia alimenta corrumpuntur. Quod fere fit, quom longo morbo vitiata corpora, etiamsi illo vacant, refectionem tamen non accipiunt; aut cum malis medicamentis corpus adfectum est; aut cum diu necessaria defuerunt; aut cum inusitatos aut inutiles cibos aliquis adsumpsit, aliquidve simile incidit. Hic praeter tabem illud quoque nonnumquam accidere solet, ut per adsiduas pusulas aut ulcera summa cutis exasperetur, vel aliquae corporis partes intumescant.
3 Tertia est longeque periculosissima species, quam Graeci pthisin nominarunt. Oritur fere a capite, inde in pulmonem destillat; huic exulceratio accedit; ex hac febricula levis fit, quae etiam cum quievit, tamen repetit; frequens tussis est, pus excreatur,

---

[a] By *tabes* Celsus means wasting or malnutrition, whether due to disease or want of food. In modern medicine the term is restricted to the special disease of the spinal cord, *tabes dorsalis* (Nomenclature of Diseases, 6th edn., 1931).

BOOK III. 22. 1-3

22. Longer and more dangerous is the illness which follows when wasting disease[a] attacks a patient. This also has several species. There is one in which the body is not nourished enough, and as there is some natural loss going on all the time without replacement, extreme emaciation ensues, which unless countered, kills. The Greeks call it atrophia. It proceeds commonly from two causes: for either from excessive dread the patient consumes less, or from excessive greed more, than he ought: thus either the deficiency weakens, or the superfluity undergoes decomposition. There is another species which the Greeks call cachexia, in which owing to bad habit of body all the aliments undergo decomposition. This occurs generally in those whose bodies have become vitiated by prolonged disease, and even if they have rid themselves of this, yet they do not regain health; or when the system has become affected by bad medicaments; or for a long while necessaries of life have been deficient; or unusual or unserviceable food has been consumed; or something similar has happened. In this case besides the wasting, it not unfrequently occurs that the skin surface undergoes irritation by persistent pustulation or ulceration, or else some parts of the body become swollen. The third species, which the Greeks call phthisis,[b] is the most dangerous by far. The malady usually arises in the head, thence it drips into the lung; there ulceration supervenes, from this a slight feverishness is produced, which even after it has become quiescent nevertheless returns; there is frequent

[b] Chronic pulmonary tuberculosis: catarrhal affections are intimately associated with tubercle bacilli in spreading the disease.

CELSUS

interdum cruentum aliquid. Quicquid excreatum est, si in ignem impositum est, mali odoris est. Itaque qui de morbo dubitant, hac nota utuntur.

4 Cum haec genera tabis sint, animadvertere primum oportet, quid sit, qui laboretur; deinde, si tantum non ali corpus apparet, causam eius attendere, et si cibi minus aliquis quam debet adsumpsit, adicere, sed paulatim, ne, si corpus insuetum subita multitudine oneraverit, concoctionem impediat. Si vero plus iusto qui adsumere solitus est, abstinere uno die, deinde ab exiguo cibo incipere, cotidie adicere, 5 donec ad iustum modum perveniat. Praeter haec convenit ambulare locis quam minime frigidis, sole vitato; per manus quoque exerceri: si infirmior est, gestari, ungui, perfricari, si potest, maxime per se ipsum, saepius eodem die, et ante cibum et post eum, sic ut interdum oleo quaedam adiciantur calfacientia, donec insudet. Prodestque in ieiuno prendere per multas partes cutem et adtrahere, ut relaxetur; aut inposita resina et abducta subinde idem 6 facere. Utile est etiam interdum balneum, sed post cibum exiguum. Atque in ipso solio recte cibi aliquid adsumitur, aut si sine hoc frictio fuit, post eam protinus. Cibi vero esse debent ex iis, quae facile concoquantur: qui maxime alunt. Ergo vini quoque, sed austeri, necessarius usus est; movenda urina.

^a This indicates destruction of the bronchiole cartilages, which include sulphur in their composition. Hipp. IV. 160 (*Aph*. V. 11).
^b Picatio (πίττωσις), also III. 27. 1 D, IV. 2. 8.

cough, pus is expectorated, sometimes blood-stained. When the sputum is thrown upon a fire, there is a bad odour,[a] hence those who are in doubt as to the disease employ this as a test.

As these are the species of wasting, the first consideration should be, which the patient is suffering from: next, if it is only that the body is not being sufficiently nourished, we must look to the cause of this, and if the patient has been consuming less food than he ought, addition is to be made, but only a little at a time, lest if the system becomes overloaded suddenly by an unaccustomed quantity, it may hinder digestion. On the other hand, if the patient has been consuming more than he ought, he should first fast for a day, then begin with a scanty amount of food, increasing daily until he reaches the proper amount. Further, he should walk in places as little cold as possible, whilst avoiding the sun; he should also use the hand exercises; if he is weaker, he should be rocked, anointed and then rubbed, doing as much as possible of this himself, several times each day, before and after meals, until he sweats—sometimes adding heating agents to the oil. It is advantageous on an empty stomach to pinch up and pull on the skin in a number of places, in order to relax it, or to do the same by applying a pitch plaster[b] and at once pulling it off. The bath also is sometimes beneficial, but only after a scanty meal. And whilst actually in the solium, some food may properly be taken, also immediately after a rubbing, when applied without the bath. The food too should be of the kinds easily digested, which are most nutritious. Hence also the use of wine is necessary, but it should be dry; urination is to be stimulated.

CELSUS

7. At si malus corporis habitus est, primum abstinendum est, deinde alvus ducenda, tum paulatim cibi dandi, adiectis exercitationibus, unctionibus, frictionibus. Utilius his frequens balineum est, sed ieiunis, etiam usque sudorem. Cibis vero opus est copiosis, variis, boni suci, quique etiam [si] minus facile corrumpantur, vino austero. Si nihil reliqua proficiunt, sanguis mittendus est, sed paulatim cotidieque pluribus diebus, cum eo ut cetera quoque eodem modo serventur.

8. Quod si mali plus est et vera pthisis est, inter initia protinus occurrere necessarium est: neque enim facile is morbus, cum inveteravit, evincitur. Opus est, si vires patiuntur, longa navigatione, caeli mutatione, sic ut densius quam id est, ex quo discedit aeger, petatur: ideoque aptissime Alexandriam ex Italia itur. Fereque id posse inter principia corpus pati debet, cum hic morbus aetate firmissima maxime oriatur, id est ab anno xiix ad annum xxxv. Si id inbecillitas non sinit, nave tamen non longe
9. gestari commodissimum est. Si navigationem aliqua res prohibet, lectica vel alio modo corpus movendum est; tum a negotiis abstinendum est omnibusque rebus, quae sollicitare animum possunt; somno indulgendum; cavendae destillationes, ne, si quid cura levarit, exasperent; et devitanda cruditas, simulque et sol et frigus; os obtegendum; fauces

---

[a] See Hipp. IV. 158 (*Aph.* V. 9).

But if there is a bad habit of body, the patient should abstain at first, next have the bowels moved by a clyster, then take food a little at a time, with exercise, anointing and rubbing. A frequent bath is useful for these cases, but on an empty stomach, prolonged till there is sweating. Abundant and varied and succulent food is necessary, such as will less readily decompose, and dry wine. If there is no relief from anything else, blood should be let, but only a little each day for several days; with this proviso, that the other remedies also should be employed as described.

But if there is more serious illness and a true phthisis, it is necessary to counter it forthwith at the very commencement; for when of long standing it is not readily overcome. If the strength allows of it a long sea voyage is requisite with a change of air, of such a kind that a denser climate should be sought than that which the patient quits; hence the most suitable is the voyage to Alexandria from Italy. And the body ought generally to be able to bear this in the early stages, since this disease arises especially during the most stable part of life, namely, between eighteen[a] and thirty-five years of age. If the patient's weak state does not allow of the above, the best thing for him is to be rocked in a ship without going far away. If anything prevents a sea voyage, the body is to be rocked in a litter, or in some other way. Further, the patient should keep away from business, and everything disturbing to his spirit; he should indulge in sleep; he is to be warned against catarrh, lest that should make worse what the treatment is relieving; indigestion should be avoided, also the sun and cold; the mouth should be covered, the neck wrapped up,

velandae; tussicula suis remediis finienda; et quamdiu quidem febricula incursat, huic interdum abstinentia, interdum etiam tempestivis cibis me-
10 dendum; eoque tempore bibenda aqua. Lac quoque, quod in capitis doloribus et in acutis febribus et per eas facta nimia siti, ac, sive praecordia tument, sive biliosa urina est, sive sanguis fluxit, pro veneno est, in pthisi tamen, sicut in omnibus longis difficilibusque febriculis recte dari potest. Quod si febris aut nondum incursat, aut iam remisit, decurrendum est ad modicas exercitationes, maximeque ambulationes, item lenes frictiones. Balineum alienum est.
11 Cibus esse debet primo acer, ut alium, porrum, idque ipsum ex aceto, vel ex eodem intubus, ocimum, lactuca, dein lenis, ut sorbitio ex tisana vel ex halica vel ex amulo, lacte adiecto. Idem oriza quoque et, si nihil aliud est, far praestat. Tum in vicem modo his cibis, modo illis utendum est; adiciendaque quaedam ex media materia, praecipueque vel ex pruna cerebellum vel pisciculus et his similia. Farina etiam cum sebo ovillo caprinove mixta, deinde incocta pro medicamento est. Vinum adsumi debet leve, austerum. Hactenus non magna mole pugna-
12 tur. Si vehementior noxa est, ac neque febricula neque tussis quiescit, tenuarique corpus apparet, validioribus auxiliis opus est. Exulcerandum est ferro candenti, uno loco sub mento, altero in gutture, duobus ad mammam utramque, item sub imis ossibus scapularum, quas ὠμοπλάτας Graeci vocant, sic, ne sanescere ulcera sinas nisi tussi finita; cui

---

a By milk Celsus means goat or ewe milk, which is not easily digested in stomach affections. Cf. Hipp. IV. 176 (*Aph.* V. 64).

any cough put a stop to by its appropriate remedies; and whenever there is an intercurrent fever, it is countered, sometimes by abstinence, sometimes by timely meals, at which water is to be drunk. Milk[a] also, which in headaches, in acute fevers and for the excessive thirst they occasion, also when the chest swells, or there is bilious urine, or a flux of blood, is as bad as a poison, can nevertheless be given appropriately in phthisis, as also in all prolonged feverishness. But if there has either been no intercurrent fever yet, or if it has already remitted, recourse should be had to moderate exercise, walking in particular, also to gentle rubbing. The bath is unsuitable. The food should at first be acrid, such as garlic and leeks, also this latter or endive, basil or lettuce after soaking in vinegar; later the food should be bland, such as a gruel made with pearl barley, or spelt flour, or starch to which milk is added. Rice also, and if there is nothing else, parched groats of spelt answer. Subsequently use is to be made of the above foods in turn, with some additions from food of the middle class, especially grilled brains, small fish and such like. Flour mixed with mutton or goat-fat and then boiled serves for a medicament. The wine taken ought to be light and dry. So far there is no great difficulty in countering the disease. But if it is more severe, and neither the feverishness nor the cough quiets down, and the body is evidently wasting, stronger remedies are required. Ulceration is to be set up by cautery, at one spot under the chin, at another on the neck, two upon each breast, and the same below the shoulder-blades which the Greeks call omoplatae. The ulcerations are not to be allowed to heal until the cough has stopped, and

per se quoque medendum esse manifestum est. Tum ter quaterve die vehementer extremae partes perfricandae; thorax levi manu pertractandus; post cibum intermittenda hora, et perfricanda crura
13 brachiaque. Interpositis denis diebus demittendus aeger in solium, in quo sit aqua calida et oleum. Ceteris diebus bibenda aqua; tum vinum; si tussis non est, potui frigidum dandum; si est, egelidum. Utile est etiam cibos in remissionibus cotidie dari, frictiones gestationesque similiter adhibere, eadem acria quarto aut quinto die adsumere, interdum herbam sanguinalem ex aceto vel plantaginem esse.
14 Medicamentum etiam est vel plantaginis sucus per se, vel marrubii cum melle incoctus, ita ut illius cyathus sorbilo sumatur huius cocleare plenum paulatim delingatur, vel inter se mixta et incocta resinae terebenthinae pars dimidia, buturi et mellis pars altera. Praecipua tamen ex his omnibus sunt victus, vehiculum et navis, et sorbitio. Alvus cita utique vitanda est. Vomitus in hoc morbo frequens perniciosus est, maximeque sanguinis. Qui meliusculus esse coepit, adicere debet exercitationes, frictiones, cibos, deinde ipse se suppresso spiritu perfricare, diu abstinere a vino, balneo, venere.

23. Inter notissimos morbos est etiam is, qui comitialis vel maior nominatur. Homo subito conci-

---

^a See Hipp. IV. 160 (*Aph.* V. 14).

^b Epilepsy. Celsus nowhere uses the Greek name perhaps because it was held to be ill-omened. The name *comitialis morbus* was given to the disease because a meeting of the *comitia* was adjourned if anyone was attacked by it, since it was looked upon as a divine manifestation; cf. the title of the Hippocratic treatise, περὶ ἱερῆς νούσου (Hipp. II. 129). Celsus mentions the premonitory symptoms, II. 8. 11 (termed *aura* by Galen VIII. 194), and does not confuse the

for this there must clearly be a special treatment also. Then three or four times a day the extremities should be rubbed vigorously, the chest being merely stroked with the hands; an hour after food both legs and arms are to be rubbed. At intervals of ten days the patient should be immersed in the solium containing oil with the hot water. On other days he should drink first water, then wine; if there is no cough, the drink should be cold, if there is a cough lukewarm. It is also of advantage to give food every day during the remission; rubbing and rocking should be employed likewise. On the fourth or fifth day he should take the above mentioned acrid food, now and then polygonum or plantain juice in vinegar. A further remedy is either plantain juice by itself, or horehound juice cooked with honey; of the former a cupful may be sipped, of the latter a spoonful, a little at a time, put upon the tongue, or one half part of turpentine resin, and another part of butter and honey may be mixed together and cooked. But of all these measures the principal ones are the diet, rocking in a litter or on a ship, and the gruel. Loose motions must be especially obviated.[a] Frequent vomiting in this affection is a sign of danger, especially when blood is vomited. A patient who is beginning to improve a little should resume exercises, rubbing, and increase of food, next rub himself while holding his breath, but for a long while abstain from wine, the bath and venery.

23. That malady which is called comitialis, or the greater, is one of the best known.[b] The man

disease with apoplexy since he does not describe the fits as followed by paralysis. While 'focal epilepsy' and minor epilepsy (*petit mal*) are now better understood, no fundamental understanding of general epilepsy has yet been reached.

dit, ex ore spumae moventur, deinde interposito tempore ad se redit, et per se ipse consurgit. Id genus saepius viros quam feminas occupat. Ac solet quidem etiam longum esse usque mortis diem et vitae non periculosum: interdum tamen cum recens est, hominem consumit. Et saepe eum, si remedia non sustulerunt, in pueris veneris, in puellis men-
2 struorum initium tollit. Modo cum distentione autem nervorum prolabitur aliquis, modo sine illa. Quidam hos quoque isdem quibus lethargicos excitare conantur; quod admodum supervacuum est, et quia ne lethargicus quidem his sanatur, et quia, cum possit ille numquam expergisci atque ita fame interire, hic ad se utique revertitur. Ubi concidit aliquis, si nulla nervorum distentio accessit, utique sanguis mitti non debet: si accessit, non utique
3 mittendus est, nisi alia quoque hortantur. Necessarium autem est ducere alvum, vel nigro veratro purgare, vel utrumque facere, si vires patiuntur; tunc caput tondere oleoque et aceto perunguere; cibum post diem tertium, simul transit hora, qua concidit, dare. Neque sorbitiones autem his aliique molles et faciles cibi neque caro, minimeque suilla, convenit, sed media materia: nam et viribus opus est et cruditates cavendae sunt; cum quibus fugere oportet solem, balneum, ignem omniaque calfacientia; item frigus, vinum, venerem, loci praecipitis conspectum omniumque terrentium, vomitum, lassitudines, sollicitudines, negotia omnia.

suddenly falls down and foam issues out of his mouth; after an interval he returns to himself, and actually gets up by himself. This kind affects men oftener than women. And usually it persists even until the day of death without danger to life; nevertheless occasionally, whilst still recent, it is fatal to the man. And often if remedies have been ineffectual, in boys the commencement of puberty, in girls of menstruation, has removed it. Now sometimes there is a spasm of the sinews when the man falls down, sometimes there is none. Some try to rouse the patients as is done in the case of those affected by lethargy; which is quite useless, both because not even the lethargic patient is cured by this method, and because, though it may be impossible to awaken him and he may thus die of starvation, the epileptic, on the other hand, returns to himself. If a man falls in a fit without the addition of spasms, certainly he should not be bled; if there are spasms, at any rate he should not be bled unless there are other indications for the bleeding. But it is necessary to move the bowels by a clyster, or by a purge of black hellebore, or by both if the strength allows of it. Next the head should be shaved and oil and vinegar poured over it, the patient should be given food on the third day, as soon as the hour has passed at which he had a fit. But neither gruels, nor other soft and easily digested food, nor meat, least of all pork, are suitable for such patients, but food materials of the middle class: for there is need to give strength and indigestion is to be avoided; in addition he should avoid sunshine, the bath, a fire, all heating agents; also cold, wine, venery, overlooking a precipice, and everything terrifying, vomiting, fatigue, anxiety, and all business.

CELSUS

4 Ubi tertio die cibus datus est, intermittere quartum et invicem alterum quemque, eadem hora cibi servata, donec quattuordecim dies transeant. Quos ubi morbus excessit, acuti vim deposuit; at si manet, curandus iam ut longus est. Quod si non, quo die primum incidit, medicus accessit, sed is, qui cadere consuevit, ei traditus est, protinus eo genere victus adhibito, qui supra (§ 3, 4) conprehensus est, expectandus est dies, quo prolabatur; utendumque tum vel sanguinis missione vel ductione alvi vel nigro
5 veratro, sicut praeceptum est (§ 3). Insequentibus deinde diebus per eos cibos, quos proposui, vitatis omnibus, quae cavenda dixi (§ 3), nutriendus. Si per haec morbus finitus non fuerit, confugiendum erit ad album veratrum, ac ter quoque aut quater eo utendum non ita multis interpositis diebus, sic tamen ne iterum umquam sumat, nisi conciderit. Mediis autem diebus vires eius erunt nutriendae, quibusdam praeter ea, quae supra (§ 3) scripta sunt, adiectis.
6 Ubi mane experrectus est, corpus eius leviter ex oleo vetere, cum capite, excepto ventre, permulceatur; tum ambulatione quam maxime recta et longa utatur; post ambulationem loco tepido vehementer et diu, ac non minus ducenties, nisi infirmus erit, perfricetur; deinde per caput multa aqua frigida perfundatur; paulum cibi adsumat; conquiescat; rursus ante noctem ambulatione utatur; iterum vehementer perfricetur, sic ut neque venter neque

BOOK III. 23. 4–6

When food has been given upon the third day, it should be omitted on the fourth, and then on alternate days, observing the same hour for the meal, until fourteen days have elapsed. When the malady lasts beyond this period, it loses its acute character, and if it persists, it is now to be treated as chronic. But if the practitioner has not been in attendance from the day of the first fit, but a patient who is liable to fits has been handed over to him, the class of diet given above should straightway be adhered to, and the day awaited upon which the patient may have a fit; then there is to be used either blood-letting, or clystering, or purgation by black hellebore, as prescribed above. Next on the following days the patient is to be supported by those foods I have mentioned, avoiding everything which I have said must be avoided. If the malady has not been brought to an end by these measures recourse should be had to white hellebore, administering it three or four times, without many days between, never, however, repeating it unless he has had a fit. Moreover, on intermediate days his strength must be supported by additions to what has been prescribed above. On awakening in the morning, his body should be lightly rubbed with old oil, including the head, but excluding the stomach; he should then walk as straight and as far as he can; after the walk he should be rubbed vigorously for a long while in a warm place, and with not less than two hundred handstrokings, unless he is weak (II. **14**); next plenty of cold water should be poured over his head; he should take a little food; rest; again before night take a walk; and once more be vigorously rubbed, yet without touching either his stomach or his head,

caput contingatur; post haec cenet, interpositisque tribus aut quattuor diebus uno aut altero acria 7 adsumat. Si ne per haec quidem fuerit liberatus, caput radat; unguatur oleo veteri, adiecto aceto et nitro; perfundatur aqua salsa; bibat ieiunus ex aqua castoreum; nulla aqua nisi decocta potionis causa utatur. Quidam iugulati gladiatoris calido sanguine epoto tali morbo se liberarunt; apud quos miserum auxilium tolerabile miserius malum fecit. Quod ad medicum vero pertinet, ultimum est iuxta talum ex utroque crure paulum sanguinis mittere, occipitium incidere et cucurbitulas admovere, ferro candenti in occipitio quoque et infra, qua summa vertebra cum capite committitur, adurere duobus locis, ut per ea 8 perniciosus umor evadat. Quibus si finitum malum non fuerit, prope est, ut perpetuum sit. Ad levandum id tantummodo utendum erit exercitatione, multa frictione, cibisque, qui supra (§ 3) conprehensi sunt, praecipueque vitanda omnia, quae ne fierent excepimus (§ 3).

24. Aeque notus est morbus, quem interdum arquatum, interdum regium nominant. Quem Hippocrates ait, si post septimum diem febricitante aegro supervenit, tutum esse, mollibus tantummodo praecordiis substantibus: Diocles ex toto, si post febrem oritur, etiam prodesse; si post hunc febris, occidere. Color autem eum morbum detegit, maxime oculorum, in quibus quod album esse debet,

---

a Pliny *N.H.* XXVIII. 4. 10.

b Jaundice; a general disturbance preventing the evacuation of the yellow bile residue after blood formation in the liver; compare the black bile residue and melancholia, note III. **19.** 17.

after this he may have dinner, and at intervals of three or four days he should eat for a day or two acrid foods. If not freed by these measures, his head should be shaved; anointed with old oil, to which vinegar and nitre have been added; have salt water poured over it; next upon an empty stomach he should take castoreum in water; no water should be used for drinking unless it has been boiled. Some have freed themselves from such a disease by drinking the hot blood from the cut throat of a gladiator:[a] a miserable aid made tolerable by a malady still more miserable. But as to what is really the concern of the practitioner, the last resources are: to let a little blood from both legs near the ankle, to incise the back of the scalp and apply cups, to burn in two places with a cautery, at the back of the scalp and just below where the highest vertebra joins the head, in order that pernicious humour may exude through the burns. If the disease has not been brought to an end by the foregoing measures, it is probable that it will be lifelong. To mitigate it to some extent all you can do is to use exercise, plenty of rubbing, and the food which has been mentioned above, particularly avoiding what we have declared to be harmful.

24. Equally well-known is the disease which they name sometimes the rainbow-hued, sometimes the royal.[b] Of this Hippocrates said that if it comes on after the patient has been suffering from fever for seven days, the patient is safe, provided that the parts under the ribs remain soft. Diocles stated positively that if it arise after a fever, it is even favourable, but deadly if fever follows it. The colour reveals this malady, particularly of the eyes; they become yellow in the parts which should be

2 fit luteum. Soletque accedere et sitis et dolor capitis et frequens singultus et praecordiorum dextra parte durities et, ubi corporis vehemens motus est, spiritus difficultas membrorumque resolutio; atque ubi diutius manet morbus, totum corpus cum pallore quodam inalbescit.

Primo die abstinere aegrum oportet, secundo ducere alvum, tum, si febris est, eam victus genere discutere; si non est, scamoniam potui dare, vel cum aqua betam albam contritam, vel cum aqua mulsa nuces amaras, absinthium, anesum, sic ut pars 3 huius minima sit. Asclepiades aquam quoque salsam, et quidem per biduum, purgationis causa bibere cogebat, iis quae urinam movent reiectis. Quidam superioribus omissis per haec et per eos cibos, qui extenuant, idem se consequi dicunt. Ego ubique, si satis virium est, validiora; si parum, 4 inbecilliora auxilia praefero. Si purgatio fit, post eam triduo primo modice cibum oportet adsumere ex media materia, et vinum bibere Graecum salsum, ut resolutio ventris maneat; tum altero triduo validiores cibos, et carnis quoque aliquid esse, intra aquam manere; deinde ad superius genus victus reverti, cum eo ut magis satietur; omisso Graeco vino bibere integrum austerum; atque ita per haec variare, ut interdum acres quoque cibos interponat, 5 interdum ad salsum vinum redeat. Per omne vero tempus utendum est exercitatione, fricatione, si hiemps est, balneo; si aestas, frigidis natationibus;

white. And it is usually accompanied by thirst and headache and frequent hiccough and induration under the ribs on the right side, and when a sharp movement of the body is made there is difficulty in breathing and laxness of the limbs; and when the disease persists for a long while, the whole body whitens with a sort of pallor.

On the first day the patient should fast, on the second day have the bowels moved by a clyster, then if there is fever, it is dispersed by appropriate diet; if not, scammony is given in a draught, or white beet pounded up in water, or bitter almonds, wormwood, and a very little aniseed in hydromel. Asclepiades used also to make the patient drink salted water, even for a couple of days, in order to purge, and rejected diuretics. Some, omitting the remedies given above, say they gain the same end through diuretics and those foods which cause thinness. For myself, if there is sufficient strength, I prefer the stronger remedies, the milder if there is but little. After purgation, for the first three days a moderate amount of food of the middle-class should be taken with salted Greek wine to drink, in order to keep the bowels loose; then on the ensuing three days, food of the stronger class with some meat, keeping to water for drink; next there is a return to the middle class of food, but in such a way that he may be more satisfied therewith, and for drink an undiluted dry wine in place of the Greek; and this diet is varied so that sometimes acrid foods are put in, sometimes the salted wine is again given. But throughout the whole time use is to be made of exercise, of rubbing, in winter of the bath, in summer of swimming in cold water, the patient should enjoy

lecto etiam et conclavi cultiore, lusu, ioco, ludis, lascivia, per quae mens exhilaretur; ob quae regius morbus dictus videtur. Malagma quoque, quod digerat, super praecordia datum, prodest, vel arida ibi ficus superimposita, si iecur aut lienis est adfectus.

25. Ignotus autem paene in Italia, frequentissimus in quibusdam regionibus is morbus est, quem ἐλεφαντίασιν Graeci vocant; isque longis adnumeratur: quo totum corpus adficitur, ita ut ossa quoque vitiari dicantur. Summa pars corporis crebras maculas crebrosque tumores habet; rubor harum paulatim in atrum colorem convertitur. Summa cutis inaequaliter crassa, tenuis, dura mollisque, quasi squamis quibusdam exasperatur; corpus emacrescit; os, surae, pedes intumescunt. 2 Ubi vetus morbus est, digiti in manibus pedibusque sub tumore conduntur; febricula oritur, quae facile tot malis obrutum hominem consumit.— Protinus ergo inter initia sanguis per biduum mitti debet, aut nigro veratro venter solvi. Adhibenda tum, quanta sustineri potest, inedia; paulum deinde vires reficiendae et ducenda alvus; post haec ubi corpus levatum est, utendum exercitatione praeci- 3 pueque cursu; sudor primum labore ipsius corporis, deinde etiam siccis sudationibus evocandus; frictio adhibenda, moderandumque inter haec, ut vires conserventur. Balneum rarum esse debet; cibus sine pinguibus, sine glutinosis, sine inflantibus;

---

[a] This description (based on hearsay) combines the symptoms seen in Egypt then and now as the result of filiariasis and leprosy infections.

BOOK III. 24. 5–25. 3

a specially good bed and room, also dicing, jesting, play-acting and jollification, whereby the mind may be exhilarated; on account of this treatment the disease seems to have been termed royal. In addition a dispersive poultice, applied under the ribs, is beneficial, or if the liver or spleen has become affected a dried fig is put on.

25. The disease which the Greeks call elephantiasis,[a] whilst almost unknown in Italy, is of very frequent occurrence in certain regions; it is counted among chronic affections; in this the whole body becomes so affected that even the bones are said to become diseased. The surface of the body presents a multiplicity of spots and of swellings, which, at first red, are gradually changed to be black in colour. The skin is thickened and thinned in an irregular way, hardened and softened, roughened in some places with a kind of scales; the trunk wastes, the face, calves and feet swell. When the disease is of long standing, the fingers and toes are sunk under the swelling: feverishness supervenes, which may easily destroy a patient overwhelmed by such troubles. At once, therefore, at the commencement, he should be bled for two days, or the bowels loosened by black hellebore, then a scanty diet is to be adopted so far as can be borne; after that the strength should be a little reinforced and the bowels clystered; subsequently, when the system has been relieved, exercise and especially running is to be used. Sweating should be induced primarily by the patient's own exertion, afterwards also by dry sweatings, rubbing is to be employed with moderation so that strength is preserved. The bath should be seldom used; neither fatty nor glutinous nor flatulent food; wine

vinum praeterquam primis diebus recte datur. Corpus contrita plantago et inlita optime tueri videtur.

26. Attonitos quoque raro videmus, quorum et corpus et mens stupet. Fit interdum ictu fulminis, interdum morbo: ἀποπληξίαν hunc Graeci appellant.—His sanguis mittendus est; veratro quoque albo, vel alvi ductione utendum; tunc adhibendae frictiones, et ex media materia minime pingues cibi, quidam etiam acres: a vino abstinendum.

27. 1. At resolutio nervorum frequens ubique morbus est: interdum tota corpora, interdum partes infestat. Veteres auctores illud ἀποπληξίαν hoc παράλυσιν nominarunt: nunc utrumque παράλυσιν appellari video. Solent autem qui per omnia membra vehementer resoluti sunt, celeriter rapi; ac si correpti non sunt, diutius quidem vivunt, sed raro tamen ad sanitatem perveniunt; plerumque miserum spiritum B trahunt, memoria quoque amissa. In partibus vero numquam acutus, saepe longus, fere sanabilis morbus est. Si omnia membra vehementer resoluta sunt, sanguinis detractio vel occidit vel liberat. Aliud curationis genus vix umquam sanitatem restituit, saepe mortem tantum differt, vitam interim infestat. Post sanguinis missionem si non redit et motus et mens, nihil spei superest; si redit, sanitas quoque prospicitur. C At ubi pars resoluta est, pro vi et mali et corporis vel sanguis mittendus vel alvus ducenda

---

[a] Hipp. IV. pp. 119, 135, 193 (*Aph.* II. 42, III. 31, VI. 56, 57).

[b] Medical writers, following Hippocrates II. 246 (*Breaths* 13) ascribed all forms of paralysis to collections of flatus and phlegm in the head, until Wepfer in 1658 discovered cerebral haemorrhage to be a cause.

is properly given except on the first days. Plantain crushed and smeared on seems to protect the body best.

26. We also see occasionally some who have been stunned, in whom both the body and the mind are stupefied. This is produced sometimes by lightning stroke, sometimes by disease; the Greeks call this latter apoplexia. In these cases, blood is to be let, and either white hellebore (**II. 13**. 2) or a clyster administered; then rubbings are applied, and food of the middle class given, and that the least fatty; also some which is acrid; there is to be abstinence from wine.

27. Relaxing of the sinews, on the other hand, is a frequent disease everywhere. It attacks at times the whole body, at times part of it. Ancient writers named the former apoplexy,[a] the latter paralysis: I see that now both are called paralysis.[b] Those who are gravely paralyzed in all their limbs are as a rule quickly carried off, but if not so carried off, some may live a long while, yet rarely however regain health. Mostly they drag out a miserable existence, their memory lost also. The disease, when partial only, is never acute, often prolonged, generally remediable. If all the limbs are gravely paralyzed withdrawal of blood either kills or cures. Any other kind of treatment scarcely ever restores health, it often merely postpones death, and meanwhile makes life a burden. If after blood-letting, neither movement nor the mind is recovered, there is no hope left; if they do return, health also is in prospect. But when a particular part is paralyzed, in accordance with the force of the disease, and the strength of the body, either blood is to be let, or the bowel

est. Cetera eadem in utroque casu facienda sunt, siquidem vitare praecipue convenit frigus; paulatimque ad exercitationes revertendum est, sic ut ingrediatur ipse protinus, si potest. Si id crurum inbecillitas prohibet, vel gestetur, vel motu lecti concutiatur; tum id membrum, quod deficit, si potest, per se, si minus, per alium moveatur, et vi D quadam ad consuetudinem suam redeat. Prodest etiam torpentibus membris summam cutem exasperare vel urticis caesam vel inposito sinapi, sic ut, ubi rubere coeperit corpus, haec removeantur. Scilla quoque contrita bulbique contriti cum ture recte inponuntur. Neque alienum est resina cutem tertio quoque die diutius vellere; pluribus etiam locis aliquando sine ferro cucurbitulas admovere. Unctioni vero aptissimum est vetus oleum vel nitrum aceto et oleo mixtum. Quin etiam fovere aqua calida marina, vel si ea non est, tamen salsa magno-E pere necessarium est. Ac si quo loco vel naturales vel etiam manu factae tales natationes sunt, iis potissimum utendum est; praecipueque in iis agitanda membra, quae maxime deficiunt; si id non est, balneum tamen prodest. Cibus esse debet ex materia media, maximeque ex venatione: potio sine vino aquae calidae. Si tamen vetus morbus est, interponi quarto vel quinto die purgationis causa vinum Graecum salsum potest. Post cenam utilis vomitus est.

2. Interdum vero etiam nervorum dolor oriri solet. In hoc casu non vomere, non urinam medicamentis movere, non sine exercitatione sudorem, ut quidam

---

[a] Because in the bath the body is supported by the water.
[b] See note on *nervus*, p. 152.

clystered. The rest that has to be done is the same in both conditions: in particular cold should be avoided; and the patient should return to exercise a little at a time, in such a way that he should begin to walk at once, if he can. If the weakness of the legs prevent this, he should be carried about in a litter or rocked in his bed, then, if possible, his defective limb should be moved by himself, failing that by someone else, and by a form of compulsion, it should be restored to its customary state. It is also beneficial to stimulate the skin of the torpid limb, either by whipping with nettles, or by applying mustard plasters, these latter being removed as soon as the skin becomes red. Appropriate applications also are crushed squills, and onions pounded up with frankincense. Nor is it amiss to pluck on the skin for some time by the aid of a pitch plaster every third day (III. 22. 6) and sometimes to apply dry cups in several places. Again for anointing, old olive oil is most suitable, or soda mixed with oil and vinegar. Further, it is also highly necessary to foment with warm sea water, or failing that with salt and water. And if there are at hand swimming baths, whether natural or artificial, they should be used as much as possible;[a] especially the defective limb should be moved in them; if there are none such the ordinary bath is of service. The food should be of the middle class, particularly game, the drink hot water without wine. If, however, the disease is of long standing, every fourth or fifth day Greek salted wine may be given, in order to purge. An emetic after supper is of use.

At times also there occurs pain in the sinews.[b] In that case it is not expedient to excite vomiting nor urination as some prescribe, nor indeed sweating

praecipiunt, expedit: bibenda aqua est, bis die in lectulo leniter satis diu corpus perfricandum est, deinde retento spiritu ab ipsa exercitatione potius superiores partes movendae; balneo raro utendum; mutandum subinde peregrinationibus caelum. Si dolor est, ea ipsa pars sine oleo nitro ex aqua perunguenda est, deinde involvenda, et subicienda pruna lenis et sulpur, atque ita diu subfumigandum; idque aliquandiu faciendum, sed ieiuno, cum bene iam concoxerit. Cucurbitulae quoque saepe dolenti parti admovendae sunt, pulsandusque leviter inflatis vesicis bubulis is locus est. Utile est etiam sebum miscere cum hyoscyami et urticae contritis seminibus, sic ut omnium par modus sit, idque inponere; fovere aqua, in qua sulpur decoctum sit. Utriculi quoque recte inponuntur aqua calida repleti, aut bitumen cum hordeacia farina ... iactum.[1] Atque in ipso potissimum dolore utendum gestatione vehementi est; quod in aliis doloribus pessimum est.

3. Tremor autem nervorum aeque vomitu medicamentisque urinam moventibus intenditur. Inimica etiam habet balinea assasque sudationes. Bibenda aqua est; acri ambulatione utendum, itemque unctionibus frictionibusque, maxime per se ipsum; pila similibusque superiores partes dimovendae; cibo quolibet utendum, dum concoctioni utique studeatur. Secundum cibum curis abstinendum; rarissima venere utendum est. Si quando quis in eam pro-

---

[1] iactum. *The reading of the best MSS. is corrupt: Daremberg following one MS. reads* mixtum, *Marx conjectures* calefactum.

---

[a] τρόμος, distinguished from tremor due to cold (I. **9.** 3), or to heat (II. **8.** 16).

otherwise than through exercise; water should be drunk; twice a day in bed the body should be rubbed gently and for some time, and then whilst holding the breath, the limbs, preferably the upper, are to be moved in the course of exercise. The bath should be seldom used; from time to time there should be a change of air by travel. If there is pain, the part should be wetted with water containing soda, but not oil, then wrapped up, and under it should be placed a brazier containing some glowing charcoal with sulphur, so that it may be fumigated for a while; this should be repeated from time to time, but only on an empty stomach and after digestion is completed. Cups also may be applied at frequent intervals to the painful part, and this place lightly beaten with inflated ox bladders. It is also of service to mix fat with pounded henbane and nettle seeds, equal parts of each, and put this on, also to foment with a decoction of sulphur. Further, it is a good plan to apply leather bottles filled with hot water, or bitumen mixed with barley meal. And for the actual pain the best remedy is forceful rocking; which in other kinds of pain is the worst.

Tremor[a] of sinews again is likewise made worse by an emetic, and by medicaments causing urination. Inimical also are baths and dry sweatings. Water is to be drunk; the patient should take a smart walk and be anointed and rubbed as well, especially by himself; the upper limbs are to be exercised by ball games and the like; he may take what food he likes provided that he studies his digestion. He should avoid worry after meals; make the rarest use of venery. If at any time he has given way to it, then

lapsus est, tum oleo leviter diuque in lectulo perfricari manibus puerilibus quam virilibus debet.

4. Suppurationes autem, quae in aliqua interiore parte oriuntur, ubi notae fuerint, primum id agere oportet per ea cataplasmata, quae reprimunt, ne coitus inutilis materiae fiat; deinde, si haec victa sunt, per ea malagmata, quae digerunt, dissipentur. Quod si consecuti non sumus, sequitur ut evocetur, deinde, ut maturescat. Omnis tum vomicae finis est, ut rumpatur; indiciumque est pus vel alvo vel B ore redditum. Sed nihil facere oportet, quo minus, quicquid est puris, excedat. Utendum maxime sorbitione est, et aqua calida. Ubi pus ferri desiit, transeundum ad faciles quidem sed tamen validiores et frigidos cibos frigidamque aquam, sic ut ab egelidis tamen initium fiat. Primoque cum melle quaedam edenda, ut nuclei pinei vel Graecae nuces vel Abellanae; postea summovendum id ipsum, quo C maturius induci cicatrix possit. Medicamentum eo tempore ulceri est sucus adsumptus vel porri vel marrubii, et omni cibo porrum ipsum adiectum. Oportebit autem uti in iis partibus, quae non adficientur, frictionibus, item ambulationibus lenibus; vitandumque erit, ne vel luctando vel currendo vel alia ratione sanescentia ulcera exasperentur: in hoc enim morbo perniciosus ideoque omni modo cavendus sanguinis vomitus est.

he ought to be rubbed, with oil, gently and for some time, whilst in bed, by the hands of boys rather than men.

Now suppurations[a] which arise in some interior part, when they become noticeable, first should be acted upon by those poultices which repress, lest there is produced a harmful collection of the material of disease; next if these remedies are unsuccessful, the suppurations may be dissipated by dispersive poultices. If we are not successful in that, it follows that the suppuration should be drawn outwards, next that it should mature. The ending of every abscess is to rupture; the indication is pus discharged either from the bowels or mouth. But nothing ought to be done to diminish the discharge of the pus. Broth and hot water are chiefly to be given. When pus ceases to be discharged, then there should be a transition to digestible yet nutritious food consumed cold, also cold water for drink, commencing, however, with lukewarm. To begin with, things such as pine kernels, or almonds, or hazel nuts, may be eaten along with honey; afterwards these make way for whatever can make the scar form earlier. At this stage as a medicament for the ulceration there is to be taken either leek or horehound juice, and whatever the food, leeks should be added. Rubbing is required also for parts unaffected, so also gentle walks; to be avoided are wrestling and running and other things tending to irritate healing ulcerations, for in this malady the vomiting of blood is most pernicious and to be guarded against in every way.

[a] Commentators have regarded this section as misplaced; but it is a summary referring both backwards and forwards in respect to treatment.

BOOK IV

LIBER IV

1. Hactenus reperiuntur ea genera morborum, quae in totis corporibus ita sunt, ut is certae sedes adsignari non possint: nunc de iis dicam, quae sunt in partibus. Facilius autem omnium interiorum morbi curationesque in notitiam venient, si prius eorum sedes breviter ostendero.

2   Caput igitur, eaque, quae in ore sunt, non lingua tantummodo palatoque terminantur, sed etiam quatenus oculis nostris exposita sunt. In dextra sinistraque circa guttur venae grandes, quae sphagitides nominantur, itemque arteriae, quas carotidas vocant, sursum procedentes ultra aures feruntur. At in ipsis cervicibus glandulae positae sunt, quae interdum cum dolore intumescunt.

3   Deinde duo itinera incipiunt: alterum asperam arteriam nominant, alterum stomachum. Arteria exterior ad pulmonem, stomachus interior ad ventriculum fertur; illa spiritum, hic cibum recipit. Quibus cum diversae viae sint, qua coeunt exigua in arteria sub ipsis faucibus lingua est; quae, cum spiramus, attollitur, cum cibum potionemque adsu-

---

[a] This brief description of the internal parts in relation to their treatment for disease is the first outline of Human Anatomy from the Clinical Standpoint; Book VIII. begins with a description of the skull and other bones in relation

BOOK IV

1. THUS far I have dealt with those classes of diseases which so affect bodies as a whole, that fixed situations cannot be assigned to them: I will now speak of diseases in particular parts. Diseases of all the internal parts and their treatment, however, will come under view more readily if I first describe briefly their situations.[a]

The head, then, and the structures within the mouth are not only bounded by the tongue and palate, but also by whatever is visible to our eyes. On the right and left sides around the throat, great blood-vessels named sphagitides,[b] also arteries called carotids, run upwards in their course beyond the ears. But actually within the neck are placed glands, which at times become painfully swollen.

From that point two passages begin: one named the windpipe, the more superficial, leads to the lung; the deeper, the gullet, to the stomach; the former takes in the breath, the latter food. Though their courses diverge, where they are joined, there is a little tongue in the windpipe, just below the fauces, which is raised[c] when we breathe, and, when we

to their treatment for injury. Cf. Vol. III. p. 474, and Appendix, p. 593.

[b] Latin *iugulares (venae)*.

[c] Because the larynx is raised; the error that the epiglottis turned down like a lid is avoided.

mimus, arteriam claudit. Ipsa autem arteria, dura et cartilaginosa, in gutture adsurgit, ceteris partibus residit. Constat ex circulis quibusdam, compositis ad imaginem earum vertebrarum, quae in spina sunt, ita tamen ut ex parte exteriore aspera, ex interiore stomachi modo levis sit; eaque descendens ad praecordia cum pulmone committitur.

4 Is spongiosus, ideoque spiritus capax, et a tergo spinae ipsi iunctus, in duas fibras ungulae bubulae modo dividitur. Huic cor adnexum est, natura musculosum, in pectore sub sinistriore mamma situm; duosque quasi ventriculos habet. At sub corde atque pulmone traversum ex valida membrana saeptum est, quod praecordiis uterum diducit; idque nervosum, multis etiam venis per id discurrentibus; a superiore parte non solum intestina, sed iecur quoque lienemque discernit. Haec viscera proxuma sed infra tamen posita dextra sinistraque sunt.

5 Iecur a dextra parte sub praecordiis ab ipso saepto orsum, intrinsecus cavum, extrinsecus gibbum; quod prominens leviter ventriculo insidet, et in quattuor fibras dividitur. Ex inferiore vero parte ei fel inhaeret: at lienis sinistra non eidem saepto sed intestino innexus est; natura mollis et rarus, longitudinis crassitudinisque modicae; isque paulum costarum regione in uterum excedens ex maxima

---

^a Here and below, sections 4 and 5, praecordia is used in the third sense mentioned on p. 100, note *a*.

^b Celsus considers the lung as a single organ with two main divisions. So Galen (K), XIV. 714.

^c Ventriculus, "small stomach", is used to describe the divisions of the heart and kidneys (cf. section 5) because of their similarity in shape to the stomach.

^d The diaphragm.

BOOK IV. i. 3-5

swallow food and drink, closes the windpipe. Now the actual windpipe is rigid and gristly; in the throat it is prominent, in the remaining parts it is depressed. It consists of certain little rings, arranged after the likeness of those vertebrae which are in the spine, but in such a way that whilst rough on the outer surface, the inside is smooth like the gullet; descending to the praecordia,[a] it makes a junction with the lung.

The lung[b] is spongy, and so can take in the breath, and at the back it is joined to the spine itself, and it is divided like the hoof of an ox into two lobes. To the lung is attached the heart, which, muscular in nature, is placed under the left breast, and has two small stomach-like pockets.[c] Now, under the heart and lung is a transverse partition of strong membrane,[d] which separates the belly from the praecordia; it is sinewy, and many blood-vessels also take their course through it; it separates from the parts above not only the intestines but also the liver and spleen. These organs are placed against it but under it, on the right and left sides respectively.

The liver, which starts from the actual partition under the praecordia on the right side, is concave within, convex without; its projecting part rests lightly on the stomach, and it is divided into four[e] lobes. Outside its lower part the gall-bladder adheres to it: but the spleen to the left is not connected to the same partition, but to the intestine;[f] in texture it is soft and loose, moderately long and thick; and it hardly projects at all from beneath the ribs into the belly, but is hidden under them for the most

[e] The human liver is divided into four lobes, right, left, quadrate, caudate.

[f] *i.e.* to the colon rather than to the diaphragm.

parte sub his conditur. Atque haec quidem iuncta sunt. Renes vero diversi; qui lumbis summis coxis inhaerent, a parte earum resimi, ab altera rotundi; qui et venosi sunt, et ventriculos habent, et tunicis super conteguntur.

6 Ac viscerum quidem hae sedes sunt. Stomachus vero, qui intestinorum principium est, nervosus: a septima spinae vertebra incipit, circa praecordia cum ventriculo committitur. Ventriculus autem, qui receptaculum cibi est, constat ex duobus tergoribus; isque inter lienem et iecur positus est, utroque ex his paulum super eum ingrediente. Suntque etiam membranulae tenues, per quas inter se tria ista conectuntur, iungunturque ei saepto, quod transversum esse supra (§ 4) posui.

7 Inde ima ventriculi pars paulum in dexteriorem partem conversa, in summum intestinum coartatur. Hanc iuncturam πυλωρόν Graeci vocant, quoniam portae modo in inferiores partes ea, quae excreturi sumus, emittit.

Ab ea ieiunum intestinum incipit, non ita inplicitum; cui tale vocabulum est, quia numquam quod accepit, continet, sed protinus in inferiores partes transmittit.

8 Inde tenuius intestinum est, in sinus vehementer inplicitum: orbes vero eius per membranulas singuli cum interioribus conectuntur; qui in dexteriorem partem conversi et e regione dexterioris coxae finiti, superiores tamen partes magis complent.

Deinde id intestinum cum crassiore altero transverso committitur; quod a dextra parte incipiens,

---

^a Pelves and calices.   ^b From πύλη, a gate.
^c The mesentery.

part. Now the foregoing are joined together. The kidneys on the other hand are different; they adhere to the loins above the hips, being concave on one surface, on the other convex; they are both vascular, have ventricles,[a] and are covered by coats.

These then are the situations of the viscera. Now the gullet, which is the commencement of the intestines, is sinewy; beginning at the seventh spinal vertebra, it makes a junction in the region of the praecordia with the stomach. And the stomach, which is the receptacle of the food, consists of two coats; and it is placed between the spleen and liver, both overlapping it a little. There are also fine membranes by which these three are interconnected, and they are joined to that partition, which I have described above as transverse.

Thence the lowest part of the stomach, after being directed a little to the right, is narrowed into the top of the intestine. This juncture the Greeks call pylorus,[b] because, like a gateway, it lets through into the parts below whatever we are to excrete.

From this point begins the fasting intestine, not so much infolded; it has this name because it does not hold what it has received, but forthwith passes it on into the parts below.

Beyond is the thinner intestine, infolded into many loops, its several coils too being connected with the more internal parts by fine membranes;[c] these coils are directed rather to the right side, to end in the region of the right hip; however, they occupy mostly the upper parts.

After that spot this intestine makes a junction crosswise with another, the thicker intestine; which, beginning on the right side, is long and pervious

359

CELSUS

in sinisteriorem pervium et longum est, in dexteriorem non est, ideoque caecum nominatur.

9 At id, quod pervium est, late fusum atque sinuatum, minusque quam superiora intestina nervosum, ab utraque parte huc atque illuc volutum, magis tamen in sinisteriores inferioresque partes, contingit iecur atque ventriculum; deinde cum quibusdam membranulis a sinistro rene venientibus iungitur, atque hinc dextra recurvatum in imo derigitur, qua excernit; ideoque id ibi rectum intestinum nominatur.

10 Contegit vero universa haec omentum, ex inferiore parte leve et strictum, ex superiore mollius; cui adeps quoque innascitur, quae sensu, sicut cerebrum quoque et medulla, caret.

At a renibus singulae venae, colore albae, ad vesicam feruntur: ureteras Graeci vocant, quod per eas inde descendentem urinam in vesicam destillare concipiunt.

11 Vesica autem in ipso sinu nervosa et duplex, cervice plena atque carnosa, iungitur per venas cum intestino eoque osse, quod pubi subest. Ipsa soluta atque liberior est, aliter in viris atque in feminis posita: nam in viris iuxta rectum intestinum est, potius in sinistram partem inclinata: in feminis super genitale earum sita est, supraque elapsa ab ipsa vulva sustinetur.

12 Tum in masculis iter urinae spatiosius et conpressius a cervice huius descendit ad colem: in feminis brevius et plenius super vulvae cervicem se ostendit. Vulva autem in virginibus quidem admodum exigua est; in mulieribus vero, nisi ubi gravidae

---

[a] Here the reference is to the muscular coat.
[b] The flexures of the colon, right, left, and sigmoid.
[c] So it seemed at the operation of lithotomy, VII. 26.

towards the left, but not towards the right, which is therefore called the blind intestine.

But that one which is pervious being widespread and winding, and less sinewy[a] than the upper intestines, has a flexure[b] on both sides, right and left, especially on the left side and in the lower parts and touches the liver and stomach, next it is joined to some fine membranes coming from the left kidney, and thence bending backwards and to the right, it is directed straight downwards to the place where it excretes; and so it is there named the straight intestine.

The omentum too, which overlies all these, is at its lower part smooth and compact, softer at its upper part; fat also is produced in it, which like the brain and marrow is without feeling.

Again from the kidneys, two veins, white in colour, lead to the bladder; the Greeks call them ureters, because they believe that through them the urine descending drops into the bladder.

Now the bladder, sinewy and in two layers at its bag, is at its neck bulky and fleshy; it is connected by blood-vessels with the intestine, and with that bone which underlies the pubes. The bladder itself is loose and rather free, and situated differently in men and women: for in men it is close to the straight intestine, being inclined rather to the left[c] side; in women it is situated over the genitals, and whilst free above, is supported actually by the womb.

Again, in males, a longer and narrower urinary passage descends from the neck of the bladder into the penis; in women, a shorter and wider one presents itself over the neck of the womb. Now the womb in virgins is indeed quite small; in women, unless they are

sunt, non multo maior, quam ut manu conprehendatur. Ea, recta tenuataque cervice, quem canalem vocant, contra mediam alvum orsa, inde paulum ad dexteriorem coxam convertitur; deinde super rectum intestinum progressa, iliis feminae latera sua innectit.
13 Ipsa autem ilia inter coxas et pubem imo ventre posita sunt. A quibus ac pube abdomen sursum versus ad praecordia pervenit: ab exteriore parte evidenti cute, ab interiore levi membrana inclusum, quae omento iungitur; peritonaeos autem a Graecis nominatur.

2. His velut in conspectum quendam, quatenus scire curanti necessarium est, adductis, remedia singularum laborantium partium exsequar, orsus a capite; sub quo nomine nunc significo eam partem, quae capillis tegitur: nam oculorum, aurium, dentium dolor, et si qui similis est, alias erit explicandus (*libb.* VI. **6–9**, VII. **7–12**).

2 In capite autem interdum acutus et pestifer morbus est, quem κεφαλαίαν Graeci vocant; cuius notae sunt horror calidus, nervorum resolutio, oculorum caligo, mentis alienatio, vomitus, sic ut vox supprimatur, vel sanguinis ex naribus cursus, sic ut corpus frigescat, anima deficiat. Praeter haec dolor intolerabilis, maxime circa tempora vel occi-
3 pitium. Interdum autem in capite longa inbecillitas, sed neque gravis neque periculosa, per hominis aetatem est: interdum gravior dolor sed brevis, neque tamen mortiferus, qui vel vino vel cruditate

---

[a] Celsus here describes the headache set up by malarial infection before the discovery of cinchona bark and quinine enabled it to be relieved.

BOOK IV. 1. 12–2. 3

pregnant, it is not really much larger than a handful. Beginning over against the middle of the rectum by a straight narrow neck, which they call canalis, it is then turned a little towards the right hip joint; next, as it rises above the right intestine, its sides are fastened into the woman's ilia. Again, these ilia are situated between the hip joints and the pubes at the bottom of the abdomen. From them and from the pubes the abdominal wall extends upwards to the praecordia; it is covered visibly upon its outside by skin, inside by a smooth membrane which makes a junction with the omentum; and it is named by the Greeks peritoneal membrane.

2. Having made a sort of survey as it were of these organs, so far as it is necessary for a practitioner to know them, I shall follow out the remedies for the several parts when diseased, starting with the head; under that term I now mean that part which is covered with hair; for pain in the eyes, ears and teeth and the like will be elsewhere explained (VI. **6–9**, VII. **7–12**).

In the head, then, there is at times an acute and dangerous disease, which the Greeks call cephalaia;[a] the signs of which are hot shivering, paralysis of sinews, blurred vision, alienation of the mind, vomiting, so that the voice is suppressed, or bleeding from the nose, so that the body becomes cold, vitality fails. In addition there is intolerable pain, especially in the region of the temples and back of the head. Again, there is sometimes a chronic weakness in the head, which, although neither severe nor dangerous, lasts through life; sometimes there is more severe pain, but of short duration, and not fatal, which is brought about

vel frigore vel igne aut sole contrahitur. Hique omnes dolores modo in febre, modo sine hac sunt; modo in toto capite, modo in parte, interdum sic 4 ut oris quoque proximam partem excrucient. Praeter haec etiamnum invenitur genus, quod potest longum esse; ubi umor cutem inflat, eaque intumescit et prementi digito cedit: ὑδροκέφαλον Graeci appellant. Ex his id, quod secundo loco positum est, dum leve est, qua sit ratione curandum, dixi, cum persequerer ea, quae sani homines in inbecillitate partis alicuius facere deberent (I. 4). Quae vero auxilia sunt capitis, ubi cum febre dolor est, eo loco explicitum est, quo febrium curatio exposita est (III. 3-17). Nunc de ceteris dicendum est.

5 Ex quibus id, quod acutum est, idque, quod supra consuetudinem intenditur, idque, quod ex subita causa etsi non pestiferum tamen vehemens est, primam curationem habet, qua sanguis mittatur. Sed id, nisi intolerabilis dolor est, supervacuum est, satiusque est abstinere a cibo; si fieri potest, etiam potione; si non potest, aquam bibere. Si postero die dolor remanet, alvum ducere, sternumenta evocare, nihil adsumere nisi aquam. Saepe enim dies unus aut alter totum dolorem hac ratione discutit, 6 utique si ex vino vel cruditate origo est. Si vero in his auxilii parum est, tonderi oportet ad cutem; dein considerandum est, quae causa dolorem excitarit. Si calor, aqua frigida multa perfundere caput expedit,

---

[a] Tic douloureux—facial neuralgia.

by wine or indigestion or cold or heat or the sun. And all these pains occur, sometimes with fever, sometimes without fever; sometimes they affect the whole head, sometimes a part only; at times so as to cause excruciating pain also in the adjacent part of the face.[a] Besides the foregoing there is a class which may become chronic, in which a humour inflates the scalp, so that it swells up and yields to the pressure of the fingers. The Greeks call it hydrocephalus. Of these forms, that mentioned second, while it is slight, is to be treated by the regimen I have stated when I was describing what healthy men should do in the case of weakness of any part (I. **4**). For pain in the head accompanied by fever the remedies have been detailed when describing the treatment of fevers in general (III. **3–17**). Now to speak of the rest.

Of these the case that is acute, also that which surpasses ordinary limits, and that which is of sudden causation and although not deadly, is yet violent, has its primary remedy in blood-letting. But this measure is unnecessary, unless the pain is intolerable, and it is better to abstain from food; also from drink, when possible; if not possible, then to drink water. If, on the day following, pain persists, the bowels should be clystered, sneezing provoked, and nothing but water taken. For often, in this way, all the pain is dispersed within one or two days, especially if it has originated from wine or indigestion. But if there is little benefit from the above, the head should be shaved down to the scalp; then it should be considered what cause excited the pain. If the cause was hot weather, it is well to pour cold water freely over the head, to put on the

CELSUS

spongiam concavam inponere subinde in aqua
frigida expressam; unguere rosa et aceto, vel potius
his tinctam lanam sucidam imponere aliave refriger-
7 antia cataplasmata. At si frigus nocuit, caput
oportet perfundere aqua calida marina vel certe
salsa, aut in qua laurus decocta sit; tum caput
vehementer perfricare; deinde calido oleo implere,
veste velare. Quidam id etiam devinciunt; alii
cervicalibus vestimentisque onerant, et sic levantur;
alios calida cataplasmata adiuvant. Ergo etiam ubi
causa incognita est, videre oportet, refrigerantia
magis an calfacientia leniant, et iis uti, quae experi-
8 mentum adprobarit. At si parum causa discernitur,
perfundere caput, primum calida aqua, sicut supra
(§ 7) praeceptum est, vel salsa, vel ex lauro decocta,
tum frigida posca. Illa in omni vetusto capitis
dolore communia sunt: sternumenta excitare, inferiores partes vehementer perfricare; gargarizare
iis, quae salivam movent; cucurbitulas temporibus
et occipitio admovere; sanguinem ex naribus extra-
here; resina subinde tempora revellere, et imposito
sinapi exulcerare ea, quae male habent ante linteolo
subiecto, ne vehementer adrodat; candentibus ferramentis, ubi dolor est, ulcera excitare; cibum permodi-
cum cum aqua sumere; ubi levatus est dolor, in
balineum ire, ibi multa aqua, prius calida, deinde
frigida per caput perfundi: si discussus ex toto dolor
est, etiam ad vinum reverti, sed postea semper
antequam quicquam aliud aquam bibere.

head a concave sponge now and again wrung out of cold water; to anoint the head with rose oil and vinegar, or better to put on unscoured wool saturated with the same, or else other refrigerant plasters. But if cold has done the harm, the head should be bathed with warm sea water, or at any rate salt and water, or with a laurel-leaf decoction, after which the head should be rubbed smartly, have warm oil poured on it, and then be covered up. Some even bandage up the head, some load it with neck-wraps and mufflers, and so get relief; warm plasters give help in other cases. Hence, even when the cause is unknown, it should be observed whether cooling or heating methods afford the more relief, and to make use of those which experience has approved. But if the cause is not known, the head should be bathed, first in warm water as noted above, or in salt and water, or in the laurel decoction, next in cold vinegar and water. For all long-standing pain in the head, the following are the general measures: to provoke sneezing; to rub the legs smartly; to gargle things which provoke salivation; to apply cups to the temples and occiput; to draw blood from the nostrils; to pluck upon the skin of the temples frequently by the aid of pitch plasters; to apply mustard in order to cause ulcers over the site of the pain, after having put a layer of linen over the skin to prevent violent erosion; to excite ulcerations by cautery, applied over the seat of the pain; to take food in great moderation, with water; after the pain has been relieved, to go to the bath, and there to have much water poured over the head, first hot, then cold; if the pain has been quite dispersed, the patient may even return to wine, but should always before anything else drink some water.

CELSUS

9 Dissimile est id genus, quod umorem in caput contrahit. In hoc tonderi ad cutem necessarium est; deinde inponere sinapi sic ut exulceret; si id parum profuit, scalpello utendum est. Illa cum hydropicis communia sunt: exerceatur, insudet, vehementer perfricetur, cibis potionibusque utatur urinam praecipue moventibus.

3 (II. 2).[1] Circa faciem vero morbus innascitur, quem Graeci κυνικὸν σπασμὸν nominant. Isque cum acuta fere febre oritur; os cum motu quodam pervertitur [ideoque nihil aliud est quam distentio oris].[2] Accedit crebra coloris in facie totoque in
2 corpore mutatio; somnus in promptu est.—In hoc sanguinem mittere optimum est: si finitum eo malum non est, ducere alvum: si ne sic quidem discussum est, albo veratro vomitum movere. Praeter haec necessarium est vitare solem, lassitudinem, vinum. Si discussum his non est, utendum est cursu, frictione in eo, quod laesum est, leni et multa; in
3 reliquis partibus breviore sed vehementi. Prodest etiam movere sternumenta; caput radere, idque perfundere aqua calida vel marina vel certe salsa, sic ut ei sulpur quoque adiciatur; post perfusionem iterum perfricare; sinapi manducare, eodemque tempore adfectis oris partibus ceratum, integris idem

[1] *Figures in brackets mark the division into chapters adopted by v. d. Linden and older editors.*
[2] *Targa deleted these words as a gloss.*

---

[a] This is a description of the Cephalhaematoma of the newly born (*Caput succedaneum*), this condition follows upon birth or may occur after a blow; the blood clots, the serum separates

BOOK IV. 2. 9–3. 3

The class in which humour collects upon the head[a] is different. In that case it is necessary to shave the head to the scalp; then to apply mustard until it causes ulcers; if this is of little avail, recourse must be had to the scalpel. The following measures are the same as for dropsical patients: exercise, sweating, smart rubbing, and such food and drink as will specially promote urination.

3. Again, about the face there originates an affection which the Greeks call " dog spasm."[b] And it begins along with acute fever; the mouth is drawn to one side by a peculiar movement, [and so it is nothing else than a distortion of the mouth]. In addition there is frequent change of colour in the face as well as over all the body, also an inclination to sleep. In this case blood-letting is the best thing; if that does not end the disorder, the bowels are moved with a clyster; when not even thus dispersed, vomiting is provoked by white hellebore. It is necessary besides to avoid the sun, fatigue and wine. If it is not dispersed by these measures, use running, rubbing of the affected part gently and repeatedly, also rub other parts for less time, but smartly. It is also useful to provoke sneezing; to shave the head, to pour over it hot sea water, or at any rate salt and water, provided that sulphur is also added; after this affusion the patient should again be rubbed; should chew mustard, applying at the same time to the parts of the mouth affected a wax salve, likewise to the

out, and should normally be absorbed, but may become infected and suppurate.

[b] Compare IV. 2. 3. As no pain is mentioned, it may be conjectured that a "functional" disorder is meant here by cynic spasm and the treatment advised corresponds.

sinapi, donec adrodat, imponere. Cibus aptissimus ex media materia est.

4 (II. 3). At si lingua resoluta est, quod interdum per se, interdum ex morbo aliquo fit, sic ut sermo hominis non explicetur, oportet gargarizare ex aqua, in qua vel thymum vel hysopum vel nepeta decocta sit; aquam bibere; caput et os et ea, quae sub mento sunt, et cervicem vehementer perfricare; lasere linguam ipsam linere; manducare quae sunt acerrima, id est in sinapi alium, cepam; magna vi luctari, ut verba exprimantur; exerceri retento spiritu; caput saepe aqua frigida perfundere; nonnumquam multam esse radiculam, deinde vomere.

5 (II. 4). Destillat autem de capite interdum in nares, quod leve est; interdum in fauces, quod peius est; interdum etiam in pulmonem, quod pessimum est. Si in nares destillavit, tenuis per has pituita profluit; caput leviter dolet, gravitas eius sentitur, frequentia sternumenta sunt; si in fauces, has exasperat, tussiculam movet; si in pulmonem, praeter sternumenta et tussim est etiam capitis gravitas, lassitudo, sitis, aestus, biliosa urina.

2 Aliud autem quamvis non multum distans malum gravedo est. Haec nares claudit, vocem obtundit, tussim siccam movet; sub eadem salsa est saliva, sonant aures, venae moventur in capite, turbida urina est. Haec omnia κόρυζας Hippocrates nominat: nunc video apud Graecos in gravedine hoc nomen servari, destillationem καταστἀγμόν appellari. Haec autem et brevia et, si neglecta sunt, longa esse

---

[a] A common cold.
[b] Hipp. II. 23 (*Prog.* 14).

BOOK IV. 3. 3–5. 2

unaffected parts mustard until it produces erosion. Food of the middle class is most suitable.

4. But if there is paralysis of the tongue, which sometimes occurs of itself, sometimes is produced by some disease, so that the man's speech is not distinct, he should gargle a decoction of thyme, hyssop or mint; drink only water; have the head, face, the parts under the chin and the neck smartly rubbed; the tongue itself smeared with laser; chew very acrid materials, mustard, onion, garlic, and strive with all his force to pronounce words; hold his breath at exercise; frequently pour cold water over his head; on occasion eat a quantity of radish and then vomit.

5. Again there is dripping from the head sometimes into the nose, which is a mild affair; sometimes into the throat, which is worse, sometimes into the lung, which is worst of all. When the drip is into the nostrils, a thin phlegm is discharged from them; there is slight pain, and a feeling of weight in the head, with frequent sneezing; if the drip is into the throat, it irritates and excites a slight cough; if the drip is into the lung, besides the sneezing, cough and even weight in the head, there is lassitude, thirst, a feeling of heat, and bilious urine.

Another although not very different affection is gravedo.[a] This closes up the nostrils, renders the voice hoarse, excites a dry cough; in it the saliva is salt, there is ringing in the ears, the blood-vessels in the head throb, the urine is turbid. Hippocrates[b] named all the above coryza; I note that now the Greeks reserve this term for gravedo, the dripping they call catastagmus. These affections are commonly of short duration, but if neglected may last a

consuerunt. Nihil pestiferum est, nisi quod pulmonem exulceravit.

3 Ubi aliquid eiusmodi sensimus, protinus abstinere a sole, balneo, vino, venere debemus; inter quae unctione et adsueto cibo nihilo minus uti licet. Ambulatione tantum acri sed tecta utendum est; post eam caput atque os supra quinquagiens perfricandum. Raroque fit ut, si biduo vel certe triduo 4 nobis temperavimus, id vitium non levetur. Quo levato, si in destillatione crassa facta pituita est, vel in gravedine nares magis patent, balneo utendum est, multaque aqua prius calida, post egelida fovendum os caputque; deinde cum cibo pleniore vinum bibendum. At si aeque tenuis quarto die pituita est, vel nares aeque clausae videntur, adsumendum est vinum Aminaeum austerum, dein rursus biduo aqua; post quae ad balneum et ad consuetudinem 5 revertendum est. Neque tamen illis ipsis diebus, quibus aliqua omittenda sunt, expedit tamquam aegros agere, sed cetera omnia quasi sanis facienda sunt, praeterquam si diutius aliquid et vehementius ista sollicitare consuerunt: huic enim quaedam curiosior observatio necessaria est.

6 Igitur huic, si in nares vel in fauces destillavit, praeter ea, quae supra (§ 3, 4) rettuli, protinus primis diebus multum ambulandum est; perfricandae vehementer inferiores partes, levior frictio adhibenda thoraci ori capiti; demenda adsueto cibo pars

---

[a] Cf. section 3 init.

long while. None is fatal, except that which causes ulcers in the lung.

Whenever we feel anything of the sort, we should forthwith keep out of the sun, and abstain from the bath, wine and coition; but the use meanwhile of anointing and of customary food is allowable. The patient should walk, but only briskly and under cover; after that the head and face should be rubbed for more than fifty strokes. This complaint is generally relieved, provided that we take care of ourselves for a couple of days, or for three at the most. When the disease has been relieved so that the drip of phlegm becomes thick, or the gravedo so that the nostrils are more open, the bath may be resumed, much water, at first hot, then lukewarm, being used to foment the face and head; next, along with more food, wine may be taken. But if on the fourth day the phlegm is still thin, or the nostrils still stuffed up, the patient should take dry Aminaean wine, then for a couple of days water; after which he can return to the bath and his usual habits. Nevertheless, even during those days, when some things are to be avoided,[a] it is not expedient to treat the patients as sick men, but they are to do everything as in health, unless these symptoms have been liable to cause more prolonged and severe trouble; for then a somewhat more careful attention is needed.

Therefore in such a case if there is a drip into the nose or into the throat, besides the treatment described above, the patient from the start should walk a good deal during the first days: have the lower limbs smartly rubbed, together with more gentle rubbing of the chest, face and head; his accustomed food should be reduced by one-half; he

dimidia; sumenda ova, amylum similiaque, quae pituitam faciunt crassiorem: siti quanta maxime sustineri potest, pugnandum. Ubi per haec idoneus aliquis balneo factus eoque usus est, adiciendus est cibo pisciculus aut caro, sic tamen ne protinus iustus modus cibi sumatur; vino meraco copiosius utendum est.

7 At si in pulmonem quoque destillat, multo magis et ambulatione et fricatione opus est eademque adhibita ratione in cibis, si non satis illi proficiunt, acrioribus utendum est; magis somno indulgendum, abstinendumque a negotiis omnibus; aliquando sed serius balineum temptandum.

8 In gravedine autem primo die quiescere, neque esse neque bibere, caput velare, fauces lana circumdare; postero die surgere, abstinere a potione, aut si res coegerit, non ultra heminam aquae adsumere; tertio die panis non tam multum ex parte interiore cum pisciculo vel levi carne sumere; aquam bibere.

9 Si quis sibi temperare non potuerit quominus pleniore victu utatur, vomere; ubi in balneum ventum est, multa calida aqua caput et os fovere usque ad sudorem; tum ad vinum redire. Post quae vix fieri potest, ut idem incommodum maneat; sed si manserit, utendum erit cibis frigidis, aridis, levibus, umore quam minimo, servatis frictionibus exercitationibusque, quae in omni tali genere valetudinis necessariae sunt.

6 (III). A capite transitus ad cervicem est, quae gravibus admodum morbis obnoxia est. Neque

may take eggs, also starchy and such-like foods. which thicken phlegm; thirst should be resisted as far as he can bear it. When by these measures a patient has been prepared for the bath, and has used it, there may be added to the diet small fish or meat, provided that at first he should not take the full quantity of food; undiluted wine should be taken more freely.

But if the drip is into the lung also, there is even more need for walking and rubbing and the same regimen as to diet, and if that diet is not effective, more acrid food is to be employed; he should allow himself more sleep, and abstain from all business; but the bath should be tried at a somewhat later stage.

In the case of gravedo, he should lie in bed on the first day, neither eat nor drink, cover the head, and wrap wool around the throat; on the next day he should get up, and still abstain from drink, or, if he must have some, take not more than one tumbler-full of water; on the third day he may eat the crumb of bread, but not much, with some small fish, or light meat, and water for drink. Should the patient be unable to restrain himself from using a fuller diet, he is to provoke a vomit; when he gets to the bath, he should foment freely his head and face with hot water until he sweats, and then have recourse to wine. After the above measures it is scarcely possible for the same discomfort to persist; but if it does so, use cold, dry, light food with the least possible fluid, whilst continuing the rubbings and the exercises, such as are needed in all such sorts of illness.

6. From the head we pass to the neck, which is liable to harm from diseases of considerable

CELSUS

tamen alius inportunior acutiorque morbus est, quam is, qui quodam rigore nervorum modo caput scapulis, modo mentum pectori adnectit, modo rectam et inmobilem cervicem intendit. Primum Graeci ὀπισθότονον, insequentem ἐμπροσθότονον, ultimum τέτανον appellant, quamvis minus subtiliter quidam indiscretis his nominibus utuntur. Ea saepe intra
2 quartum diem tollunt: si hunc evaserunt, sine periculo sunt.—Eadem omnia ratione curantur idque convenit . . .[1] Sed Asclepiades utique emittendum sanguinem credidit; quod quidam utique vitandum esse dixerunt, eo quod maxime tum corpus calore egeret, isque esset in sanguine. Verum hoc quidem falsum est: neque enim natura sanguinis est, ut utique caleat, sed ex iis, quae in homine sunt, hic celerrime vel calescit vel refrigescit. Mitti vero necne debeat, ex iis intellegi potest, quae de sanguinis missione praecepta sunt (II. **10, 11**).
3 Utique autem recte datur castoreum, et cum hoc piper vel laser; deinde opus est fomento umido et calido. Itaque plerique aqua calida multa cervices subinde perfundunt. Id in praesentia levat, sed oportuniores nervos frigori reddit, quod utique vitan-

[1] *Probably some words describing the general treatment which is agreed on have fallen out.*

---

[a] For tetanus see Hippocrates IV. 153 (*Aph.* V. 6). This disease continued unchanged and uncontrolled by treatment until the discovery that it is caused by special bacilli, and that it is a chronic disease in old horses. Active spores pass out in the manure to infect roads and the soil of gardens and fields and to enter any open blood-vessel. The relative re-

BOOK IV. 6. 1-3

gravity. There is, however, no disease more distressing, and more acute, than that which by a sort of rigor of the sinews, now draws down the head to the shoulder-blades, now the chin to the chest, now stretches out the neck straight and immobile. The Greeks call the first opisthotonus, the next emprosthotonus, and the last tetanus,[a] although some with less exactitude use these terms indiscriminately. These diseases are often fatal within four[b] days. If the patients survive this period, they are no longer in danger. They are all treated by the same method and this is agreed upon, but Asclepiades in particular believed in blood-letting, which some said should be particularly avoided, because the body was then especially in need of that heat which was in the blood. But this is false; for it is not in the nature of the blood to be especially hot, but of all that composes man, the blood[c] most quickly turns, now hot, now cold. Still, whether or no it ought to be let, can be learnt from the instructions concerning blood-letting (II. **10, 11**). But anyhow it is right to give castory, and with it pepper or laser; further, a warm and moist fomentation is needed. For this purpose most pour hot water freely at intervals over the neck. This affords temporary relief, but renders the sinews more susceptible to cold, a thing certainly

---

sistance of the horse gives rise to an antitoxin in the animal's serum, which being used as a vaccine immediately following upon inoculation, arrests the progress of the infection.

[b] The disease usually shows itself by stiffness of the jaw, more rarely by stiffness near the infected spot; in the last stages tetanic spasms bend the patient so that he rests on the back of his head and his heels, Opisthotonos; or he is doubled up forwards, his chin to his chest, Emprosthotonos.

[c] Hipp. *De Corde* 12 (Littré, IX. 92).

CELSUS

dum est. Utilius igitur est cerato primum liquido cervicem perunguere; deinde admovere vesicas bubulas vel utriculos oleo calido repletos, vel ex farina calidum cataplasma, vel piper rotundum cum 4 ficu contusum. Utilissimum tamen est umido sale fovere; quod quomodo fieret iam ostendi (II. **17**. 9, 10; **33**. 1). Ubi eo aliquid factum est, admovere ad ignem; vel si aestas est, in solem aegrum oportet; maximeque oleo veteri, si id non est, Syriaco; si ne id quidem est, adipe quam vetustissima cervicem et scapulas et spinam perfricare. Frictio cum omnibus in homine vertebris utilis sit, tum is praecipue, quae in collo sunt. Ergo die nocteque, interpositis tamen quibusdam temporibus, hoc remedio utendum est. Dum intermittitur, imponendum malagma aliquod 5 ex calfacientibus. Cavendum vero praecipue frigus; ideoque in eo conclavi, quo cubabit aeger, ignis continuus esse debebit, maximusque tempore antelucano, quo praecipue frigus intenditur. Neque inutile erit caput adtonsum habere idque irino vel cyprino calido madefacere et superinposito pilleo velare; nonnumquam etiam in calidum oleum totum descendere, vel in aquam calidam, in qua faenum Graecum decoctum sit et adiecta olei pars tertia. Alvus quoque ducta saepe superiores partes resolvit. Si vero etiam vehementius dolor crevit, admovendae cervicibus cucurbitulae sunt, sic ut cutis incidatur; eadem aut ferramentis aut sinapi adurenda. 6 Ubi levatus est dolor moverique cervix coepit, scire

---

[a] Nard from Syria.

to be avoided. It is, therefore, more beneficial, first to anoint the neck with a liquid wax-salve, then to apply ox-bladders or leathern bottles filled with hot oil, or else a hot meal plaster, or a pod of round pepper crushed up in a fig. The best thing, however, is to foment with moistened salt according to the method already described (II. **17.** 9, 10; **33.** 1). Whatever meanwhile is being done, the patient should be brought near a fire, or into the sun in hot weather, and old oil in particular should be rubbed into his neck, shoulder-blades and spine; or if that is not at hand, Syriac[a] oil, or if not even that, oldest lard. Rubbing applied to the whole length of the vertebrae is beneficial, but especially so to those of the neck. Therefore, with certain intervals however, this procedure should be carried out both by day and by night. During such intervals some kind of an emollient composed of heating substances should be put on. Cold is especially to be guarded against; and so there ought to be a fire kept burning constantly in the room in which the patient is lying, especially during the hours before dawn, when the cold is particularly intense. It is not unserviceable to keep the head closely clipped, moistened with hot iris or cyprus oil, and covered by putting on a cap; sometimes even to submerge the patient either in hot oil, or in hot water in which fenugreek has been boiled and a third part of oil added. If the bowels also have been moved by a clyster, this often relaxes the upper parts. Should the pain grow even still more severe, cups should be applied to the neck after the skin has been incised; or the same spot is to be burnt either with the cautery, or by mustard. When the pain has been relieved and the neck begins to be

CELSUS

licet cedere remediis morbum. Sed diu vitandus cibus, quisquis mandendus est: sorbitionibus utendum itemque ovis sorbilibus aut apalis [mollibus]; ius aliquod adsumendum. At si bene processerit iamque ex toto recte se habere cervices videbuntur, incipiendum erit a pulticula vel intrita bene madida. Celerius tamen etiam panis mandendus quam vinum gustandum, siquidem huius usus praecipue periculosus ideoque in longius tempus differendus est.

7 (IV. 1). Ut hoc autem morbi genus circa totam cervicem, sic alterum aeque pestiferum acutumque in faucibus esse consuevit. Nostri anginam vocant: apud Graecos nomen, prout species est. Interdum enim neque rubor neque tumor ullus apparet, sed corpus aridum est, vix spiritus trahitur, membra solvuntur: id συνάγχην vocant. Interdum lingua faucesque cum rubore intumescunt, vox nihil significat, oculi vertuntur, facies pallet, singultus est: id κυνάγχην vocant. Illa communia sunt: aeger non cibum devorare, non potionem potest, 2 spiritus eius intercluditur. Levius est, ubi tumor tantummodo ruborque est, cetera non secuntur: id παρασυνάγχην appellant.—Quicquid est, si vires patiuntur, sanguis mittendus est; si non abundat, secundum est ducere alvum. Cucurbitula quoque recte sub mento et circa fauces admovetur, ut id, quod strangulat, evocet. Opus est deinde

---

[a] Angina (ἄγχειν—throttle), an acute infection of the throat by a mixture of virulent micro-organisms producing choking, oedema, suppuration and gangrene. The three Greek terms (συνάγχη, κυνάγχη and παρασυνάγχη) were later used indiscriminately. κυνάγχη (? panting like a dog) became *quinantia*, quinsy, now restricted in application to peritonsillar abscess.

moved, it can be recognized that the disease is yielding to treatment. But for a long while food which has to be chewed should be avoided; sops and eggs, raw or soft boiled, are to be used; any kind of soup may be taken. But if the patient has done well, and the neck appears to be all right, then will be the time to begin with pulse porridge, or well-moistened crumbled bread. He is to chew bread, however, earlier than to drink wine, because the use of wine is particularly risky, and so ought to be deferred for a longer time.

7. Whilst this kind of disease involves the region of the neck as a whole, another equally fatal and acute has its seat in the throat. We call it angina;[a] the Greeks have names according to its species. For sometimes no redness or swelling is apparent, but the skin is dry, the breath drawn with difficulty, the limbs relaxed; this they call synanche. Sometimes the tongue and throat are red and swollen, the voice becomes indistinct, the eyes are deviated, the face is pallid, there is hiccough; that they call cynanche: the signs in common are, that the patient cannot swallow food nor drink, and his breathing is obstructed. It is a slighter case when there is merely redness and swelling, not followed by the other symptoms; this they call parasynanche. Whichever form occurs blood must be let if strength permits; if there is no surplus strength, then move the bowels by a clyster. Cups also may be applied with benefit under the chin, also outside the throat, so as to draw out the matter which is suffocating.

Celsus considered tonsillar affections separately (IV. **9**, VI. **10**, VII. **12**. 2).

fomentis umidis: nam sicca spiritum elidunt. Ergo admovere spongias oportet, quae melius in calidum oleum quam in calidam aquam subinde demittuntur; efficacissimumque est hic quoque salis cali-
3 dus sucus. Tum commodum est hysopum vel nepetam vel thumum vel apsinthium vel etiam furfures aut ficus aridas aut mulsam aquam decoquere, eaque gargarizare; post haec palatum unguere vel felle taurino vel eo medicamento, quod ex moris est. Polline etiam . . . si[1] piperis id recte respergitur. Si per haec parum proficitur, ultimum est incidere satis altis plagis sub ipsis maxillis supra collum, vel in palato citra uvam, vel eas venas, quae sub lingua sunt, ut per ea vulnera morbus erumpat. Quibus si non fuerit aeger adiutus, scire licet malo victum
4 esse. Si vero his morbus levatus est, iamque fauces et cibum et spiritum capiunt, facilis ad bonam valetudinem recursus est. Atque interdum natura quoque adiuvat, si ex angustiore sede vitium transit in latiorem; itaque rubore et tumore in praecordiis orto scire licet fauces liberari. Quicquid autem eas levavit, incipiendum est ab umidis, maximeque aqua mulsa decocta: deinde adsumendi molles et non acres cibi sunt, donec fauces ad pristinum habitum
5 revertantur. Vulgo audio, si quis pullum hirundininum ederit, anginam toto anno non periclitari; servatumque eum ex sale, cum is morbus urget, comburi, carbonemque eius contritum in aquam

[1] *As it stands the text is corrupt. Marx suggests reading* tusi *for* si *and this is translated.*

[a] In disease, a change of place, metastasis, is deemed a favourable sign.
[b] Cf. Hipp. IV. 186, 204 (*Aph.* VI. 27–VII. 49).
[c] Pliny, *N.H.* XXX. 33.

BOOK IV. 7. 2–5

Next, moist foments are needed, for dry ones hinder the breath. Consequently sponges, dipped into hot oil at intervals, should be put on; that is better than hot water; but most efficacious here too is hot moistened salt. Moreover, it is useful: to make a decoction with hydromel of hyssop, catmint, thyme, wormwood, or even of bran, and dried figs, and to gargle with it; afterwards to smear the palate with ox-gall, or with the medicament made of mulberries. It is also appropriate for a cough to dust the palate with pounded pepper. If there is little effect from these remedies, the last resource is to make sufficiently deep incisions into the upper part of the neck under the lower jaw, or into the palate in front of the uvula, or into the veins under the tongue, in order that the disease may discharge through the incisions. If the patient is not benefited by all this, it must be recognized that he has been overcome by the disease. But if these measures have relieved the disease, and the throat again admits both food and breath, a return to health is easy. And sometimes nature also assists when the disease moves from a more restricted to a more widespread seat; so when redness and swelling have arisen over the praecordia,[a] it may be recognized that the throat is becoming free.[b] But whatever has relieved it, the patient should begin with fluids, especially with the hydromel decoction; next soft and unacrid food should be taken until the throat has returned to its original condition. I hear it commonly said that if a man eat a nestling swallow,[c] for a whole year he is not in danger from angina; and that when the disease attacks anyone it is also beneficial to burn a nestling which has been preserved in salt and to crumble the powdered ash into

mulsam, quae potui datur, infriari et prodesse. Id cum idoneos auctores ex populo habeat, neque habere quicquam periculi possit, quamvis in monumentis medicorum non legerim, tamen inserendum huic operi meo credidi.

8 (IV. 2). Est etiam circa fauces malum, quod apud Graecos aliud aliudque nomen habet, prout se intendit. Omne in difficultate spirandi consistit; sed haec dum modica est neque ex toto strangulat δύσπνοια appellatur; cum vehementior est, ut spirare aeger sine sono et anhelatione non possit, ἄσθμα: at cum accessit id quoque, quod aegre[1] nisi recta cervice spiritus trahitur, 2 ὀρθόπνοια. Ex quibus id, quod primum est, potest diu trahi: duo insequentia acuta esse consuerunt. His communia sunt, quod propter angustias, per quas spiritus evadit, sibilum exit; dolor in pectore praecordiisque est, interdum etiam scapulis, isque modo decedit, modo revertitur; ad haec tussicula accedit.—Auxilium est, nisi aliquid prohibet, in sanguinis detractione. Neque id satis est, sed lacte venter quoque solvendus est, liquanda alvus, interdum etiam ducenda; quibus extenuatum corpus 3 incipit spiritum trahere commodius. Caput autem etiam in lecto sublime habendum est; thorax fomentis cataplasmatisque calidis aut siccis aut etiam umidis adiuvandus est, et postea vel malagma superimponendum, vel certe ceratum ex cyprino vel irino unguento. Sumenda deinde ieiuno potui mulsa aqua, cum qua vel hysopus cocta vel contrita capparis

---

[1] quod aegre *added by Marx.*

---

[a] The three Greek terms are not used elsewhere in Celsus.

BOOK IV. 7. 5–8. 3

hydromel which is administered as a draught. Since this remedy has considerable popular authority, and cannot possibly be a danger, although I have not read of it in medical authorities, yet I thought that it should be inserted here in my work.

8. There is also in the region of the throat a malady which amongst the Greeks has different names according to its intensity. It consists altogether in a difficulty of breathing; when moderate and without any choking, it is called dyspnoea; when more severe, so that the patient cannot breathe without making a noise and gasping, asthma; but when in addition the patient can hardly draw in his breath unless with the neck outstretched, orthopnoea.[a] Of these, the first can last a long while, the two following are as a rule acute. The signs common to them are: on account of the narrow passage by which the breath escapes, it comes out with a whistle; there is pain in the chest and praecordia, at times even in the shoulder-blades, sometimes subsiding, then returning; to these there is added a slight cough. Bloodletting is the remedy unless anything prohibits it. Nor is that enough, but also the bowels are to be relaxed by milk, the stool being rendered liquid, at times even a clyster is given; as the body becomes depleted by these measures the patient begins to draw his breath more readily. Moreover, even in bed the head is to be kept raised; the chest movement assisted by hot foments and plasters, dry or even moist, and later either emollients are to be applied or at any rate a wax-salve made with cyprus, or iris ointment. Next, on an empty stomach the patient should take a draught of hydromel, in which either hyssop or crushed caper root has been boiled.

radix sit. Delingitur etiam utiliter aut nitrum aut nasturcium album frictum, deinde contritum et cum melle mixtum; simulque coquntur mel, galbanum, resina terebenthina, et ubi coierunt, ex his quod fabae magnitudinem habet cotidie sub lingua liquatur: aut sulpuris ignem non experti P. ✳ = — habrotoni P. ✳ = in vini cyatho teruntur, idque
4 tepefactum sorbetur. Est etiam non vana opinio vulpinum iecur, ubi siccum est et aridum factum, contundi oportere polentamque ex eo potioni aspergi; vel eiusdem pulmonem quam recentissimum assum, sed sine ferro coctum, edendum esse. Praeter haec sorbitionibus et lenibus cibis utendum est, interdum vino tenui, austero, nonnumquam vomitu. Prosunt etiam quaecumque urinam augent, sed nihil magis quam ambulatio lenta paene usque ad lassitudinem; frictio multa praecipue inferiorum partium, vel in sole vel ad ignem, et per se ipsum et per alios usque ad sudorem.

9 (IV. 3). In interiore vero faucium parte interdum exulceratio esse consuevit. In hac plerique extrinsecus cataplasmatis calidis fomentisque umidis utuntur: volunt etiam vaporem calidum ore recipi. Per quae molliores alii partes eas fieri dicunt oportunioresque vitio iam haerenti. Sed si bene vitari frigus potest, tuta illa praesidia; si metus eius est, supervacua sunt. Utique autem perfricare fauces
2 periculosum est: exulcerat enim. Neque utilia sunt quae urinae movendae sunt, quia possunt, dum transeunt, ibi quoque pituitam extenuare [quam supprimi melius est]. Asclepiades multarum rerum,

It is also of use to suck either soda or white nasturtium seed, parched, crushed and then mixed with honey; and for the same purpose, galbanum and turpentine resin are boiled together to a coherent mass, and a bit of this, the size of a bean, is sucked every day, or unfused sulphur 1 grm. and 66 c.grm. of southernwood are pounded up in a cupful of wine and sipped lukewarm. It is also not a foolish idea that the liver of a fox should be dried, pounded and the mash sprinkled into the above, or that the lung of that animal, as fresh as possible, roasted without touching iron in the cooking, should be eaten. In addition to the above, gruels and light food are to be used, at intervals also a light dry wine, occasionally an emetic. Some kind of diuretic is also beneficial, but there is nothing better than a walk until almost fatigued, also frequent rubbings, especially of the lower extremities, either in the sun, or before a fire, done by the patient himself or others, until he sweats.

9. But in the interior parts of the throat there is sometimes ulceration. For this most employ plasters and hot foments externally; they also order hot steam to be inhaled by the mouth. Others say that by these measures the parts are rendered more soft and more liable to the complaint already existing there. But these applications are salutary if cold can be completely avoided; if cold is to be apprehended, they are useless. But anyhow to rub the throat is dangerous; for it provokes ulceration. Nor are diuretics useful, because in the course of being swallowed they can also make thin the phlegm there, [which is better suppressed]. Asclepiades, who wisely advises many things, which we

quas ipsi quoque secuti sumus, auctor bonus, acetum ait quam acerrimum esse sorbendum: hoc enim sine ulla noxa comprimi ulcera. Sed id supprimere sanguinem, ulcera ipsa sanare non potest. Melius huic rei lycium est, quod idem quoque aeque probat, vel porri vel marrubii sucus, vel nuces Graecae cum tracanto contritae et cum passo mixtae, vel lini semen contritum et cum dulci vino mixtum. 3 Exercitatio quoque ambulandi currendique necessaria est, frictio a pectore vehemens toti inferiori parti adhibenda. Cibi vero esse debent neque nimium acres neque asperi, mel, lenticula, tragum, lac, tisana, pinguis caro, praecipueque porrum et quicquid cum hoc mixtum est. Potionis quam minimum esse convenit; aqua dari potest vel pura, vel in qua malum Cotoneum palmulaeve decoctae sint. Gargarizationes quoque lenes, sin hae parum proficiunt, reprimentes, utiles sunt. Hoc genus neque acutum est et potest esse non longum: curationem tamen maturam, ne vehementer et diu laedat, desiderat.

10 (IV. 4). Tussis vero fere propter faucium exulcerationem est. Quae multis modis contrahitur: itaque illis restitutis ipsa finitur. Solet tamen interdum per se quoque male habere, et vix, cum vetus facta est, eliditur. Et modo arida est, modo pituitam citat.—Oportet hysopum altero quoque die bibere; spiritu retento currere, sed minime in pulvere; ac lectione uti vehementi, quae primo impedita tussi, post eam vincit; tum ambulare; deinde per manus

also ourselves practise, said that very sour vinegar should be sipped; for by this the ulcers are constricted without doing harm. But whilst vinegar can suppress bleeding, it cannot heal the actual ulcerations. For that purpose lycium is better, and Asclepiades approved equally of it, or leek or marrubium juice, or almonds pounded up with tragacanth and mixed with raisin wine, or linseed pounded and mixed with sweet wine. Exercise also by walking and by running is necessary, and smart rubbing from the chest downwards should be applied to the whole of the lower part of the body. The food too should be neither very acrid nor rough, honey, lentils, wheat porridge, milk, pearl barley gruel, fat meat and especially a leek decoction and anything mixed with it. Of drink the least possible is proper; water can be given either by itself, or when quince or dates have been boiled in it. Bland gargles are of service also, or when ineffectual then repressants. This sort of affection is not acute, and cannot last long; nevertheless, it requires timely treatment, lest it should become a severe and chronic complaint.

10. Cough again is generally owing to ulceration of the throat. This is incurred in many ways: and so when the throat has healed the cough is ended. Nevertheless, at times cough is a trouble by itself, and when it has become chronic, is difficult to get rid of. Sometimes the cough is dry, sometimes it excites phlegm. Hyssop should be taken every other day; the patient should run whilst holding the breath, but not where there is dust; he should practise reading loudly, which may at first be impeded by the cough, but later overcomes it; next walking; then manual exercises also, and the chest should

CELSUS

quoque exerceri, et pectus diu perfricari; post haec quam pinguissimae ficus uncias tres super prunam 2 incoctas esse. Praeter haec, si umida est, prosunt frictiones validae cum quibusdam calfacientibus, sic ut caput quoque siccum vehementer perfricetur; item cucurbitulae pectori admotae; sinapi ex parte exteriore faucibus inpositum, donec leviter exulceret; potio ex menta nucibusque Graecis et amylo; primoque adsumptus panis aridus, deinde aliquis 3 cibus lenis. At si sicca tussis est, cum vehementissime urget, adiuvat vini austeri cyathus adsumptus, dum ne amplius id interposito tempore aliquo quam ter aut quater fiat; item laseris quam optimi paulum devorare opus est; porri vel marrubii sucum adsumere; scillam delingere, acetum ex ea vel certe acre sorbere; aut cum spica alii contriti duos vini 4 cyathos. Utilis etiam in omni tussi est peregrinatio, navigatio longa, loca maritima, natationes; cibus interdum mollis, ut malva, ut urtica, interdum acer; lac cum alio coctum; sorbitiones, quibus laser sit adiectum, aut in quibus porrum incoctum tabuerit; ovum sorbile sulpure adiecto; potui primum aqua calida, deinde in vicem aliis diebus haec, aliis vinum.

11 (IV. 5). Magis terreri potest aliquis, cum sanguinem expuit: sed id modo minus, modo plus periculi habet. Exit modo ex gingivis, modo ex ore et quidem ex hoc interdum etiam copiose, sed sine tussi, sine ulcere, sine gingivarum ullo vitio, ita ut nihil excreetur. Verum ut ex naribus aliquando,

---

[a] 84 grms.
[b] On this subject see Aretaeus, *Causes and Symptoms of Acute Diseases*, Book II. 2; περὶ αἵματος ἀναγωγῆς (pub. Adams, Sydenham Society). This work of unknown date has many similar passages to Celsus, Book IV.

be rubbed for a long while   After such exercises he should eat three ounces[a] of very juicy figs, cooked over charcoal. Besides the above, when the cough is moist, smart rubbings with some kind of heating substance are good, provided that the head too is briskly rubbed when dry; in addition, cups are applied to the chest; mustard put on outside over the throat until there is slight excoriation; and a draught taken, composed of mint, almonds, and starch; first of all dry bread should be eaten, then any kind of bland food.   But if the cough is dry and very troublesome, it is relieved by taking a cup of dry wine, provided that this is done only three or four times at rather long intervals; further, there is need to swallow a little of the best laser, to take juice of leeks or horehound; to suck a squill, to sip vinegar of squills, or at any rate sharp vinegar; or two cupfuls of wine with a bruised clove of garlic. In every case of cough it is of use to travel, take a long sea voyage, live at the seaside, swim, sometimes to take bland food, such as mallows, or nettle-tops, sometimes acrid; milk cooked with garlic; gruels to which laser has been added, or in which leeks have been boiled to pieces; a raw egg to which sulphur has been added; at first warm water to drink, then, in turn, one day water, the next day wine.

11. When blood is spat up there is more cause for alarm,[b] although that presents at one time less, at another more of danger. Blood sometimes comes from the gums, sometimes from the mouth, and that at times copiously, yet without cough, without ulceration, without any disease of the gums, so that there is no expectoration.   But just as there is on occasion bleeding from the nostrils, so also does it

sic ex ore prorumpit. Atque interdum sanguis profluit, interdum simile aquae quiddam, in qua caro
2 recens lota est. Nonnumquam autem is a summis faucibus fertur, modo exulcerata ea parte, modo non exulcerata, sed aut ore venae alicuius adaperto, aut tuberculis quibusdam natis, exque his sanguine erumpente. Quod ubi incidit, neque laedit potio aut cibus neque quicquam ut ex ulcere excreatur. Aliquando vero gutture et arteriis exulceratis frequens tussis sanguinem quoque extundit: interdum etiam fieri solet, ut aut ex pulmone aut ex pectore aut ex latere aut ex iocinere feratur. Saepe feminae, quibus sanguis per menstrua non respondit, hunc expuunt.
3 Auctoresque medici sunt vel exesa parte aliqua sanguinem exire, vel rupta, vel ore alicuius venae patefacto: primam διάβρωσιν, secundam ῥῆξιν [σχασμόν],¹ tertiam ἀναστόμωσιν appellant. Ultima minime nocet, prima gravissime. Ac saepe quidem evenit, uti sanguinem pus sequatur. Interdum autem qui sanguinem ipsum suppressit, satis ad valetudinem profuit: sed si secuta ulcera sunt, si pus, si tussis est, prout sedes ipsa est, ita varia et periculosa genera morborum sunt. Si vero sanguis tantum fluit, expeditius et remedium et
4 finis est. Neque ignorari oportet eis, quibus fluere sanguis solet aut quibus dolet spina coxaeve aut post cursum vehementem vel ambulationem, dum febris absit, non esse inutile sanguinis mediocre profluvium,

¹ σχασμός = σχάσις (scarifying) appears to be a gloss on ῥῆξις (rupture) and is bracketed by Marx.

ᵃ Hipp. IV. 196 (Aph. VII. 15).

burst out from the mouth. And sometimes it is blood which flows, sometimes something resembling water in which fresh meat has been washed. On the other hand, blood may come from the uppermost part of the throat, at one time when there is ulceration in that part, at another without ulceration, but either the mouth of some blood-vessel has opened, or the blood breaks out of certain tubercles which have originated there. When this happens, neither food nor drink does harm, nor is there any expectoration as from an ulcer. When, however, the throat and air tubes are ulcerated, the frequent cough also forces out blood; at times it is even brought up out of the lung or out of the chest or out of the sides or out of the liver. Often women, in whom the blood is not being given out through the menses, expectorate blood.

According to medical authorities blood gains exit either from some part eroded, or ruptured, or from the opened mouth of some blood-vessel; the first they call diabrosis, the second rhexis, the third anastomosis. The last is the least harmful, the first, the worst. And often indeed it happens that pus follows the blood.[a] Now at times to stop the bleeding suffices to promote recovery; but if there follow ulcerations, or pus, or a cough, according to the situation there arise various and dangerous classes of diseases. But if only blood flows out, both the remedy and the ending are the quicker. Nor ought we to ignore that in those who are in the habit of bleeding or in whom the back or hips ache whether after hard running or walking, a limited flow of blood is not disadvantageous as long as fever is absent, and when blood is passed by the

idque per urinam redditum quoque ipsam lassitudinem solvere; ac ne in eo quidem terribile esse, qui ex superiore loco decidit, si tamen in eius urina nihil novavit; neque vomitum huius adferre periculum, etiam cum repetit, si ante confirmare et implere corpus licuit et ex toto nullum nocere, qui in corpore robusto neque nimius est neque tussim aut calorem movet.

5 Haec pertinent ad universum: nunc ad ea loca, quae praeposui (§ 1, 2), veniam. Si ex gingivis exit, portulacam manducasse satis est; si ex ore, continuisse eo merum vinum; si parum id proficit, acetum. Si inter haec quoque graviter erumpit, quia consumere hominem potest, commodissimum est impetum eius admota occipitio cucurbitula, sic ut cutis quoque incidatur, avertere: si id mulieri, cui menstrua non feruntur, evenit, eandem cucur-
6 bitulam incisis inguinibus eius admovere. At si ex faucibus interioribusve partibus processit, et metus maior est et cura maior adhibenda. Sanguis mittendus est, et si nihilo minus ex ore processit, iterum tertioque, et cotidie paulum aliquid. Protinus autem debet sorbere vel acetum vel cum ture plantaginis aut porri sucum, imponendaque extrinsecus supra id, quod dolet, lana sucida ex aceto est, et id spongia subinde refrigerandum. Erasistratus horum crura quoque et femora brachiaque pluribus locis deligabat. Id Asclepiades adeo non prodesse, etiam
7 inimicum esse proposuit. Sed id saepe commode

---

^a *i.e.* owing to some bruising of the kidneys.

^b This was done for nose- and mouth-bleeding after head injury to divert blood away from the head.

^c This method was exceptionally employed to raise blood-pressure in the heart and prevent immediate death; but it usually results in persistence of the haemorrhage.

urine it even relieves this very lassitude; nor indeed, in the case of a fall from a height, is there anything alarming if blood comes with the urine,[a] so long as there is nothing else unusual in the urine; nor does vomiting of blood bring about danger, even when repeated, if before it recurs the body is allowed to regain strength and fill up; and it does no harm at all in a robust man, if not excessive, and when it excites neither cough nor fever.

The foregoing are general remarks: now I come to the particular points mentioned above. If blood escapes from the gums, it suffices to chew purslane; if from the mouth, undiluted wine should be held in it; if this does no good, then vinegar. If in spite of these remedies there is a severe outburst, since this may be the death of the patient, its attack is best diverted by applying a cup to the occipital region,[b] after first incising the skin; when this happens in a woman whose menses are not forthcoming, a cup is applied to each groin, likewise after making incisions. But if the bleeding comes from the throat, or from more internal parts, there is more to fear, and a more active treatment is to be adopted. Blood should be let, and if the flow from the mouth is not lessened, the venesection should be repeated a second or third time, and every day a little. From the first also the patient should sip either vinegar or plantain or leek juice with frankincense, and outside over the seat of pain there is to be applied unscoured wool soaked in vinegar, cooled at intervals by means of a sponge. Erasistratus used also to bind up the legs and thighs and the forearms of such patients in several places.[c] This constricting Asclepiades declared far from being beneficial, to be even harmful. But that it often

CELSUS

respondere experimenta testantur. Neque tamen pluribus locis deligari necesse est, sed satis est infra inguina et super talos, summosque umeros, etiam brachia. Tum si febris urget, danda est sorbitio, et potui aqua, in qua aliquid ex iis, quae alvum adstringunt, decoctum sit. At si abest febris, vel elota halica vel panis ex aqua frigida et molle quoque ovum dari potest, potui vel idem, quod supra scriptum est, vel vinum dulce vel aqua frigida: sed sic bibendum erit, ut sciamus huic morbo sitim prodesse.
8 Praeter haec necessaria sunt quies, securitas, silentium. Caput huius quoque cubantis sublime esse debet, recteque tondetur; facies saepe aqua frigida fovenda est. At inimica sunt vinum, balneum, venus, in cibo oleum, acria omnia, item calida fomenta, conclave calidum et inclusum, multa vestimenta corpori iniecta. Etiam frictiones[1] . . . ubi bene etiam sanguis conquievit. Tum vero incipiendum est a brachiis cruribusque, a thorace abstinendum. In hoc casu per hiemem locis maritumis, per aestatem . . .[2] mediterraneis opus est.

12 (V). Faucibus subest stomachus; in quo plura longa vitia incidere consuerunt. Nam modo ingens calor, modo inflatio hunc, modo inflammatio, modo exulceratio adficit: interdum pituita, interdum bilis oritur; frequentissimumque eius malum est, quo resolvitur, neque ulla re magis aut adficitur aut

---

[1] *After* frictiones *Marx inserts* utiles sunt, *which is translated*.
[2] *Marx suggests* uti *to fill the lacuna*.

---

[a] στομάχος originally meant the oesophagus including the cardiac orifice of the stomach (Galen, *Natural Faculties* III. 8, 260). Here Celsus with Aretaeus, *Chronic Diseases* (περὶ στο-

answers well experience bears witness. Yet there is no necessity to bind the patient in many places; it is enough to do it below the groins and above the ankles and at the upper part of the arms, also the forearms. Further, if fever is troublesome, gruel must be given, and for drink water in which has been boiled any one of the intestinal astringents. But if fever is absent, there may be given: either washed spelt or bread soaked in cold water and also soft eggs, and for drink either that mentioned above, or sweet wine or cold water; but drink must be given with the knowledge that in this disease thirst is an advantage. Besides these, there are needed rest, freedom from care, and silence. The head also of the patient whilst in bed should be kept raised, and well shaved; the face is often to be bathed with cold water. On the other hand, there is danger in wine, the bath, coition, oil in the food, all acrid food; the same applies to hot foments, a hot close room, much bedclothes piled on the patient. Rubbings also are useful after the bleeding has quite stopped. Then indeed a beginning can be made on the arms and legs, avoiding the chest. A patient in this case should live through the winter by the sea, during the summer inland.

12. Below the throat is placed the stomach,[a] in which there tend to occur many chronic complaints. For sometimes great heat affects it, sometimes flatulence, sometimes inflammation, sometimes ulceration; at times phlegm collects, at times bile; but the most frequent malady is that in which it undergoes paralysis, nor does anything else so affect it, or,

μαχικῶν, Book II. Ch. 6) refers to the cardiac orifice as well as to what he elsewhere terms ventriculus (IV. 1. 3, 6).

CELSUS

corpus adficit. Diversa autem ut vitia eius, sic etiam remedia sunt.

Ubi exaestuat, aceto cum rosa extrinsecus subinde fovendus inponendusque pulvis cum oleo et ea cataplasmata, quae simul et reprimunt et emolliunt. Potui, nisi quid obstat, egelida aqua praestanda.

2 Si inflatio est, prosunt admotae cucurbitulae neque incidere cutem necesse est: prosunt sicca et calida fomenta, sed non vehementissima. Interponenda abstinentia est; utilis in ieiunio potio est apsinthi vel hysopi vel rutae. Exercitatio primo levis, deinde maior adhibenda est, maximeque quae superiores partes moveat; quod genus in omnibus stomachi
3 vitiis aptissimum est. Post exercitationem opus est unctione, frictione; balneo quoque nonnumquam, sed rarius; interdum alvi ductione; cibis deinde calidis neque inflantibus, eodemque modo calidis potionibus, primo aquae, post, ubi resedit inflatio, vini austeri. Illud quoque in omnibus stomachi vitiis praecipiendum est, [ut] quo modo se quisque aeger refecerit, eo ut sanus utatur: nam redit huic inbecillitas sua, nisi eisdem defenditur bona valetudo, quibus reddita est.

4 At si inflammatio aliqua est, quam fere tumor et dolor sequitur, prima sunt quies, abstinentia, lana sulpurata circumdata, in ieiuno absinthium. Si ardor stomachum urget, aceto cum rosa subinde

---

[a] Cf. III. **19. 2**, *ex via pulvis*.
[b] These were believed to extract flatus—φύσα; the resulting oedematous subcutaneous swelling being termed emphysema, see p. 166, note *a*.

through it, the whole body. As diverse as are its complaints, so are the remedies.

When heated, it should be bathed externally at intervals with vinegar and rose oil, and road dust [a] applied with oil, and those plasters which simultaneously repress and soothe. For drink, unless there is anything against it, lukewarm water is the best.

If there is flatulence, it is beneficial to apply cups,[b] but there is no need to incise the skin; dry and hot foments do good, but not the strongest kind. At intervals there should be enjoined abstinence from food; a draught of wormwood or of hyssop or of rue on an empty stomach is useful. Exercise at first should be light, then more is to be taken, especially such as moves the upper limbs; the kind most appropriate in all complaints of the stomach.[c] After exercise there is need of anointing, rubbing, occasionally also the bath, yet less often than usual; now and then a clyster; later, food which is hot but not flatulent, and similarly hot drinks, first water and after the flatulence has subsided, dry wine. In all complaints of the stomach this also is to be prescribed, that each should adopt in health that regimen which has cured him; for his weakness will recur unless his health is protected by the same measures as those by which it was restored.

But if there is inflammation of any kind, which is generally followed by swelling and pain, the primary remedies are rest, abstinence, a belt of sulphurated wool, and the wormwood draught upon an empty stomach. If a burning heat troubles the stomach, it should be fomented at intervals with vinegar and

[c] Because the arms and stomach derive nerves from the same level of the spinal cord.

fovendus est: deinde cibis quidem utendum modicis; inponenda vero extrinsecus quae simul et reprimunt et emolliunt: postea deinde his detractis utendum calidis ex farina cataplasmatis, quae reliquias digerant; interdum alvus ducenda, adhibenda exercitatio et cibus plenior.

5 At si ulcera stomachum infestant, eadem fere facienda sunt, quae in faucibus exulceratis praecepta sunt (IV. 9). Exercitatio, frictio inferiorum partium adhibenda; adhibendi lenes et glutinosi cibi, sed citra satietatem: omnia acria atque acida removenda. Vino, si febris non est, dulci, aut si id inflat, certe leni utendum: sed neque praefrigido neque nimis calido.

Si vero pituita stomachus impletur, necessarius modo in ieiuno, modo post cibum vomitus est: utilis exercitatio, gestatio, navigatio, frictio. Nihil edendum bibendumque nisi calidum vitatis tantum iis, quae pituitam contrahere consuerunt.

6 Molestius est, si stomachus bile vitiosus est. Solent autem ii, qui sic temptantur, interpositis quibusdam diebus hanc, et quidem, quod pessimum est, atram vomere.—His recte alvus ducitur, potiones ex absinthio dantur: necessaria gestatio, navigatio est; si fieri potest, ex nausia vomitus [1] . . . vitanda cruditas, sumendi cibi faciles et stomacho non alieni, vinum austerum.

7 Vulgatissimum vero pessimumque stomachi vitium est [resolutio est], cum cibi non tenax est, soletque

[1] *After* vomitus Marx *inserts* evocandus.

rose oil; next food should be given in moderation, external applications are also to be made which simultaneously both repress and soothe; next after that, when these are taken off, hot meal plasters are put on to disperse the remnants of the disease: now and again a clyster must be given, exercise must be taken, and a fuller diet.

But if ulcers attack the stomach, generally the same treatment should be applied as has been prescribed in the case of ulcerated throat (IV. **9**). Exercise, also rubbing of the lower extremities, is to be practised; bland and glutinous foods taken short of satiety; and all pungent and sharp food withdrawn. Sweet wine is to be used if there is no fever, or if that causes flatulence at any rate light wine, but neither very cold nor too hot.

If the stomach becomes filled with phlegm an emetic is needed, sometimes on an empty stomach, sometimes after food: there is benefit in exercise, rocking, a sea-voyage, rubbing. Nothing should be eaten or drunk unless hot, whilst such things must be avoided as have tended to collect phlegm.

There is worse trouble when the stomach is vitiated by bile. Patients who are troubled with this, vomit up bile at intervals of some days, and worst of all, vomit black bile. For such a clyster is appropriate, and draughts of wormwood should be given; rocking and a sea-voyage are necessary; vomiting when possible is induced by sea-sickness; indigestion must be avoided, the food should be such as is readily swallowed, and not repugnant to the stomach, the wine must be dry.

But the commonest and worst complaint of the stomach is [paralysis], when it does not retain food,

CELSUS

desinere ali corpus ac sic tabe consumi.—Huic generi inutilissimum balineum est: lectiones exercitationesque superioris partis necessariae, item unctiones frictionesque; his perfundi frigida atque in eadem natare, canalibus eiusdem subicere et stomachum ipsum et magis etiam a scapulis id, quod contra stomachum est, consistere in frigidis medicatisque fontibus, quales Cutiliarum Sumbruinarumque sunt, salutare est. Cibi quoque adsumendi sunt frigidi, qui potius difficulter coquuntur, quam facile vitiantur. Ergo plerique, qui nihil aliud concoquere possunt, bubulam coquunt. Ex quo colligi potest neque avem neque venationem neque piscem dari debere nisi
8 generis durioris. Potui quidem aptissimum est vinum frigidum, vel certe bene calidum meracum, potissimum Raeticum vel Allobrogicum aliudve, quod et austerum et resina conditum est: si id non est, quam asperrimum maximeque Signinum. Si cibus non continetur, danda aqua et eliciendus plenior vomitus est, iterumque dandus cibus; et tum admovendae duobus infra stomachum digitis cucurbitulae ibique duabus aut tribus horis con-
9 tinendae sunt. Si simul et vomitus et dolor est, imponenda supra stomachum est lana sucida vel spongia ex aceto vel cataplasma, quod refrigeret. Perfricanda vero non diu sed vehementer brachia et crura et calfacienda. Si plus doloris est, infra[1] praecordia quattuor digitis cucurbitula utendum est, et protinus dandus panis ex posca frigida: si non continuit, post vomitum lene aliquid ex iis, quae

---

[1] *So Marx and other editors for the MSS.* in qua.

[a] Cf. II. 28. 2.

BOOK IV. 12. 7–9

and the nutrition of the body is wont to cease, and so it is consumed by wasting. In this sort of disease the bath is most harmful; reading aloud and exercise of the upper limbs are needed, as also anointing and rubbings; it is good for the patient to have cold water poured over him, and to swim in cold water, also to submit his stomach to jets of it, especially at the back of the stomach from the shoulder-blades downwards, to bathe in cold medicinal springs, such as those at Cutilia and Simbruvium. Food should be also taken cold, rather than that which is digested with difficulty than that which readily decomposes. Hence many who can digest nothing else, digest beef,[a] and therefore it may be inferred that neither poultry nor venison, nor fish except the harder kinds, should be given. The most suitable drink is wine cold, or else undiluted and well heated, particularly Rhaetic or Allobrogic wine, or any other which is both dry and seasoned with resin; if there is none of the above at hand, then the harshest possible, especially Signine wine. If food is not retained, water must be given and a more copious vomit elicited, after which food is to be given again, and then cups are to be applied two fingers' breadth below the stomach, and they are to be kept on two or three hours. If simultaneously there is both vomiting and pain, there should be placed over the stomach unscoured wool or sponge soaked in vinegar, or a refrigerant plaster. The arms and legs too are to be rubbed sharply, but not for long, and to be kept warm. If pain is more severe, a cup is to be put on four fingers' breadth below the praecordia, and following that bread in cold vinegar and water is to be given; should this not be retained, after the vomiting,

non aliena stomacho sint: si ne id quidem tenuit, singuli cyathi vini singulis interpositis horis, donec stomachus consistat. Valens etiam medicamentum est radiculae sucus, valentius acidi Punici mali cum pari modo suci, qui ex dulci Punico malo est, adiecto etiam intubi suco et mentae, sed huius minima parte; quibus tantundem, quantum in his omnibus est, aquae frigidae quam optime miscetur: id enim plus quam vinum ad conprimendum stomachum potest. Supprimendus autem vomitus est, qui per se venit. Sed si nausia est aut si coacuit intus cibus aut computruit, quorum utrumlibet ructus ostendit, eiciendus est; protinusque, cibis adsumptis iisdem, quos proxime (§ 9) posui, stomachus restituendus. Ubi sublatus est praesens metus, ad ea redeundum est, quae supra (§ 7) praecepta sunt.

13 (VI). Stomachus lateribus cingitur, atque in his quoque vehementes dolores esse consuerunt. Et initium vel ex frigore vel ex ictu vel ex nimio cursu vel ex morbo est. Sed interdum omne[1] malum intra dolorem est isque modo tarde, modo celeriter solvitur: interdum ad perniciem quoque procedit oriturque acutus morbus, qui pleuriticus a Graecis nominatur. Huic dolori lateris febris et tussis accedit; et per hanc excreatur, si tolerabilis morbus est, pituita; si gravis, sanguis. Interdum etiam sicca tussis est, quae nihil emolitur, idque primo vitio gravius, secundo tolerabilius est.—Remedium vero est magni et recentis doloris sanguis missus: at sive levior sive vetustior casus est, vel super-

---

[1] omne, *so Stangl, followed by Marx for the* ne *of the MSS.*

[a] Cf. Aretaeus, Book I. 10, περὶ πλευρίτιδος.

BOOK IV. 12. 9–13. 2

anything light or not unsuitable for the stomach can be given; if even that is not retained, give a cupful of wine every hour until the stomach settles down. Radish juice also is an active remedy; more active still is a mixture of the juice of sour and sweet pomegranates in equal parts, with the addition of endive and mint juice, the least quantity of the last; the whole of the above may be mixed thoroughly well with an equal quantity of cold water, which is better than wine for tightening up the stomach. Now, vomiting when spontaneous should be arrested; but if there is nausea, or if food turns acid inside, or decomposes, both of which are manifested by eructations, the food should be evacuated, and the stomach forthwith replenished by the taking of food of the kind just noted (§ 9). When immediate apprehension has been removed a return should be made to the prescriptions given above (§ 7).

13. The stomach is girt about by the ribs, and in these also severe pains occur. And the commencement either is from a chill, or from a blow, or from excessive running, or from disease. But at times pain is all there is the matter, and this is recovered from be it slowly or quickly; at times it goes on until it is dangerous, and the acute disease arises which the Greeks call pleurisy.[a] To the aforesaid pain in the side is added fever and cough; and by means of the cough, phlegm is expectorated when the disease is less serious, but blood when it is grave. At times also there is a dry cough without expectoration, which is worse than the former condition, and better than the latter. The appropriate remedy for severe and recent pain is blood-letting; but if the case is either of a slighter or of a more chronic kind,

vacuum vel serum id auxilii est; confugiendumque ad cucurbitas est, ante summa cute incisa. Recte etiam sinapi ex aceto super pectus inponitur, donec ulcera pusulasque excitet, et tum medicamentum, quod 3 umorem illuc citet. Praeter haec circumdare primum oportet latus abso lanae sulpuratae; deinde, cum paululum inflammatio se remisit, siccis et calidis fomentis uti. Ab his transitus ad malagmata est. Si vetustior dolor remanet, novissime resina inposita discutitur. Utendum cibis potionibusque calidis, vitandum frigus. Inter haec tamen non alienum est extremas partes oleo et sulpure perfricare; si levata tussis est, leni lectione uti, iamque et acres cibos et vinum meracius adsumere. Quae ita a medicis praecipiuntur, ut tamen sine his rusticos nostros 4 epota ex aqua herba trixago satis adiuvet. Haec in omni lateris dolore communia sunt: plus negotii est, si acutus quoque morbus is factus est. In hoc praeter ea, quae supra posita sunt, haec animadvertenda sunt; ut cibus sit quam maxime tenuis et lenis, praecipueque sorbitio eaque ex tisana potissimum, aut ius, in quo porrus cum pullo gallinaceo coctus sit, idque non nisi tertio quoque die detur, si tamen per vires licebit: potui vero aqua mulsa, 5 in qua hysopum aut ruta decocta sit. Quae quibus temporibus danda sint, ex ratione vel adiectae vel levatae febris apparebit, sic ut in remissione quam maxima dentur, cum eo tamen, ut sciamus non esse

then this remedy becomes either unnecessary or belated; and recourse is to be had to cupping after incising the skin. It is also appropriate to apply vinegar and mustard upon the chest until this raises ulcerations and pustulations, and then a medicament to draw out the humour that way. Besides the above the side should be first surrounded with a sheet of sulphurated wool; next, after the inflammation has subsided somewhat, have dry and hot foments applied to it. From these transition is made to emollients. If the pain persists for a longer time, it may finally be dispersed by resin plaster. Food and drink should be taken hot, avoiding cold. Along with the above treatment, however, it is not unfitting to rub the lower limbs with oil and sulphur. If the cough has been relieved, the patient should read a little out loud, and now take both sharp food and undiluted wine. Though such are what medical practitioners prescribe, yet our country people, lacking these remedies, find help enough in a draught of germander. The foregoing are the remedies common to all cases of pain in the side: there is more to do if this affection has also become acute. In such cases, besides what has been described above, attention must be given to the following: that the food be as thin and bland as possible, and gruel is most suitable, especially that made with pearl barley, or soup made by boiling a chicken with leeks, and this may be given, but only every third day, if the patient's strength permits of this; the drink should be hydromel in which hyssop or rue has been boiled. The times at which these should be given will become apparent from the way the fever increases or diminishes, so that it should be given when there is least fever, not forgetting,

eius generis tussi aridas fauces committendas: nam saepe, ubi nihil est, quod excreetur, continuatur et strangulat. Ob quam causam dixi (§ 2) etiam peius id genus esse tussis, quod nihil quam quod pituitam moveret. Sed hic vinum sorbere, ut supra (**10,** 3) praecepimus, morbus ipse non patitur: in vicem
6 eius cremor tisanae sumendus est. Ut his autem in ipso morbi fervore sustinendus aeger est, sic, ubi paululum is se remisit, alimenta pleniora et vini quoque aliquid dari potest, dum nihil detur, quod aut refrigeret corpus aut fauces asperet. Si in refectione quoque manserit tussis, intermittere oportebit uno die, posteroque cum cibo vini paulo plus adsumere. Atque incipiente quoque tussi, tum non erit alienum, ut supra (**10,** 3) quoque positum est, vini cyathos sorbere, sed id in hoc genere valetudinis dulce vel certe lene commodius est. Si malum inveteravit, athletico victu corpus firmandum est.

14 (VII). A compagine corporis ad viscera transeundum est, et inprimis ad pulmonem veniendum; ex quo vehemens et acutus morbus oritur, quem peripleumoniacon Graeci vocant. Eius haec condicio est: pulmo totus adficitur; hunc casum eius subsequitur tussis bilem vel pus trahens, praecordiorum totiusque pectoris gravitas, spiritus difficultas, magnae febres, continua vigilia, cibi fastidium, tabes. Id genus morbi plus periculi quam doloris habet.—Oportet, si satis validae vires sunt, sanguinem mittere; si minores, cucurbitulas sine ferro praecordiis admovere.
2 Tum si satis valet, gestando aegrum digerere; si

---

^a Celsus is describing pneumonia. Cf. Aretaeus, *Acute Diseases*, Book II. Ch. 1, περὶ περιπνευμονίης, Hipp. IV. 158, 160, 194 (*Aph.* V. 8, 15; VII. 11).
^b Cf. II. 4. 1, 8. 2.

however, that a dry throat must not be combined with that kind of cough; for often when there is no expectoration, the cough is incessant and chokes the patient. On this account I stated above that a cough which brings up nothing is of a worse kind than that causing phlegm to be expectorated. But here the disease does not allow of wine being sipped as prescribed above (**10.** 3); pearl barley gruel is to be taken instead. As these have to sustain the patient during the hot stage of the disease, as soon as there is a little remission, the diet can be increased and also some wine given, as long as nothing is given that will either chill the body or irritate the throat. If the cough persists in convalescence, it will be well on one day to omit the wine, and on the next to take a little extra wine with the food. And also at the beginning of a cough, as stated above, it is not amiss to sip cupfuls of wine; but sweet or at any rate light wine, is the more suitable in this kind of illness. If the malady has become inveterate the body must be strengthened by food fit for an athlete.

14. Passing from the framework of the body to the viscera, we come first to the lung, where a grave and acute disease arises, which the Greeks name peripleumoniacon.[a] The conditions are these: the lung is attacked as a whole; this is followed by a cough which draws up bile or pus; there is a feeling of weight over the praecordia and all the chest; there is difficult breathing, high fever, persistent insomnia, loss of appetite, wasting. This sort of disease has in it more of danger[b] than of pain. Blood should be let if there is strength enough; if not, dry cups should be applied over the praecordia. Then if the patient is strong enough he should be rocked to disperse the

parum, intra domum tamen dimovere: potionem autem hysopi ...[1] cum quo ficus arida sit incocta, aut aquam mulsam, in qua vel hysopum vel ruta decocta sit: frictione uti diutissime in scapulis, proxime ab his in brachiis et pedibus et cruribus, leviter contra pulmonem, idque bis cottidie facere. Quod ad cibum vero pertinet, huic nec salsis opus est neque acribus neque amaris neque alvum adstringentibus, 3 sed paullo lenioribus. Ergo primis diebus danda est sorbitio tisanae vel halicae vel oryzae, cum qua recens adeps cocta sit: cum hac sorbile ovum, nuclei pinei ex melle, panis vel elota halica ex aqua mulsa; potui deinde non solum pura aqua sed etiam mulsa egelida, aut si aestas est, etiam frigida, nisi quid obstat. Haec autem altero quoque die increscente morbo dare satis est. Ubi incrementum constitit, quantum res patitur, ab omnibus abstinendum est, praeterquam aqua egelida. Si vires desunt, adiuvandae sunt aqua mulsa. Prosuntque adversus dolores imposita calida fomenta vel ea, quae simul et repri4 munt et emolliunt. Prodest impositus super pectus sal bene contritus cum cerato mixtus, quia leviter cutem erodit, eoque impetum materiae, quo pulmo vexatur, evocat. Utile etiam aliquod malagma est ex iis, quae materiam trahunt. Neque alienum est, dum premit morbus, clausis fenestris aegrum continere: ubi paullum levatus est, ter aut quater die fenestris aliquis apertis parvum aerem recipere. Deinde in refectione pluribus diebus a vino abstinere; gestatione, frictione uti, sorbitionibus; et prioribus cibis adicere ex holeribus porrum, ex

---

[1] *Daremberg and other editors inserted* dare *to fill the lacuna.*

[a] V. 18. 6, 7.

disease; if not, he should yet be moved about in the house: his drink should then be a decoction of hyssop with a dried fig, or hydromel in which hyssop or rue has been boiled; he should be rubbed twice daily, longest between the shoulder-blades, then the arms, feet and legs, but lightly over the lung. As regards food too, in this instance it should be neither salted nor acrid nor bitter nor constipating, but of the rather blander kinds. Therefore on the first days pearl barley or spelt or rice gruel in which fresh lard has been boiled are to be given; with this raw eggs, pine kernels in honey, bread or washed groats of spelt in hydromel; then he may drink not only water by itself but also lukewarm hydromel, or even this cold in summer, unless there is some objection. But whilst the disease is on the increase, it is enough to give these every other day. When the increase has come to a stand, he should abstain, so far as is practicable, from everything except lukewarm water. If the strength begins to fail, hydromel is to be added. For the relief of pain it is helpful to apply foments hot, or those which both repress and soothe. The application to the chest of salt, well rubbed up and mixed with wax-salve, is beneficial because it slightly erodes the skin, and thereby draws out that flood of the matter by which the lung is being oppressed. Useful also is any one of the emollients [a] which draw out the matter. During the pressure of the disease it is not wrong to keep the patient with the windows shut: when he is somewhat better, some windows should be opened three or four times a day to let in a little air. Next during recovery he should for several days abstain from wine, use rocking, rubbing and gruels; to the previous foods add: of vegetables,

carne ungulas et summa trunculorum, atque pisciculos, sic ut diu nihil nisi molle et lene sumatur.

15 (VIII). Alterius quoque visceris morbus, iocineris, aeque modo longus, modo acutus esse consuevit: ἡπατικὸν Graeci vocant. Dextra parte sub praecordiis vehemens dolor est, idemque ad latus dextrum et ad iugulum umerumque partis eiusdem pervenit: nonnumquam manus quoque dextra torquetur, horror calidus est. Ubi male est, bilis evomitur; interdum singultus prope strangulat. Et haec quidem acuti morbi sunt: longioris vero, ubi suppuratio in iocinere est, dolorque modo finitur, modo intenditur, dextra parte praecordia dura sunt et tument; post cibum maior spiritus difficultas est;
2 accedit maxillarum quaedam resolutio. Ubi inveteravit malum, venter et crura pedesque intumescunt, pectus atque umeri circaque iugulum utrumque extenuatur.—Initio sanguinem mittere optimum est: tum venter solvendus est, si non potest aliter, per nigrum veratrum. Inponenda extrinsecus cataplasmata, primum quae reprimant, deinde calida quae diducant, quibus recte iris vel apsinthium
3 adicitur; post haec malagma. Dandae vero sorbitiones sunt, omnesque cibi et calidi et qui non multum alant, ei fere qui pulmonis quoque dolori conveniunt, praeterque eos, qui urinam movent, potionesque ad id efficaces. Utilia in hoc morbo sunt thymum, satureia, hysopum, nepeta, amulum,

---

[a] ἡπατικὸν πάθος, *hepaticus morbus* (liver abscess) was distinguished by Celsus from jaundice (*morbus arquatus*), cf. III. 24; Aretaeus, *Chronic Diseases*, Book I., Ch. 13 and 15, also distinguished between the two.

[b] The nerves of the adjacent surface of the liver and diaphragm, and those of the upper limbs, are derived from the same level of the spinal cord in the neck.

BOOK IV. 14. 4–15. 3

leeks, of meat, trotters and tit-bits, also small fish, so long as for a while nothing but what is soft and bland is consumed.

15. Further a disease of another of the viscera, the liver, is also sometimes chronic, sometimes acute: the Greeks call it hepaticon.[a] There is severe pain in the right part under the praecordia, which spreads to the right side, to the clavicle and arm of that side;[b] at times there is also pain in the right hand, there is hot shivering. In a grave case there is vomiting of bile; sometimes the patient is nearly choked by hiccough. Such are the signs when acute; but in a more chronic case, where there is suppuration within the liver, and the pain at times ceases, at times is intensified, the praecordia on the right side become hard and swollen; after a meal there is greater difficulty in breathing; then supervenes a sort of paralysis of the lower jaws.[c] When the disease has become inveterate, the abdomen and legs and feet swell; there is wasting of the chest and arms and about the clavicle on both sides. It is best to begin by letting blood; then the bowel is to be moved, if nothing else takes effect, by black hellebore. Externally plasters are to be applied, first repressants, then hot ones to disperse; appropriate additions are iris or wormwood unguents; after these emollients. Gruels, moreover, are to be given, all food hot and not too nourishing, generally that kind which is also suitable to pleurisy (IV. **13**, 4), and in addition such food and drink as promote urination. Beneficial in this disease are: thyme, savory, hyssop, catmint, starch, sesamum

[c] In the dying.

sesamum, lauri bacae, pini flos, herba sanguinalis, menta, ex malo Cotoneo medium, columbae iecur recens et crudum. Ex quibus quaedam per se esse, quaedam adicere vel sorbitioni vel potioni licet, sic tamen, ut parce adsumantur. Neque alienum est absinthium contritum ex melle et pipere, eiusque catapotium cotidie devorare. Abstinendum utique est ab omnibus frigidis: neque enim res ulla magis 4 iecur laedit. Frictionibus utendum in extremis partibus: vitandus omnis labor, omnis vehementior motus; ne spiritus quidem diutius continendus est. Ira, trepidatio, pondus, ictus, cursus inimica sunt. Perfusio corporis multa prodest ex aqua, si hiemps est, calida, si aestas, tepida; item liberalis unctio et in balneo sudor. Si vero iecur vomica laborat, eadem facienda sunt, quae in ceteris interioribus suppurationibus. Quidam etiam contra id scalpello aperiunt et ipsam vomicam adurunt.

16 (IX). At lienis, ubi adfectus est, intumescit, simulque cum eo pars sinistra; eaque dura est et prementi renititur: venter intentus est; aliquis etiam cruribus tumor est. Ulcera aut omnino non sanescunt, aut certe cicatricem vix recipiunt: in intenta ambulatione cursuque dolor et quaedam difficultas est.—Hoc vitium quies auget: itaque exercitatione et labore opus est, habita tamen ratione, ne febrem ista, si nimium processerint, excitent. Unctiones frictionesque et sudores neces-

---

[a] The same recommendation was made by Aretaeus, *op. cit.*, except that Aretaeus omitted the incision through the skin; he used the cautery at white heat, and pushed it down to the pus, for at the same time it cuts and burns. Recovery might follow if the flow of pus was pure and white—Hipp. IV. 202 (*Aph*. VII. 45).

BOOK IV. 15. 3–16. 1

seeds, laurel berries, young pine-cone tips, knotgrass, mint, quince pulp, the fresh raw liver of a pigeon. Some of the above may be eaten alone, some can be added to the gruel or draughts, so long as they are taken sparingly. There is no objection to wormwood rubbed up in honey and pepper, of which a dose is taken daily. All cold things must be especially avoided; for nothing is more harmful to the liver. Rubbings of the extremities should be employed; all manual work should be avoided, and all more active movement; the patient should never even hold his breath for long together. Anger, hurry, weight-lifting, boxing, running are harmful. A copious affusion of the body with water, hot in winter, tepid in summer, is beneficial, also free anointing and sweating at the bath. But if the liver suffers from an abscess, the same is to be done as in other internal suppurations. Some even with a scalpel make an incision over the liver, and burn through into the actual abscess with the cautery.[a]

16. Now the spleen[b] when affected swells, and with it simultaneously the left side; and this becomes hard and resists pressure. The abdomen is tense: there is even some swelling of the legs. Ulcerations either do not heal at all, or at any rate form a scar with difficulty: there is also pain and some difficulty in walking fast or running. Rest increases this complaint, and so there is need for exercise and work; nevertheless, care must be taken lest if carried too far fever be excited. Anointings and rubbings and

---

[b] Hippocrates and Aretaeus had already noted the connection between marshy districts and enlargement of the spleen, Hipp. I. 82, *Airs, Waters, and Places*, Ch. 8, Aretaeus, *Chronic Diseases*, Bk. I., Ch. 14,

CELSUS

sarii sunt. Dulcia omnia inimica sunt, item lac et
2 caseus: acida autem maxime conveniunt. Ergo acetum acre per se sorbere, et magis etiam quod scilla conditum est, expedit. Edenda sunt salsamenta vel oleae ex muria dura, tinctae in aceto lactucae, intubique ex eodem, betae ex sinapi, asparagus, armoracia, pastinaca, ungulae, rostra, aves macrae, eiusdem generis venatio. Potui vero ieiuno dari debet apsinthium incoctum: at post cibum aqua a ferrario fabro, in qua candens ferrum subinde tinctum sit: haec enim vel praecipue lienem coercet. Quod animadversum est in iis animalibus, quae aput hos fabros educata exiguos lienes habent.
3 Potest etiam dari vinum tenue, austerum; omniaque in cibis et potionibus, quae urinae movendae sunt. Praecipueque ad id valet vel trifolii semen vel cuminum vel apium vel serpullum vel cytisus vel portulaca vel nepeta vel thymum vel hysopum vel satureia: haec enim inde commodissime videntur umorem deducere. Lienis quoque bubulus utiliter esui datur; praecipueque eruca et nasturcium lienem
4 extenuant. Inponenda quoque extrinsecus sunt quae levent. [Fit ex unguento et palmulis, quod myrobalanon Graeci vocant; fit ex lini et nasturci semine, quo vinum et oleum adicitur; fit ex cupresso viridi et arida ficu; fit ex sinapi, cui sebi hirquini a renibus quarta pars ponderis adicitur, teriturque in sole et protinus imponitur.][1] Multisque modis huic rei cappari aptum est: nam et ipsum cum cibo

---

[1] *The words in brackets were deleted by Targa as a gloss.*

[a] Possibly because of an absorbable ferric hydrate which relieved anaemia.
[b] A homoeopathic remedy. [c] IV. 8. 3.

BOOK IV. 16. 1–4

sweatings are necessary. All sweet things are hurtful, also milk and cheese; but sour things are the most suitable. Therefore sharp vinegar may be sipped by itself, vinegar of squills is even better. Such patients should eat salt fish or olives preserved in strong brine, lettuce dipped in vinegar, also endive in the same, beet with mustard, asparagus, horse-radish, parsnip, trotters, chaps, poultry not fatted, and similar game. The drink too, when taken on an empty stomach, should be wormwood decoction; after food, water in which a blacksmith has from time to time dipped his red-hot irons; since this water especially reduces the spleen. For it has been observed that animals reared by our blacksmiths, have small spleens.[a] Dry thin wine can also be given: and everything, whether food or drink, which causes urination. Of particular value in this respect are: trefoil seeds or cummin or celery or creeping thyme or broom tops or purslane or catmint or thyme or hyssop or savory: for these seem best adapted to draw out humour from the spleen. Ox-spleen[b] may be usefully given to eat; rocket and nasturtium in particular render the spleen smaller. Palliatives must also be applied externally: [there is one made of ointment and dates which the Greeks call myrobalanon, or that made of linseed and nasturtium[c] seeds, to which wine and oil have been added; or that made of green cypress and a dried fig; or that made with mustard to which is added a fourth part by weight of he-goat's kidney fat, and which is rubbed up in the sun and applied forthwith.] Moreover, capers[d] may be employed in several ways; for they may be both

---

[d] Capparis spinoza, buds and root, yield a volatile oil resembling rue.

CELSUS

adsumere et muriam eius cum aceto sorbere commodum est. Quin etiam extrinsecus radicem contritam vel corticem eius cum furfuribus aut ipsum cappari cum melle contritum imponere expedit. Malagmata quoque huic rei aptantur.

17 (X). At renes ubi adfecti sunt, diu male habent. Peius est, si frequens biliosus vomitus accedit.—Oportet conquiescere, cubare molliter, solvere alvum, si aliter non respondet, etiam ducere; saepe desidere in aqua calida; neque cibum neque potionem frigidam adsumere; abstinere ab omnibus salsis, acidis, acribus, pomis; bibere liberaliter; adicere modo cibo modo potioni piper, porrum, ferulam, album papaver; quae maxime inde urinam movere 2 consuerunt. Auxilio quoque his exulceratis sunt, si adhuc ulcera purganda sunt, cucumeris semina detractis corticibus sexaginta, nuclei ex pinu silvestri XII, anesi quod tribus digitis sumi possit, croci paulum, contrita et in duas mulsi potiones divisa: si vero dolor tantum levandus est, eiusdem cucumeris semina xxx, idem nuclei xx, nuces Graecae v, croci paululum, contrita et cum lacte potui data. Ac super quoque recte quaedam malagmata iniciuntur, maximeque ea, quae umori extrahendo sunt.

18 (XI). A visceribus ad intestina veniendum est, quae sunt et acutis et longis morbis obnoxia.

taken with the food, and the brine and vinegar in which they have been soaked may be sipped. They may be even applied externally, the root or bark having been rubbed up with bran or the capers themselves with honey. There are also emollients suitable for this affection.

17. As regards the kidneys,[a] these when they have become affected, continue diseased for a long while. It is worse if bilious vomiting is added. The patient should rest, sleep on a soft bed, keep the bowels loose even using a clyster when they do not act otherwise; he should sit frequently in a hot bath; take neither food nor drink cold, abstain from everything salted, acid, acrid, and from orchard fruit; drink freely; add whether to the food or to the drink pepper, leeks, fennel, white poppy; which are the most active in causing a discharge of urine. As an additional remedy when there is ulceration of the kidneys, if the ulcerations are still in need of being cleaned, sixty cucumber seeds stript of the husk, twelve pine kernels, of aniseed as much as can be taken up by three fingers, and a little crocus, are rubbed up together, and divided between two draughts of honey wine: but if it is merely pain which has to be relieved, thirty of the cucumber seeds, twenty pine kernels, five almonds, and a little crocus are rubbed up together and given in milk. And besides it is right to apply certain emollients, and especially such as extract humour.

18. From the viscera we proceed to the intestines, which are subject to diseases, both acute and chronic.

[a] Hippocrates had distinguished four forms of Calculous disease of the kidneys, *Internal Diseases* 14–17 (Littré VII, 203) and presumably Celsus dealt with this subject in the lost portion of book IV; cf. p. 448, note 1. See also Aretaeus *Chronic Diseases*, Bk. II. Ch. 3.

CELSUS

Primoque facienda mentio est cholerae, quia communne id stomachi atque intestinorum vitium videri potest: nam simul et deiectio et vomitus est, praeterque haec inflatio est. Intestina torquentur, bilis supra infraque erumpit, primum aquae similis, deinde ut in ea recens caro lota esse videatur, interdum alba, nonnumquam nigra vel varia. Ergo eo nomine
2 morbum hunc choleram Graeci nominarunt. Praeter ea vero, quae supra conprehensa sunt, saepe etiam crura manusque contrahuntur, urget sitis, anima deficit; quibus concurrentibus non mirum est, si subito quis moritur: neque tamen ulli morbo minore momento succurritur.—Protinus ergo, ubi ista coeperunt, aquae tepidae quam plurimum bibere oportet et vomere. Vix umquam fit ne vomitus sequatur: sed etiamsi incidit, miscuisse tamen novam materiam corruptae prodest; parsque sanitatis est vomitum
3 esse subpressum. Si id incidit, protinus ab omni potione abstinendum est: si vero tormina sunt, oportet frigidis et umidis fomentis stomachum fovere, vel si venter dolet, iisdem egelidis, sic ut venter ipse mediocriter calentibus iuvetur. Quod si vehementer et vomitus et deiectio et sitis vexant, et adhuc subcruda sunt, quae vomuntur, nondum vino maturum tempus est: aqua, neque ea ipsa frigida, sed potius egelida danda est; admovendumque naribus est

---

[a] Cholera nostras—native sporadic cholera especially common in children, a severe and fatal bilious vomiting with diarrhoea or with obstinate constipation, intimately related to season and food. Hipp. *Epid.* V. 10, 79—Aretaeus, *Acute*

BOOK IV. 18. 1–3

And in the first place mention is to be made of cholera,[a] because it appears to be a complaint common to the stomach and intestines: for there occur simultaneously diarrhoea and vomiting, and in addition flatulence. The intestines are griped, bile bursts out upwards and downwards; first it is watery, then like water in which fresh meat has been washed; at times it is white in colour, at other times black or variously coloured. Hence the Greeks term this affection by the name of cholera. Besides those symptoms which are mentioned above, often the legs and arms are also contracted, there is urgent thirst, and fainting; when such things occur together, it is not to be wondered at if the patient dies suddenly, and yet in no other disease is there less time for affording relief. Therefore immediately upon the commencement of the above signs, the patient should drink as much as he can of tepid water, and vomit. Vomiting hardly ever fails to follow; but even if it does not occur, nevertheless it is advantageous to have mixed fresh material with that which is decomposed; the cessation of vomiting is a step towards recovery. If this happens, the patient should abstain forthwith from all drink; if there are still gripings, the stomach should be treated with cold and moist foments, or if there is pain in the belly, these should be lukewarm, so that the belly itself is relieved by moderately warm applications. But if vomiting, diarrhoea and thirst give rise to severe distress, and the vomit still contains undigested food, it is not yet a fitting time for wine: water should be given, not cold but rather lukewarm: pennyroyal

Diseases, Book II. 5. Asiatic cholera was first described by Bontius, *De Medicina Indorum*, 1778.

puleium ex aceto, vel polenta vino adsparsa, vel
4 menta . . .¹ secundum naturam est. At cum discussa
cruditas est, tum magis verendum est, ne anima
deficiat. Ergo tum confugiendum est ad vinum. Id
esse oportet tenue, odoratum, cum aqua frigida mixtum, vel polenta adiecta vel infracto pane, quem ipsum
quoque adsumere expedit quotiensque aliquid aut
stomachus aut venter effudit, totiens per haec vires
restituere. Erasistratus primo tribus vini guttis aut
quinis aspergendam potionem esse dixit, deinde
paulatim merum adiciendum. Is si et ab initio
vinum dedit et metum cruditatis secutus est, non
sine causa fecit: si vehementem infirmitatem adiu-
5 vari posse tribus guttis putavit, erravit. At si inanis
est homo, et crura eius contrahuntur, interponenda
potio apsinthi est. Si extremae partes corporis
frigent, unguendae sunt calido oleo, cui cerae paulum sit adiectum, calidisque fomentis nutriendae.
Si ne sub his quidem quies facta est, extrinsecus
contra ventriculum ipsum cucurbitula admovenda
est, aut sinapi superinponendum. Ubi is constitit,
dormire oportet; postero die utique a potione
abstinere, die tertio in balneum ire; paulatim se
cibo reficere. Somno quisquis facile adquiescit . . .²
itemque lassitudine et frigore. Si post suppressam
choleram febricula manet, alvum duci necessarium
est, tum cibis vinoque utendum est.

¹ *Marx suggests that* prout *has fallen out.*
² *Targa noted a lacuna here. Marx suggests* "celerrime restituitur: redit malum imprimis cruditate," *and this text is translated.*

in vinegar should be applied to the nostrils, or wine sprinkled with polenta, or mint in its natural state. But when the indigestion has been relieved, there is then greater apprehension of fainting. Recourse therefore should then be had to wine. The wine taken should be thin, aromatic, mixed with cold water, adding either polenta or crumbled bread, and bread by itself ought also to be taken, and as often as either the stomach or intestines discharge their contents, so often should the patient recruit his strength by these means. Erasistratus said that a draught should have mixed with it at first three or five drops of wine, subsequently gradual additions of undiluted wine. If Erasistratus both gave wine at the beginning and was influenced by fear of causing indigestion, he acted not without reason; if he thought that severe weakness could be relieved by three drops of wine, he erred. But if the patient is empty and his legs are contracted, a draught of wormwood should be given at intervals. If the extremities become cold, they should be anointed with hot oil to which a little wax has been added, and stimulated by hot foments. If there is no relief even from the above remedies, outside over the actual stomach cups should be applied, or mustard laid upon it. When he has settled down, he should go to sleep. On the next day he should be sure to abstain from drinking, on the third day he should go to the bath, gradually recruit himself with food. Whoever easily gets to sleep is quickly restored; the trouble is brought back by indigestion and also by fatigue and cold. If, after the suppression of the cholera, slight fever persists, there is need for a clyster, and then to take food and wine.

19 (XII). Sed hic quidem morbus et acutus est, et inter intestina stomachumque versatur sic, ut cuius potissimum partis sit, non facile dici possit. In ipsius vero ventriculi porta consistit is, qui[1] . . . et longus esse consuevit: coeliacus a Graecis nominatur. Sub hoc venter indurescit, dolorque eius est; alvus nihil reddit, ac ne spiritum quidem transmittit; extremae partes frigescunt; difficulter spiritus redditur.—Commodissimum est inter initia calida[2] . . . et cataplasmata toto ventri imponere, ut dolorem leniant, post cibum vomere, atque ita ventrem exinanire; proxumis deinde diebus cucurbitulas sine ferro ventri et coxis admovere; ventrem ipsum liquare, dato lacte et vino salso frigido; si tempus anni patitur, etiam viridibus ficis, sic tamen, ne quis aut cibus aut umor universus detur sed paulatim. Ergo per intervalla temporis sat est cyathos binos ternosve sumere, et cibum pro portione huius; commodeque facit cyatho lactis cyathus aquae mixtus et sic datus; cibique inflantes et acres utiliores sunt, adeo ut lacti quoque recte contritum alium adiciatur. Procedente vero tempore opus est gestari, maximeque navigare; perfricari ter aut quater die, sic ut nitrum oleo adiciatur; perfundi aqua calida post cibum; deinde sinapi imponere per omnia membra, excepto capite, donec adrodatur et rubeat maximeque si corpus durum et virile est. Paulatim deinde faciendus est transitus ad ea, quae ventrem conprimunt. Assa caro danda valens, et quae non facile corrumpatur: potui vero pluvialis aqua decocta, sed

[1] *M. inserts* et acutus est *after* qui (*cf.* 15. 1), *and this is translated.*

[2] *Marx inserts* fomenta *after* calida, *and this is translated.*

19. Now the disorder just described is both acute and has its seat between the intestines and stomach, so that it is not easy to say to which part it most belongs. That which the Greeks term coeliacus[a] has its seat at the gateway of the stomach and is usually both acute and chronic. Under this affection the belly becomes hard and painful; the bowels void nothing, not even wind; the extremities become cold; the breath is passed with difficulty. To begin with it is best to apply hot foments and plasters all over the belly to relieve pain, after food to induce a vomit and thus to empty the belly; next on the following days to apply dry cups to the abdomen and hips; to loosen the bowels, by giving milk and cold salted wine; also if in season green figs, provided that neither drink nor food is given all at once but a little at a time. It is enough, therefore, to take two or three cupfuls at intervals, and food in the same proportion; a cup of milk, mixed with one of water, and so administered, is suitable; flatulent and pungent foods are more useful, hence it is well to add pounded garlic to the milk. And as time goes on there is need for: rocking, especially a sea-voyage; rubbing three or four times a day, soda being added to the oil; hot-water affusions after food; then mustard should be put upon all the extremities, omitting the head, until there is irritation and redness, especially if the body is robust and virile. Gradual transition should next be made to remedies which confine the bowels. Roast meat, such as is nutritious and does not readily decompose, is to be given; and for drink,

---

[a] Celsus appears to have in mind the effects of pyloric spasm and intestinal atony, cf. Aretaeus, *Chronic Diseases*, Book II. 7, περὶ κοιλιακῆς διαθέσιος.

quae per binos ternosve cyathos bibatur. Si vetus vitium est, oportet laser quam optimum ad piperis magnitudinem devorare, altero quoque die vinum vel aquam bibere, interdum interposito cibo singulos vini cyathos sorbere: ex inferiore parte infundere pluviatilem egelidam aquam, maximeque si dolor in imis partibus remanet.

20 (XIII). Intra[1] ipsa vero intestina consistunt duo morbi, quorum alter in tenuiore, alter in pleniore est. Prior acutus est, insequens esse longus potest. Diocles Carystius tenuioris intestini morbum χορδαψόν, plenioris εἰλεόν nominavit: a plerisque video nunc illum priorem εἰλεόν, hunc κολικόν nominari. Sed prior modo supra umbilicum, modo sub umbilico dolorem movet. Fit alterutro loco inflammatio; nec alvus nec spiritus infra transmittitur. Si superior pars adfecta est, cibus, si inferior, stercus per os redditur, si utrumlibet, vetus est.
2 Adicit periculo vomitus biliosus, mali odoris, aut varius aut niger.—Remedium est sanguinem mittere aut cucurbitulas pluribus locis admovere, non ubique cute incisa: id enim duobus aut tribus locis satis est: ex ceteris spiritum evocare abunde est. Tum animadvertere oportet quo loco malum sit: solet enim contra id tumere. Et si supra umbilicum est, alvi ductio utilis non est: si infra est, alvum ducere, ut Erasistrato placuit, optimum est, et saepe id auxilii satis est. Ducitur autem percolato tisanae cremore cum oleo et melle,
3 sic ut praeterea nihil adiciatur. Si nihil tumet, duas

[1] *So v.d. Linden, for the MSS.* inter.

---

[a] Aretaeus, *Chronic Diseases*, Book II. 8, περὶ κωλικῶν, included both eileos and colicos.   [b] IV. 12. 2.

boiled rainwater, of which two or three cupfuls should be drunk at a time. If the disorder is of longer standing the proper thing is to swallow a bit of the best laser the size of a peppercorn, to drink wine and water on alternate days, between meals at times to sip a cupful of wine; to administer a clyster of tepid rain-water, especially if pain persists in the lower bowel.

20. In the intestines proper two diseases have their seat, one in the small, the other in the large. The former is acute, the latter may become chronic. Diocles of Carystus named the disease of the small intestines chordapsos, of the large eileos. I note that by many the former is now termed eileos, the latter colicos.[a] The former excites pain, at times above, at times below the navel. At one or other of these places there is inflammation; neither motion nor wind is passed downwards. If the upper part is affected, food, if the lower, faeces is returned by the mouth, if either happens the disease is chronic. Additional signs of danger are if the vomit is bilious, malodorous, either varying in colour or black. The remedy is blood-letting or cupping in several places, the skin not being incised in all; for it is sufficient to do so in two or three places; in the others it is quite enough to extract wind.[b] Next attention should be turned to the seat of disease: for there is commonly a swelling over it. And if this is situated above the navel, there is no use in the clyster; if below, to clyster the bowels as Erasistratus advised is the best remedy and often that is all the treatment required. Now the clyster should consist of strained pearl barley gruel, together with oil and honey, nothing else being added. If there is no

manus imponere oportet supra summum ventrem, paulatimque deducere: invenietur enim mali locus, qui necesse est renitatur; et ex eo deliberari poterit, ducenda necne alvus sit. Illa communia sunt; calida cataplasmata admovere, eaque imponere a mammis usque ad inguina et spinam ac saepe mutare; brachia cruraque perfricare, demittere totum hominem in calido oleo: si dolor non quiescit, etiam in alvum ex parte inferiore tres aut quattuor cyathos calidi olei dare. Ubi per haec consecuti sumus, ut iam ex inferiore parte spiritus transmittatur, offerre potui mulsum tepidum non multum: nam ante magna cura vitandum est, ne quid bibat. Si id commode 4 cessit, adicere sorbitionem. Ubi dolor et febricula quierunt, tum demum uti cibo pleniore, sed neque inflante neque duro neque valido, ne intestina adhuc imbecilla laedantur: potui vero nihil . . .[1] praeterquam puram aquam. Nam sive quid vinulentum sive acidum est, id huic morbo alienum est. Ac postea quoque vitare oportet balneum, ambulationem, gestationem, ceterosque corporis motus: nam facile id malum redire consuevit, et, sive cum frigus subit sive aliqua iactatio, nisi bene iam confirmatis intestinis revertitur.

21 (XIV). Is autem morbus, qui in intestino pleniore est, in ea maxime parte est, quam caecam[2] esse proposui (IV. **1**, 8). Vehemens fit inflatio, vehementes dolores, dextra magis parte; intestinum, quod verti videtur[3] . . . prope spiritum elidit. In plerisque post frigora cruditatesque oritur, deinde

---

[1] *Marx inserts* praestare *after* nihil, *which is translated.*
[2] *So Caesar for the* quam certam *of the MSS.* qua caecum *v. d. Linden.*
[3] *Marx notes a lacuna here.*

swelling, the two hands should be placed upon the upper part of the belly, and little by little drawn downwards; for the seat of the trouble may be thus discovered, owing to its being necessarily resistent; and from this one can form an opinion whether the bowels should be clystered or not. The treatments common to both forms are: the application of hot plasters, put on from the breasts to the groins, and back to the spine, and often changed; rubbing of the arms and legs; immersing the patient all over in hot oil. If the pain is not relieved, there is injected into the bowels from below three or four cupfuls of hot oil. When we have brought it about by these measures that wind is now passed down and out, tepid honeyed wine, not much, is given to drink; for before that every care should be taken that nothing at all is drunk. If the honeyed wine is kept down, then give gruel. When pain and feverishness have subsided, then at length a fuller diet is adopted, but nothing flatulent nor solid nor rich, lest the intestines, whilst still weak, take harm; but for drink nothing is better than plain water, for in this disease vinous and acid drinks are objectionable. Subsequently the patient should avoid the bath, walking, rocking and other bodily movements; for this disorder is very liable to recur, and, unless the intestines have already returned to a sound state, either cold or shaking of any kind may cause a return of the trouble.

21. The disease which is in the larger intestine is situated chiefly in that part which I have described as a blind alley (IV. **1**, 8). There is extreme flatulence, violent pains especially on the right side; the intestine which appears to undergo torsion, . . . almost forces out wind. In most cases it comes on after chills and

quiescit, et per aetatem saepe repetens sic cruciat, ut vitae spatio nihil demat.—Ubi is dolor coepit, admovere sicca et calida fomenta oportet, sed primo lenia,[1] deinde validiora, simulque frictione[2] ad extremas partes [id est crura brachiaque] materiam evocare:[3] si discussus[4] non est, qua dolet, cucurbitulas sine ferro defigere. Est etiam medicamentum eius rei causa comparatum, quod colicon nominatur: id se repperisse Cassius gloriabatur. Magis prodest potui datum, sed impositum quoque extrinsecus digerendo spiritum dolorem levat. Nisi finito vero tormento recte neque cibus neque potio adsumitur. Quo victu sit utendum, qui hoc genere temptantur, iam mihi dictum est (**1**, 7). [Confectio medicamenti, quod colicon nominatur: ex his constat: costi, anesi, castorei, singulorum P⁰ ✻ III, petroselini P⁰ den̄ III, piperis longi et rotundi, singulorum P⁰ ✻ II, papaveris lacrimae, iunci rotundi, myrrae, nardi, singulorum P⁰ ✻ VI; quae melle excipiuntur. Id autem et devorari potest et ex aqua calida sumi.][5]

22 (XV). Proxima his inter intestinorum mala tormina esse consueverunt: dysenteria Graece vocatur. Intus intestina exulcerantur; ex his cruor manat isque modo cum stercore aliquo semper liquido, modo cum quibusdam quasi muccosis excernitur, interdum simul quaedam carnosa descendunt; frequens deiciendi cupiditas dolorque in ano est.

[1] *So v. d. Linden for the MSS.* lenta.
[2] *So Marx for the MSS.* frictiones.
[3] *So v. d. Linden for the* revocare *of the MSS.* which Marx follows.
[4] *So v. d. Linden for the* discussum *of the MSS.*
[5] *Targa deleted this prescription as an interpolation from Book V 25. 12.*

[a] Cp. p. 37.

fits of indigestion, then subsides, and in course of time often recurs so as to be a cause of suffering but without shortening the length of life. At the commencement of the pain, dry, hot foments should be applied, at first mild, then stronger ones, at the same time rubbing is used to draw off the matter to the extremities, [into the legs and arms]; if the pain be not so dispersed, dry cups should be applied. There is even a medicament compounded for this very purpose called colicos: Cassius[a] used to boast that he had invented it. It is of more benefit when given as a draught, but when applied externally too it relieves pain by dispersing the wind. Until griping has quite ceased it is not right to take properly either food or drink. I have already stated what kind of food should be used in this kind of disorder (1, 7). [The composition of the medicament termed colicos is as follows: costmary, anise, castor, of each 12 grms., of parsley 12 grms., of long and round peppers, a.a. 8 grms., of poppy tears, round rush, myrrh, nard, a.a. 24 grms., all mixed together with honey. This may be either swallowed by itself or taken in hot water.]

22. The most akin to the above among intestinal maladies are gripings,[b] called by the Greeks dysenteria. The insides of the intestines ulcerate; from these blood trickles and at times is excreted with some faeces which are always liquid, at times with a sort of mucus, sometimes at the same time something fleshlike comes down; there is frequent desire to stool and pain in the anus. Along with this

---

[b] Celsus uses *tormina* to translate the Greek στρόφοι; Hippocrates also uses the term δυσεντερία for this. All writers on ancient medicine agreed in ascribing the disease chiefly to unsuitable food, cf. Aretaeus, *Chronic Diseases*, Book II. 9.

CELSUS

Cum eodem dolore exiguum aliquid emittitur atque eo quoque tormentum intenditur; idque post tempus aliquod levatur exiguaque requies est; somnus interpellatur; febricula oritur; longoque tempore id malum cum inveteraverit, aut tollit hominem, 2 aut, etiamsi finitur, excruciat.—Oportet inprimis conquiescere, siquidem omnis agitatio exulcerat; deinde ieiunum sorbere vini cyathum, cui contrita radix quinquefolii sit adiecta; imponere cataplasmata super ventrem, quae reprimunt, quod in superioribus ventris morbis non expedit; quotiensque desidit, subluere aqua calida, in qua decoctae verbenae sint; portulacam vel coctam vel ex dura muria esse; 3 cibos potionesque eas, quae adstringunt alvum . . .[1] Si vetustior morbus est, ex inferioribus partibus tepidum infundere vel tisanae cremorem vel lac vel adipem liquatam vel medullam cervinam vel oleum vel cum rosa butyrum vel cum eadem album crudum ex ovis vel aquam, in qua lini semen decoctum sit, vel si somnus non accedit, vitellos cum aqua, in qua rosae floris folia cocta sint: levant enim dolorem haec et mitiora ulcera efficiunt, maximeque utilia sunt, si 4 cibi quoque secutum fastidium est. Themison muria dura quam asperrima hic[2] utendum memoriae prodidit. Cibi vero esse debent, qui leniter ventrem adstringant. At ea, quae urinam movent, si ea consecuta sunt, in aliam partem umorem avertendo prosunt: si non sunt consecuta, noxam augent; itaque nisi in quibus prompte id facere consuerunt, non sunt adhibenda. Potui, si febricula est, aqua pura calida vel ea, quae ipsa quoque adstringat, dari

[1] *Marx supplies* adsumere, *but notes that the clause* cibos . . . alvum *is suspected by Targa to be a gloss.*
[2] hic *so edd. for MSS.* sic.

BOOK IV. 22. 1–4

pain a scanty motion is discharged, and by this too the griping pain is intensified: and after a while there is some relief and a short interval of ease; sleep is broken, feverishness comes; when the disorder has continued for a long while, it either carries off the patient, or even, although it come to an end, puts him to torture. Rest must be adopted from the first, since any shaking sets up ulceration; next on an empty stomach he is to sip a cupful of wine to which has been added powdered cinquefoil root; then repressant plasters are put upon the abdomen, which in the case of disorders of the upper abdomen is not expedient; whenever the patient goes to stool, he should bathe the anus with hot water in which vervains have been boiled; purslane should be eaten, whether cooked or pickled in strong brine; also such foods and drink as are astringent to the bowel. If the distemper is of longer standing, there should be injected into the rectum either a tepid cream of pearl barley, or milk, or melted fat, or deer marrow, or olive oil, or rose oil with butter or with raw white of egg, or a decoction of linseed, or if sleep does not occur, yolk of eggs in a decoction of rose-leaves: for such remedies relieve pain and mitigate ulceration, and are of special utility if loss of appetite has ensued. Themison has stated in writing that the strongest brine should be used in these cases. Food too should be of the kind which will act as mild astringents. But diuretics if they take effect are beneficial by directing humour to another part: if they do not take effect, they increase the trouble; so unless for those on whom they act promptly, they should not be used. If there is feverishness, the drink should be hot water, either plain or with some astrin-

debet: si non est, vinum leve, austerum. Si pluribus diebus nihil remedia alia iuverunt vetusque iam vitium est, aquae bene frigidae potio adsumpta ulcera adstringit et initium secundae valetudinis facit.
5 Sed ubi venter suppressus est, protinus ad calidam potionem revertendum est. Solet autem interdum etiam putris sanies pessimique odoris descendere, solet purus sanguis profluere. Si superius vitium est, alvus aqua mulsa duci debet, tunc deinde eadem infundi, quae supra (§ 3) comprehensa sunt. [Valensque est etiam adversus cancerem intestinorum minii gleba cum salis hemina contrita, si mixta his aqua in alvum datur.][1] At si sanguis profluit, cibi potionesque esse debent, quae adstringant.

23 (XVI). Ex torminibus interdum intestinorum levitas oritur, quae . . .[2] continere nihil possunt, et, quicquid adsumptum est, imperfectum protinus reddunt. Id interdum aegros trahit, interdum praecipitat.—In hoc utique adhibere oportet comprimentia, quo facilius tenendi aliquid intestinis vis sit. Ergo et super pectus ponetur sinapi, exulcerataque cute malagma, quod umorem evocet; et ex verbenis decocta in[3] aqua desidat; et cibos potionesque adsumat, quae alvum adstringunt; et frigidis
2 utetur perfusionibus. Oportet tamen prospicere, ne simul his omnibus admotis vitium contrarium per

[1] *Deleted by Targa.*
[2] *The text as it stands is corrupt. Marx supplies the words* leienteria a Graecis nominatur. Intestina: *v. d. Linden reads* qua *for* quae.
[3] decocta in *Constantine followed by Marx for the MSS.* decoctis.

---

[a] II. 30, 1–8.
[b] For the term leienteria cf. p. 90, note *b*.

gent[a] in it; if none, then light dry wine. If for several days other remedies have done no good, and the disease is now of long standing, drinking of very cold water acts as an astringent upon the ulcerations and starts recovery. But as soon as the movement of the bowels is under control, there should forthwith be a return to warm drinks. Sometimes also there is discharged a putrid sanies having a foul odour, sometimes unmixed blood escapes. If the former occurs, a hydromel clyster should be given, and then the other things mentioned above injected. [An effective remedy even for intestinal canker is a lump of minium rubbed up with 250 grms. of salt, dissolved in water, and administered as a clyster.] But if there is a flux of blood, food and drink should be astringent.

23. From dysentery there proceeds sometimes leienteria,[b] when the intestines cannot retain anything, and whatever is swallowed is straightway excreted imperfectly digested.[c] Sometimes in the patients this drags on, sometimes it hurries them off. In this affection especially astringents are to be adopted, to give the intestines strength to retain better. For this purpose mustard should be put on the chest, and when the skin becomes ulcerated, then an emollient to draw out humour; and the patient should sit in a decoction of vervains; take both food and drink which control the bowel: and have cold water poured over him. Nevertheless, care should be taken lest with all these remedies there be an opposite trouble

[c] Hippocrates, *Prorrh.* II. 23, Littré, IX. 53. The food passed in a liquid state following cicatrization of ulcerations set up by dysentery, or from weakness of intestines without previous ulceration, cf. Aretaeus, *Chronic Diseases*, Book II.10.

immodicas inflationes oriatur. Paulatim ergo firmari intestina debebunt aliquibus cotidie adiectis. Et cum in omni fluore ventris, tum in hoc praecipue necessarium est, non quotiens libet desidere, sed quotiens necesse est, ut haec ipsa mora in consuetudinem ferendi oneris intestina deducat. Alterum quoque, quod aeque ad omnes similes adfectus pertinet, in hoc maxime servandum est, ut, cum pleraque utilia insuavia sint, qualis est plantago et rubi et quicquid malicorio mixtum est, ea potissimum ex 3 his dentur, quae maxime aeger volet. Deinde, si omnia ista fastidit, ad excitandam cibi cupiditatem interponatur aliquid minus utile, sed magis gratum. Exercitationes, frictiones huic quoque morbo necessariae sunt, et cum his sol, ignis, balneum; vomitus, ut Hippocrati visum est, etiam albo veratro, si cetera parum proficient, evocatus.

24 (XVII). Nonnumquam autem lumbrici quoque occupant alvum, hique modo ex inferioribus partibus, modo foedius ore redduntur; atque interdum latos eos, qui peiores sunt, interdum teretes videmus.—Si lati sunt, aqua potui dari debet, in qua lupinum aut cortex mori decoctus sit, aut cui adiectum sit contritum vel hysopum vel piperis acetabulum vel scamoniae paulum. Vel etiam pridie, cum multum alium ederit, vomat, posteroque die mali Punici tenues radiculas colligat, quantum manu comprehendet; eas contusas in aquae tribus sextariis decoquat, donec tertia pars

---

a Diet in Acute Diseases, Littré, II. 475.
b Taenia solium.
c Ascaris lumbricordes.
d 63 c.cm.

set up by excessive flatulence. Consequently, little by little, the intestines should be strengthend by some additions daily. As in the case of any abdominal flux, so in this, it is particularly necessary that the patient should go to stool, not as often as inclined, but as often as compelled, so that by such delay the intestines may be got into the habit of holding up their contents. There is another thing which, whilst applicable equally to all similar affections, is to be specially observed in this, that as many beneficial medicaments are disagreeable to the taste, such as the mixture containing plantain and blackberries and any mixture containing pomegranate bark, that shall be chosen which the patient likes most. Moreover, if he loathes all of them, something to excite his appetite should be interposed, less useful, perhaps, but more pleasant. Exercise and rubbing are needed in this disease also, as well as heat, whether of the sun, or a fire, and baths; and according to Hippocrates,[a] a vomit even by white hellebore, when other measures prove of little avail.

24. Again, worms also occasionally take possession of the bowel, and these are discharged at one time from the lower bowel, at another more nastily from the mouth: and we observe them sometimes to be flattened,[b] which are the worse, at times to be rounded.[c] For the flat worms there should be given as draughts, a decoction of lupins, or of mulberry bark, to which may be added, after pounding, either hyssop or a vinegar cupful[d] of pepper, or a little scammony. Alternatively on one day let him eat a quantity of garlic and vomit, then on the next day take a handful of fine pomegranate roots, crush them and boil them in a litre and half of water down to one-

CELSUS

supersit; huc adiciat nitri paulum, et ieiunus bibat.
2 Interpositis deinde tribus horis duas potiones sumat; at aquae[1] . . . vel muriae durae sit adiecta; tum desidat subiecta calida aqua in pelve. Si vero teretes sunt, qui pueros maxime exercent, et eadem dari possunt et quaedam leviora, ut contritum semen urticae aut brassicae aut cumini cum aqua, vel menta cum eadem vel absinthium decoctum vel hysopum ex aqua mulsa vel nasturcii semen cum aceto contritum. Edisse etiam et lupinum et alium prodest, vel in alvum oleum subter dedisse.

25 (XVIII). Est autem aliud levius omnibus proximis, de quibus supra dictum est, quod tenesmon Graeci vocant. Id neque acutis neque longis morbis adnumerari debet, cum et facile tollatur neque umquam per se iugulet. In hoc aeque atque in torminibus frequens desidendi cupiditas est, aeque dolor, ubi aliquid excernitur. Descendunt autem pituitae muccisque similia, interdum etiam leviter subcruenta: sed his interponuntur nonnumquam
2 ex cibo quoque recte coacta.—Desidere oportet in aqua calida saepiusque ipsum anum nutrire. Cui plura medicamenta idonea sunt: butyrum cum rosa; acacia ex aceto liquata; emplastrum id, quod τετραφάρμακον Graeci vocant, rosa liquatum; alumen lana circumdatum et ita adpositum, eademque ex inferiore parte indita, quae torminum auxilia sunt; eaedem[2] verbenae decoctae, ut inferiores partes foveantur. Alternis vero diebus aqua, alternis leve et austerum vinum bibendum est. Potio esse

---

[1] *Marx conjectures that some words have fallen out after* aquae, *such as* marinae hemina = *half a pint of sea water, and this conjecture is translated.*

[2] eaedem *Constantine for MSS.* eadem. *Marx retains*

third, to this add a little soda, and drink it on an empty stomach. At three hours' interval, let him take two further draughts; but with the addition of half a pint of sea water or strong brine; then on going to stool, sit over a basin of hot water. Again, for the round [a] worms which especially trouble children, both the same remedies may be given and some milder ones, such as pounded-up seeds of nettles or of cabbage or of cummin in water, or mint in the same or a decoction of wormwood or hyssop in hydromel or cress seeds pounded up in vinegar. It is also of service either to eat lupin or garlic, or administer into the lower bowel a clyster of olive oil.

25. There is, again, another affection which the Greeks call tenesmos, slighter than all those last spoken of. It should be counted neither with acute nor with chronic diseases, since it is readily relieved, and never by itself fatal. As in the case of dysentery, there is equally the frequent desire for stool, and equally the pain when anything is passed. There is a discharge resembling phlegm and mucus; sometimes it is even slightly bloodstained; but mingled with properly formed faeces derived from food. The patient should sit in hot water, and make application frequently to his anus. For this there are several suitable medicaments; butter in rose oil, gum acacia dissolved in vinegar; that wax-salve which the Greeks call tetrapharmacon, made liquid with rose oil; alum wrapped up in wool and so applied; the same clysters as are beneficial in dysentery; the same decoction of vervains to foment the lower parts. He should drink on alternate days water and a thin dry wine

[a] Oxyuris vermicularis, thread worm.

---

eadem, *and thinks that some words have fallen out* (adhibenda aqua calida in qua sunt).

debet egelida et frigidae propior; ratio victus talis, qualem in torminibus supra (22) praecepimus.

26 (XIX). Levior etiam, dum recens, deiectio est, ubi et liquida alvus et saepius quam ex consuetudine fertur; atque interdum tolerabilis dolor est, interdum gravissimus, idque peius est. Sed uno die fluere alvum saepe pro valetudine est, atque etiam pluribus, dum febris absit et intra septimum diem id conquiescat. Purgatur enim corpus, et quod intus laesurum erat, utiliter effunditur. Verum spatium periculosum est: interdum enim tormina ac 2 febriculas excitat viresque consumit.—Primo die quiescere satis est, neque impetum ventris prohibere. Si per se desiit, balneo uti, paulum cibi capere; si mansit, abstinere non solum a cibo sed etiam a potione. Postero die si nihilo minus liquida alvus est, aeque conquiescere, paulum astringentis cibi sumere. Tertio die in balneum ire; vehementer omnia praeter ventrem perfricare, ad ignem lumbos scapulasque admovere; cibis uti, sed ventrem contra-3 hentibus, vino non multo meraco. Si postero quoque die fluet, plus edisse, sed vomere et, ex toto donec conquiescat, contra siti, fame, vomitu niti: vix enim fieri potest, ut post hanc animadversionem alvus non contrahatur. Alia via est, ubi velis subprimere, cenare, deinde vomere; postero die in lecto conquiescere, vespere ungi, sed leniter; deinde panis circa selibram ex vino Aminaeo mero sumere; tum assum aliquid, maximeque avem, et postea vinum

---

[a] Hippocrates always uses the words διάρροια κοιλίης or ῥύσις κοιλίης, Celsus *alvi deiectio, profluvio*.

lukewarm or better cold. The diet should be the same as prescribed above (22) for dysentery.

26. Even slighter, while recent, is diarrhoea,[a] in which the stool is liquid and more frequent than ordinary; and sometimes the pain is bearable, at times very severe, when it is a worse affair. But a flux from the bowel for one day is often salutary, and even for several days, provided that fever is absent and it subsides within seven days. For the body is purged, and whatever is about to cause a complaint inside is evacuated with advantage. But persistence is the danger; for it excites at times dysentery and feverishness and exhausts strength. It is sufficient on the first day to rest, and not to check the movement of the bowels. If it stops of itself, the patient should make use of the bath, and take a little food; if it persists, he should abstain, not only from food, but even from drink. If on the day following, in spite of all, the stool is still liquid, he should rest as before and take a little astringent food. On the third day he should go to the bath; be rubbed all over vigorously except the abdomen, sit with his loins and shoulder-blades before a fire; take food of an astringent kind, and a little undiluted wine. If on the fourth day the flux persists, he should eat more but provoke a vomit afterwards, and counter in a general way the diarrhoea by thirst, hunger and vomiting, until it subsides; for it is scarcely possible that after so attending to it, the bowel will not be controlled. Another method to suppress the diarrhoea is to dine and then vomit; the next day to rest in bed, in the evening to be anointed, but lightly, then to eat about half a pound of bread soaked in undiluted Aminaean wine; after that something roasted, poultry in par-

idem bibere aqua pluviali mixtum, idque usque
4 quintum diem facere, iterumque vomere. Frigidam autem adsidue potionem esse debere contra priores auctores Asclepiades affirmavit, et quidem quam frigidissimam. Ego experimentis quemque in se credere debere existimo, calida potius an frigida utatur. Interdum autem evenit, ut id pluribus diebus neglectum curari difficilius possit. A vomitu oportet incipere; deinde postero die vespere tepido loco ungi; cibum modicum adsumere, vinum meracum quam asperrimum; impositam super ventrem
5 habere cum cerato rutam. In hoc autem affectu corporis neque ambulatione neque frictione opus est: vehiculo sedisse vel magis etiam equo prodest: neque enim ulla res magis intestina confirmat. Si vero etiam medicamentis utendum, aptissimum est id, quod ex pomis fit. Vindemiae tempore in grande vas coicienda sunt pira atque mala silvestria: si ea non sunt, pira Tarentina viridia vel Signina, mala
6 Scaudiana vel Amerina, myrapia; hisque adicienda sunt Cotonea, et cum ipsis corticibus suis Punica, sorba, et, quibus magis utimur, et torminalia, sic ut haec tertiam ollae partem teneant; tum deinde ea musto implenda est, coquendumque id, donec omnia, quae indita sunt, liquata in unitatem quandam coeant. Id gustu non insuave est, et, quandocumque opus est, adsumptum, leniter sine ulla stomachi noxa ventrem tenet. Duo aut tria coclearia

---

[a] There is no section throughout the whole work which exhibits more plainly the personal experience of a practitioner. See Introduction, p. xi.

ticular, and lastly to drink the same wine mixed with rain-water; and to do so until the fifth day, then vomit again. Now Asclepiades, against the opinion of previous writers, affirmed that the drink should be kept constantly cold, indeed as cold as possible. I[a] myself hold that each should trust in his own experiences, whether hot rather than cold drink should be made use of. It sometimes happens also that this disorder, having been neglected for several days, is more difficult to relieve. Such a patient should commence with an emetic; then the following day at evening be anointed in a warm room; take food in moderation, and the sourest wine undiluted; a wax-salve with rue should be applied to the abdomen. In this affection neither walking nor rubbing is of benefit; sitting in a carriage and even more riding on horseback is advantageous; for nothing strengthens the intestines more. But if use is to be made of medicaments as well, the most suitable is that made from orchard fruit. At the time of the vintage, pears and crab apples are thrown into a large vessel; and if the latter are not to be had, green Tarentine or Signine pears, Scaudian, or Amerian apples, sweet-scented.[b] To these are added quinces and pomegranates with their rind, service fruit, and those that are called torminalia, which we use by preference, so that these occupy one-third of the jar; then this is next to be filled up with must, and boiled until all the ingredients have become resolved into a uniform mass. It is not unpleasant to the taste, and taken as needed, it controls the bowel gently, without any harm to the stomach. It is enough to take in one day two or three

[b] See Pliny, *N.H.* XXV. 15e, clove-scented; the Italian *pera garofane*.

uno die sumpsisse satis est. Alterum valentius genus est: murtae bacas legere, ex his vinum exprimere, id decoquere, ut decima pars remaneat, eiusque
7 cyathum sorbere. Tertium, quod quandocumque fieri potest: malum Punicum excavare, exemptisque omnibus seminibus, membranas, quae inter ea fuerunt, iterum[1] . . . coicere; tum infundere cruda ova, rudiculaque miscere; dein malum ipsum super prunam imponere, quod, dum umor intus est, non aduritur: ubi siccum esse coepit, removere oportet, extractumque cocleari quod intus est esse. Aliquibus . . .[2] adiectis maius momentum habet; itaque etiam in piperatum coicitur misceturque cum sale et
8 pipere.[3] Est quid ex his edendum est. Pulticula etiam, cum qua paulum ex favo vetere coctum[4] sit, et lenticula cum malicorio cocta, rubique cacumina in aqua decocta, et ex oleo atque aceto adsumpta, efficacia sunt, atque ea aqua, in qua vel palmulae vel malum Cotoneum vel arida sorba vel rubi decocti sunt, potata. Quod genus significo, quotiens
9 potionem dandam esse dico, quae astringat. Tritici quoque hemina in vino Aminaeo austero decoquitur, idque triticum ieiuno ac sitienti datur, superque id vinum id sorbetur; quod iure valentissimis medicamentis adnumerari potest. Atque etiam potui datur vinum Signinum vel resinatum austerum vel quodlibet austerum. Contunditurque cum corticibus seminibusque suis Punicum malum vinoque tali miscetur;

---

[1] *After* iterum, *Marx supplies* in cavum, *which is translated.*
[2] *Marx supplies* acribus, *which is translated.*
[3] *Several emendations are suggested for this corrupt passage.* Marx et sic quidem ex his edendum est; *Targa (followed by Daremberg)* atque ex his edendum; *v.d. Linden,* esturque ex his. Edenda. . . . *The general sense is clear (see translation).*

spoonfuls. Another composition is stronger: myrtle berries are gathered, and wine expressed from them is boiled down to one-tenth, of which a cup[a] is sipped. A third can be prepared at any time by scooping out the inside of a pomegranate, removing all the seeds, and returning the pulp into the cavity, then raw eggs are pounded in, and stirred round with a small rod; next the fruit itself is heated over charcoal, for it does not burn so long as the inside is liquid; when the inside begins to dry the pomegranate is taken off the brazier and with a spoon the inside is scooped out and eaten. By certain acrid additions this remedy can be made more active; thus also it may be stirred up in peppered wine and mixed with salt and pepper, and so eaten. Pease porridge, with which a little of an old honeycomb has been boiled, also lentil porridge boiled with pomegranate rind, also a decoction of bramble tops eaten with oil and vinegar, are efficacious, as also draughts of a decoction of dates or quinces or dried service fruit or brambles. Such are the kind I refer to whenever I say an astringent draught should be administered. Also a half-pint of wheat is boiled in dry Aminaean wine, and first the wheat is eaten on a stomach empty both of food and drink, afterwards the wine itself is drunk and can be justly counted amongst the most active remedies. Also there can be given to drink Signian wine, or dry and resinated wine, or any other dry wine. And a pomegranate may be pounded up along with its rind and seeds, and mixed with wine of the above sort; the

[a] 42 c.cm.

[4] *So some MSS. Marx with others reads* cocti.

idque vel merum sorbet aliquis vel bibit mixtum. Sed medicamentis uti nisi in vehementibus malis supervacuum est.

27 (XX). Ex vulva quoque feminis vehemens malum nascitur proximeque ab stomacho vel adficitur haec vel corpus adficit. Interdum etiam sic exanimat, ut tamquam comitiali morbo prosternat. Distat tamen hic casus eo, quod neque oculi vertuntur nec spumae profluunt nec nervi distenduntur: sopor tantum est. Idque quibusdam feminis crebro revertens perpetuum est.—Ubi incidit, si satis virium est, sanguis missus adiuvat; si parum est, cucurbitulae tamen defigendae sunt in inguinibus,
B Si diutius aut iacet aut alioqui iacere consuevit, admovere oportet naribus extinctum ex lucerna linamentum, vel aliud ex iis, quae foedioris odoris esse rettuli (III. 20, 1), quod mulierem excitet. Idemque aquae quoque frigidae perfusio efficit. Adiuvatque ruta contrita cum melle, vel ex cyprino ceratum, vel quodlibet calidum et umidum cataplasma naturalibus pube tenus impositum. Inter haec etiam perfricare coxas et poplites oportet. Deinde ubi ad se redit, circumcidendum vinum est in totum annum, etiamsi casus idem non revertitur.
C Frictione cottidie utendum totius quidem corporis, praecipue vero ventris et poplitum. Cibus ex media materia dandus; sinapi super imum ventrem tertio quoque aut quarto die imponendum, donec corpus rubeat. Si durities manet, mollire commode videtur

---

[a] For a description of hysteria, cf. Hipp. IV. 166 (*Aph.* V. 35), and Aretaeus, Book II. 10. θεραπεία ὑστερικῆς πνιγός, cure of hysterical suffocation.

[b] κατοχή catalepsy.

patient either sips it undiluted, or drinks it mixed with water. But it is superfluous except in bad cases to make use of medicaments.

27. From the womb of a woman, also, there arises a violent malady;[a] and next to the stomach this organ is affected the most by the body, and has the most influence upon it. At times it makes the woman so insensible that it prostrates her as if by epilepsy. The case, however, differs from epilepsy, in that the eyes are not turned nor is there foaming at the mouth nor spasm of sinews; there is merely stupor.[b] In some women this attack recurs at frequent intervals and lasts throughout life. When this happens, if there is sufficient strength, bloodletting is beneficial; if too little, yet cups should be applied to the groins. If she lies prostrate for a long while, or if she has done so at other times, hold to her nostrils an extinguished lamp wick, or some other of these materials which I have referred to as having a specially foetid odour (III. 20, 1), to arouse the woman. For the same end, affusion with cold water is also effectual. And there is benefit from rue pounded up in honey, or from a wax-salve made up with cyprus oil or from hot moist plasters of some sort applied to the external genitals as far as the pubes. At the same time also the hips and the backs of the knees should be rubbed. Then when she has come to herself, she should be cut off from wine for a whole year, even if a similar attack does not recur. Friction should be applied daily to the whole body, but particularly to the abdomen and behind the knees. Food of the middle class should be given: every third or fourth day mustard is to be applied over the hypogastrium until the skin is reddened. If induration

CELSUS

solanum in lac demissum, deinde contritum, et cera alba atque medulla cervina cum irino, aut sebum taurinum vel caprinum cum rosa mixtum. Dandum etiam potui vel castoreum est vel git vel anetum.

D Si parum pura est, purgatur iunco quadrato. Si vero vulva exulcerata est, ceratum ex rosa fiet, ei recens suilla adeps et ex ovis album misceatur, idque adponatur; vel album ex ovo cum rosa mixtum, adiecto, quo facilius consistat, contritae rosae pulvere. Dolens vero ea sulpure suffumigari debet. At si purgatio nimia mulieri nocet, remedio sunt cucurbitulae cute incisa inguinibus vel etiam sub mammis admotae. Si maligna purgatio est, subicienda sunt medicamenta[1] quae evocent sanguinem: costum, puleium, violae albae, apium, nepita et satureia et hysopum. In cibum quoque accipiat quae apta sunt: porrum, rutam, cyminum, caepas, sinapi, vel omne acrum olus. Si vero sanguis, qui ex inferiore parte erumpere solet, is ex naribus eruperit, incisis inguinibus adponenda est cucurbita idque per tres vel quattuor menses tricesimo quoque die repetieris: tunc scias hoc vitium sanasse. Si vero non se sanguis ostenderit, scias ei dolores capitis surgere. Tunc ex brachio ei sanguis emittendus est et statim curasti eam.[1]

. . .

E coeuntia. Idem faciunt etiam albae olivae, et nigrum papaver cum melle adsumptum, et cummis

[1] *The words* medicamenta . . . eam *are recorded by Pseudoranus. There is a note in several MSS. that two leaves were missing here from the oldest copy. Their contents can be gathered from the list of chapter headings given by one MS. J (cf. p. 466),* De vessica, de calculis in vessica. *The MS. resumes in the middle of a chapter entitled* In omni dolore vessicae.

BOOK IV. 27. 1 C–1 E

persists, a convenient emollient appears to be bitter sweet steeped in milk, then pounded and mixed with white wax and deer marrow in iris oil, or suet of beef or goat mixed with rose oil.   Also there should be given in draught either castory, or git,[a] or dill. If the womb is not healthy, it is cleaned with square rushes; but if it is actually ulcerated a wax-salve is made with rose oil, mixed with fresh lard and white of egg, and applied to it, or else white of egg mixed with rose oil, with pounded rose-leaves added to give it consistence. When painful the womb should be fumigated from below with sulphur. But if excessive menstruation is doing harm to the woman, the remedy is to scarify and cup the groins, or even to apply cups under the breasts. If the menstrual discharge is bad, the following medicaments are to be applied to evoke blood, costmary, pennyroyal, white violet, parsley, catmint and savory and hyssop. Let her include what is suitable in her diet: leeks, rue, cummin, onion, mustard, or any other acrid vegetable. If blood bursts out from the nose at a time when it should do so from the genitals, the groins are to be scarified and cupped, repeating this every thirtieth day for three or four months, then you may be sure that this affection has been cured.   But if there is no show of blood, you may be sure that there are pains coming in the head. Then blood is to be let from the arms, and you have given relief at once.

. . . constricting remedies. White olives also produce the same effect, also black poppy seeds, taken

[a] *Nigella sativa, melanthium,* black cummin.

CELSUS

cum contrito semine apii liquatum et cum cyatho passi datum. Praeter haec in omnibus vesicae doloribus idoneae potiones sunt, quae ex odoribus fiunt, id est spica nardi, croco, cinnamo, casia, similibusque. Idemque etiam decocta lentiscus praestat. Si tamen intolerabilis dolor est et sanguis profluit, etiam sanguinis detractio apta est, aut certe coxis admotae cucurbitulae cute incisa.

2 At cum urina super potionum modum etiam sine dolore profluens maciem et periculum[1] . . . facit, si tenuis est, opus est exercitatione et frictione, maximeque in sole vel ad ignem. Balneum rarum esse debet, neque longa in eo mora, cibus conprimens, vinum austerum meracum, per aestatem frigidum, per hiemem egelidum, sed tantum, quantum minimum sitim finiat. Alvus quoque vel ducenda vel lacte purganda est. Si crassa urina est, vehementior esse debet et exercitatio et frictio, longior in balneo mora; cibis opus est tenerioribus, vino eodem. In utroque morbo vitanda omnia sunt, quae urinam movere consuerunt.

28 (XXI). Est etiam circa naturalia vitium, nimia profusio seminis; quod sine venere, sine nocturnis imaginibus sic fertur, ut interposito spatio tabe hominem consumat.—In hoc adfectu salutares sunt vehementes frictiones, perfusiones natationesque quam frigidissimae, neque cibi nec potio nisi frigida
2 adsumpta. Vitare autem oportet cruditates, omnia

---

[1] *After* periculum, *Marx supplies* tabis, *which is translated.*

[a] Celsus here describes diabetes, which was said to have been recognised by Demetrius of Apameia in Phrygia, about 100 B.C. It was well known in Ancient India, and Susruta named it honey urine and connected it with thirst.

[b] Profusio seminis, γονόρροια. Greek and Roman medical writers refused to recognise this as an infection, in opposition

with honey, and liquid gum, mixed with pounded celery seeds, and given in a cupful of raisin wine. Besides the above, draughts suited for all bladder pains are made from aromatics, such as spikenard, saffron, cinnamon, cassia, and such like, also decoction of mastic does good. If in spite of these pain becomes intolerable and there is blood in the urine, venesection is proper, or at any rate wet cupping over the hips.

But when the urine [a] exceeds in quantity the fluid taken, even if it is passed without pain, it gives rise to wasting and danger of consumption; if it is thin, there is need for exercise and rubbing, particularly in the sun and before a fire. The bath should be taken but seldom, and the patient should not stay in it for long; the food should be astringent, the wine dry and undiluted, cold in summer, lukewarm in winter, and in quantity the minimum required to allay thirst. The bowels also are to be moved by a clyster or by taking milk. If the urine is thick, exercise and rubbing should be more thorough, and the patient should stay longer in the bath; food and wine should be of the lighter kind. In both affections, everything that promotes urine should be avoided.

28. There is also a complaint about the genitals, an excessive outflow of semen;[b] which is produced without coition, without nocturnal apparitions, so that in course of time the man is consumed by wasting. Salutary remedies in this affection are: vigorous rubbings, affusions, and swimming in quite cold water; no food and drink taken unless cold. He should, moreover, avoid everything indigestible, everything

to all popular knowledge. See Leviticus XV. 2; XXII. 4; LXX.; it was placed in the same category with leprosy. See also Josephus, *B. J.* 6, 9, 3.

inflantia; nihil ex is adsumere, quae contrahere semen videntur, qualia sunt siligo, simila, ova, halica, amylum, omnis caro glutinosa, piper, eruca, bulbi, nuclei pinei. Neque alienum est fovere inferiores partes aqua decocta ex verbenis comprimentibus, ex iisdem aliqua cataplasmata imo ventri inguinibusque circumdare, praecipueque ex aceto rutam; vitare etiam, ne supinus obdormiat.

29 (XXII). Superest ut ad extremas partes corporis veniam, quae articulis inter se conseruntur. Initium a coxis faciam. Harum ingens dolor esse consuevit, isque hominem saepe debilitat et quosdam non dimittit; eoque id genus difficillime curatur, quod fere post longos morbos vis pestifera huc se inclinat; quae ut illas partes liberat, sic hanc, iam
2 ipsam quoque adfectam, prehendit.—Fovendum primum aqua calida est, deinde utendum calidis cataplasmatibus. Maxime prodesse videtur aut cum hordeacea farina aut cum ficu ex aqua decocta mixtus capparis cortex concisus, vel lolii farina ex vino diluto cocta et mixta cum acida faece; quae quia refrigescunt, imponere noctu malagmata commodius est. Inulae quoque radix contusa et ex vino austero postea cocta et late super coxam imposita inter valentissima auxilia est. Si ista non solverunt, sale calido et umido utendum est. Si ne sic quidem finitus dolor est, aut tumor ei accedit, incisa cute admovendae sunt cucurbitulae; movenda urina; alvos, si compressa est, ducenda. Ultimum

---

[a] A description of hip-joint disease, such as is common in old people, *malum coxae senile*, cf. Hipp. IV. 192 (*Aph*. VI. 60), ἰσχίας. The tuberculous disease of the hip in young people may be referred to in the latter part of the chapter.

flatulent; nothing should be taken of those things which appear to collect the semen, such things are siligo, simila, eggs, spelt, starch, all glutinous flesh, pepper, colewort, bulbs, pine kernels. It is not inexpedient to bathe the lower extremities in a decoction of astringent vervains, to cover the hypogastrium and groins with plasters prepared from the same decoction, and in particular from rue preserved in vinegar: also the patient should avoid sleeping on his back.

29. It remains for me to come to the extremities of the body which are interconnected by joints. I begin with the hips. In these severe pain is wont to occur,[a] and this often weakens the patient, and some it never leaves: and on this account it is a difficult class to treat, for it is generally after chronic diseases that a pestiferous force directs itself to the hip; which, as it releases other parts, seizes upon this, which now becomes the seat of the disease. The hip is to be first fomented with hot water, after which hot plasters are applied. Those which appear to be especially beneficial are these: caper bark chopped up and mixed either with barley meal or with fig decoction, or darnel meal boiled in diluted wine and mixed with sour wine lees: since these are apt to grow cold, by night it is better to put on emollients. Inula root also pounded and afterwards boiled in dry wine and applied widely over the hip is among the most efficacious of remedies. If these do not resolve the trouble, then hot moist salt[b] is to be employed. If even these measures do not end the pain, and a swelling supervenes, the skin is incised and cups are to be applied; diuretics are given; and the bowels if costive are to be clystered.

[b] II. **17.** 10.

est et in veteribus quoque morbis efficacissimum tribus aut quattuor locis super coxam cutem candentibus ferramentis exulcerare. Sed frictione quoque utendum est maxime in sole et eodem die saepius, quo facilius ea, quae coeundo nocuerunt, digerantur; eaque, si nulla exulceratio est, etiam ipsis coxis; si est, ceteris partibus adhibenda est. Cum vero saepe aliquid exulcerandum candenti ferramento sit, ut eo materia inutilis evocetur, illud perpetuum est, non, ut primum fieri potest, huius generis ulcera sanare, sed ea trahere, donec id vitium, cui per haec opitulamur, conquiescat.

30 (XXIII). Coxis proxima genua sunt; in quibus ipsis non numquam dolor esse consuevit.—In iisdem autem cataplasmatis cucurbitulisque praesidium est, sicut etiam cum in umeris aliisve commissuris dolor aliquis exortus est. Equitare ei, cui genua dolent, inimicissimum omnium est. Omnes autem eiusmodi dolores, ubi inveteraverunt, vix citra ustionem finiuntur.

31 (XXIV). In manibus pedibusque articulorum vitia frequentiora longioraque sunt, quae in podagris cheragrisve esse consuerunt. Ea raro vel castratos vel pueros ante femina coitum vel mulieres, nisi quibus menstrua suppressa sunt, temptant.—Ubi sentire coeperunt, sanguis mittendus est: id enim inter initia statim factum saepe annuam, nonnumquam perpetuam valetudinem tutam praestat. Quidam etiam, cum asinino lacte poto sese eluissent, in perpetuum hoc malum eva-

---

[a] Hipp. *Prorrh.* II. 42 (Littré IX. 73).
[b] See Appendix, p. 463–5.
[c] Hipp. IV. 186 (*Aph.* VI. 28–30).

The ultimate measure and the most efficacious in cases of old standing, is to set up issues in three or four places over the hip by burning the skin with cauteries. But rubbing is also to be employed, particularly in the sun and often each day, in order that the materials of the disease, which have been doing harm by collecting, may be the more readily dispersed; and the rubbing is applied actually over the hips in the absence of ulceration; if there is any, then to other parts. Since now some issue often has to be set up by the hot cautery, in order that matter may be extracted, it is the general rule not to let ulcerations of this kind heal offhand, but to let them drag on until the complaint which we aim to relieve has quieted down.

30. Next to the hips come the knees, in which pain now and again occurs, and these same plasters and cuppings are a safeguard, as also when any pain arises in the shoulder or other joints. Riding on horseback is of all things the most injurious to anyone with painful knees. All such pains, when of long standing, are hardly ever ended except by cauterization.

31. Joint troubles[a] in the hands and feet are very frequent and persistent, such as occur in cases of podagra and cheiragra.[b] These seldom attack eunuchs or boys before coition with a woman, or women except those in whom the menses have become suppressed.[c] Upon the commencement of pain blood should be let; for when this is carried out at once in the first stages it ensures health, often for a year, sometimes for always. Some also, when they have washed themselves out by drinking asses' milk, evade this disease in perpetuity; some have obtained lifelong

serunt: quidam, cum toto anno a vino, mulso, venere sibi temperassent, securitatem totius vitae consecuti sunt; idque utique post primum dolorem servandum est, etiamsi quievit. Quod si iam consuetudo eius facta est, potest quidem aliquis esse securior iis temporibus, quibus dolor se remisit: maiorem vero curam adhibere debet iis, quibus id revertitur; quod
3 fere vere autumnove fieri solet. Cum vero dolor urget, mane gestari debet; deinde ferri in ambulationem; ibi se dimovere, et, si podagra est, interpositis temporibus exiguis invicem modo sedere, modo ingredi; tum, antequam cibum capiat, sine balneo loco calido leviter perfricari, sudare, perfundi aqua egelida: deinde cibum sumere ex media materia, interpositis rebus urinam moventibus, quotiensque plenior est, vomere. Ubi dolor vehemens urget, interest sine tumore is sit, an tumor cum
4 calore, an tumores iam etiam obcalluerint. Nam si tumor nullus est, calidis fomentis opus est. Aquam marinam vel muriam duram fervefacere oportet, deinde in pelvem coicere, et, cum iam homo pati potest, pedes demittere, superque pallam dare, et vestimento tegere; paulatim deinde iuxta labrum ipsum ex eadem aqua leviter infundere, ne calor intus destituat; ac deinde noctu cataplasmata calfacientia imponere, maximeque ibisci radicem ex vino coctam.
5 Si vero tumor calorque est, utiliora sunt refrigerantia, recteque in aqua quam frigidissima articuli continentur sed neque cotidie neque diu, ne nervi indurescant.

---

[a] Hipp. IV. 192 (*Aph.* VI. 55).
[b] For *ambulatio* in this sense, cf. Varro R.R. 3. 5. 9; Cicero, *Ad Familiares*, 3. 1. 1.

security by refraining from wine, mead and venery for a whole year; indeed this course should be adopted especially after the primary attack, even although it has subsided. But if the malady has already become established, it may be possible to act with more freedom in those seasons[a] in which the pain tends to remit; but he should adopt more careful treatment at those times in which it recurs, which is generally in spring or autumn. Now when the pain requires it, in the morning the patient should be rocked; then carried to a promenade;[b] there he should move about, and in the case of podagra he should take short turns at sitting down and walking about: next before taking food and without entering the bath itself, but in a hot room, he should be gently rubbed, sweated, and then douched with lukewarm water: the food following should be of the middle class; diuretics are given with it, and an emetic whenever he is of a fuller habit. When the pain is very severe, it makes a difference whether there is an absence of swelling, or a swelling with heat, or swellings which are already hardened. For if there is no swelling, hot foments are needed. Either sea-water, or strong brine should be heated, then poured into a vessel; and as soon as he can bear it, the man puts his feet in, over the vessel is spread a cloak, and over him a blanket; after that hot water is poured over the lip of the vessel, a little at a time, to prevent the contents from losing heat: and then at night heating plasters are applied, especially mallow root boiled in wine. But if there is swelling and heat, refrigerants are more useful, and the joints may be rightly held in very cold water, but not every day, nor for long, lest the sinews become hardened. There is to be applied

CELSUS

Inponendum vero est cataplasma, quod refrigeret, neque tamen in hoc ipso diu permanendum, sed ad ea transeundum, quae sic reprimunt, ut emolliant. Si maior est dolor, papaveris cortices in vino coquendi miscendique cum cerato sunt, quod ex rosa factum sit; vel cerae et adipis suillae tantundem una liquandum, deinde his vinum miscendum; atque ubi quod ex eo impositum est incaluit, detrahendum, et 6 subinde aliud inponendum est. Si vero tumores etiam obcalluerunt et dolent, levat spongia inposita, quae subinde ex oleo et aceto vel aqua frigida exprimitur, aut pari portione inter se mixta pix, cera, alumen. Sunt etiam plura idonea manibus pedibusque malagmata. Quod si nihil superinponi dolor patitur, id, quod sine tumore est, fovere oportet spongia, quae in aquam calidam demittatur, in qua vel papaveris cortices vel cucumeris silvestris radix decocta sit; tum inducere articulis crocum cum 7 suco papaveris et ovillo lacte. At si tumor est, foveri quidem debet aqua egelida, in qua lentiscus aliave verbena ex reprimentibus decocta sit, induci vero medicamentum ex nucibus amaris cum aceto tritis, aut cerussa, cui contritae herbae muralis sucus sit adiectus. Lapis etiam, [qui carnem edit,] quem σαρκοφάγον Graeci vocant, excisus sic ut pedes capiat, demissos eos, cum dolent, retentosque ibi levare consuevit. Ex quo in Asia lapidi Assio gratia 8 est. Ubi dolor et inflammatio se remiserunt, quod intra dies quadraginta fit, nisi vitium hominis accessit, modicis exercitationibus, abstinentia, unctionibus

---

[a] Limestone highly impregnated with salt which was used for mummifying bodies; Pliny *N.H.* II. 97. (L.C.L. I, p. 341).
[b] Assos, a city in the Troas.

also a cooling plaster; this, however, is not to be kept on for long, but a change made to those which soothe as well as repress. If pain is greater, rind of poppy-heads is to be boiled in wine, and mixed with wax-salve made up with rose oil; or wax and lard, equal parts, are melted together, and then the wine mixed with these; and as soon as this application becomes hot, it is to be removed and another immediately put on. But if the swellings have grown hard and are painful, the application of a sponge frequently squeezed out of oil and vinegar, or out of cold water, or the application of pitch, wax and alum, equal parts mixed, gives relief. There are also several emollients suitable alike for the hands and feet. But if the pain does not allow of anything being put on, when there is no swelling, the joint should be fomented with a sponge which has been dipped in a warm decoction of poppy-head rind, or of wild cucumber root, next the joints are smeared with saffron, poppy-juice and ewe's milk. But if there is a swelling, this ought to be bathed with a tepid decoction of mastic or some other repressant vervain, and then covered with a medicament composed of bitter almonds pounded up in vinegar, or of white lead, to which has been added the juice of pounded pellitory. The stone, too, [which corrodes flesh], which the Greeks call sarcophagos,[a] is carved out so as to admit the feet; when these are painful, they are inserted and held there, and are usually relieved. In Asia Minor Assian[b] limestone is held in esteem for this purpose. When pain and inflammation have subsided, which should happen within forty[c] days, unless the patient is in fault, gentle exercise, spare diet, soothing anoint-

[c] Hipp. IV. 190 (*Aph.* VI. 49).

lenibus utendum est, sic ut etiam tum acopo vel liquido cerato cyprino articuli perfricentur. Equi-
9 tare podagricis quoque alienum est. Quibus vero articulorum dolor certis temporibus revertitur, hos ante et curioso victu cavere oportet, ne inutilis materia corpori supersit, et crebriore vomitu; et si quis ex corpore metus, vel alvi ductione uti vel lacte purgari. Quod Erasistratus in podagricis expulit, ne in inferiores partes factus cursus pedes repleret, cum evidens sit omni purgatione non superiora tantummodo sed etiam inferiora exinaniri.

32 (XXV). Ex quocumque autem morbo quis invalescit, si tarde confirmatur, vigilare prima luce debet; nihilo minus in lecto conquiescere; circa tertiam horam leviter unctis manibus corpus permulcere. Deinde delectationis causa, quantum iuvat, ambulare, circumcisa omni negotiosa actione; tum gestari diu, multa frictione uti, loca, caelum, cibos
2 saepe mutare. Ubi triduo quadriduove vinum bibit, uno aut etiam altero die interponere aquam. Per haec enim fiet, ne in vitia tabem inferentia incidat et ut mature vires suas recipiat. Cum ex toto vero convaluerit, periculose vitae genus subito mutabit et inordinate aget. Paulatim ergo debebit omissis his legibus eo transire, ut arbitrio suo vivat.

[a] *Gestari* may refer to riding as well as carriage exercise.
[b] Cf. I. 1. 1.

ings, are to be employed, provided that also then the joints may be rubbed with an anodyne salve or with a liquid wax-salve of cyprus oil. But riding on horseback is harmful for those with podagra. Those, too, in whom joint-pains tend to recur at certain seasons ought both to take precautions beforehand as to their diet, lest there should be a surfeit of harmful material in the body, and to use an emetic the more frequently; and those in any anxiety as to their body should make use of clystering, or of purgation by milk. This treatment for those with podagra was rejected by Erasistratus, lest a flux directed downwards might fill up the feet, though it is evident that any purgation extracts, not only from the upper parts, but also from the lower as well.

32. Now from whatever disease he is recovering, if his convalescence is slow, the patient ought to keep awake from dawn, but nevertheless stay at rest in bed: about nine o'clock he should be gently stroked over with anointed hands, after that by way of amusement, and as long as he pleases, walk, all business being omitted: then he should use conveyances[a] for a good while, be rubbed much, often change his residence, climate and diet. Having taken wine for three or four days, he should for one or two days drink water only. For thus he will ensure that he does not lapse into a complaint which causes wasting, but soon gets back his full strength. When he has, in fact, completely recovered, it will be dangerous for him suddenly to change his way of life and to act without restraint. Therefore he should only little by little leave off what has been prescribed, and pass to a way of life of his own choosing.[b]

APPENDIX

ON PODAGRA AND CHEIRAGRA

DICTIONARIES wrongly give gout as the only meaning for these two words, which, in many classical writers,[a] and sometimes in Celsus, are simply used to mean pain in the foot or hand.

It is, however, a fact that all animals, as well as man, are subject in old age to a degeneration of the cartilaginous surfaces of their joints. This is true even in prehistoric times; such changes occurred among the Dinosaurs of the chalk period, among various primitive animals and in early man. This degeneration is now met with, not only in the old, but in middle age; it is a form of precocious senility; often the tendency is clearly inherited and the condition becomes an essential primary factor in various joint diseases. It was to this condition, when occurring in the feet, that the terms podagra and cheiragra came specially to be applied. Gout was first scientifically diagnosed after the finding of urate of soda in the tophi and an excess of uric acid in the blood. But the description of Aretaeus[b] seems to show that it was known to the physicians

[a] See Aristotle, *History of Animals*, VI. 21, 575 B, and VIII. 22, 23, 24, 604 A. By podagra he meant a disease of the feet in animals, cattle, horses, etc. Cf. also Virgil, *Georgics*, III. 299.

[b] *Chronic Diseases*, II. 12 (ischiatica). His date is unknown; it has been conjectured that he flourished about 70 A.D.

APPENDIX

of the 1st century A.D. He mentions that the disease (podagra) sometimes begins by attacking the great toe and that the tophi break down into a chalky fluid.[a]

Pliny [b] and Galen,[c] writing in the 2nd century A.D., declared that a new form of podagra had appeared, definitely inheritable from parents and grandparents, which they both attributed to luxury in food and drink; this had become prevalent in the Roman world, but was hardly known in the Greek world of Hippocrates.

The development of the disease, which Pliny and Galen referred entirely to the increase of luxury, may have had, in part at least, another cause. In the 16th century an outbreak of illness occurred at Poitou [d] with similar symptoms, which was finally diagnosed as chronic lead poisoning due to the use of lead pipes or vats in the preparation of the wine. Another outbreak, due to the same cause, occurred in the 18th century in connection with the manufacture of cider in Devonshire.[e] On the other hand, the marked decrease in cases of gout during the last half-century may be related to the simultaneous reduction of cases in chronic lead poisoning.

It is, therefore, not improbable that the form of podagra described by Pliny and Galen may have been closely related to chronic lead poisoning due to the

---

[a] The earliest skeleton definitely recognized as affected by true gout was found in a Nubian cemetery belonging to the 5th century A.D.

[b] *Natural History*, XXVI., X. 64.

[c] XVII. A, 431–XVIII. A, 43 (K).

[d] Citesius, Opera 1639.

[e] Huxham, *Obs. Med. Phys.*, 1784, III. 54.

APPENDIX

great increase in the use of lead pipes owing to the extensive building of aqueducts under the Empire.[a]

To sum up, it seems clear that if the words podagra and cheiragra when used by Celsus are simply translated " gout," we are attributing to him a specialized use of the term which he probably did not intend; his podagra and cheiragra were used of any pain in the feet and hand, though such pain was often a joint pain and sometimes, no doubt, occurred in a case of true gout.[b]

[a] For an account of lead poisoning see Celsus V. 27, 12 B; for the use of lead pipes see Vitruvius VIII. 35, and Frontinus, Aqueducts I. 25.

[b] The word *gutta* from which gout is derived was used by later Latin medical writers as equivalent to Destillatio ($ῥεῦμα$), the " dripping " from the head described by Celsus. The first to use it in the sense of gout was probably Alexander of Tralles in the 6th century A.D. The term gout is wrongly used by Dryden in his translation of the *Georgics* to translate the *podagra* of Virgil (III. 299) which refers to foot-rot in sheep.

LIST OF CHAPTER HEADINGS

CAPITVLA LIBRI I IN CODICE *J* PRAESCRIPTA

Qualiter debeant languentem medicinae operare (*prohoem.*).

Qualiter eveniunt (*conf.* 22) qui usu curant et rationem non reddunt (*prohoem.* 12—19).

Si se concoctione (*sequitur spatium XX litterarum*) adhibebit abstinentia (*prohoem.* 20—41 : *conf.* 33 : *supplendum* sentiet laborare, *scribendum* abstinentiam).

De membrana † quae (*scribendum* qua inferiora) a superioribus dividuntur; diaphragma a Graecis dicitur (*prohoem.* 42).

Qualiter se sanus agere debeat (I).

Qualiter se agere debeant qui in stomacho imbecilles sunt (II).

Haec observatio si catharum est aut stomachus reumatizat (II. 3 extr.).

Qualiter debeat corpus fricari vel ungi (III. 4).

De vomitu (III. 17).

De ventre (III. 25).

De [a]escis temporum (III. 34).

De capite (IV).

De lippitudine quae gravedinem tollit (V).

De alvo soluto (VI).

De dolore intestini interioris (VII).

LIST OF CHAPTER HEADINGS

De stomacho (VIII).
De dolore nervorum vel podagra (IX).
Observatio adversus pestilentiam (X).

CAPITVLA LIBRI II IN COD *J* PRAESCRIPTA

I. De temporibus (I. 1).
II. De tempestatibus (I. 2).
III. De ventis (I. 3).
IIII. De aetatibus (I. 5).
V. De forma corporis (I. 5).
VI. De vere (I. 6).
VII. De aestate (I. 7).
VIII. De autumno (I. 8).
VIIII. De siccitatibus (I. 12).
X. De imbribus (I. 12).
XI. Muliebria corpora molliora (I. 16).
XII. Sanguinem prohibentem (profluentem *in marg. capitis*) ex quibusdam venis (I. 21).
XIII. De notis adversae valitudinis (II).
XIIII. Incipiente febre signa bona vel mala (III).
XV. Mala signa languentium (IV).
XVI. De longa valetudine signa (V).
XVII. De ultimis signis valetudinum (VI).
XVIII. De muliebri valetudine (VI. 8).
XVIIII. Plerosque post haec inditia vixisse (VI. 13).
XX. Ex quibus notis singulas possis valetudines agnoscere (VII).
XXI. Signa urinae mirifice (mirife *in marg. capitis*, mystice *in indice*) exposita [possit agnosci] (VII. 11).

LIST OF CHAPTER HEADINGS

XXII. De calculosis (VII. 14).
XXIII. Quemadmodum femina calculosa possit agnosci (VII. 15).
XXIIII. De mulieribus pregnantibus (VII. 16).
XXV. De lateribus (VII. 17).
XXVI. De dentibus (VII. 21).
XXVII. De diversis doloribus ostendendis (VII. 21).
XXVIII. De signis quae insaniam ostendunt (VII. 24).
XXVIIII. Quae valetudo ex qua oriatur (VII. 28).
XXX. Quae notae (nota *utroque loco*) spem salutis quae pericula ostendunt (VIII).
XXXI. De fluxu narium (VIII. 6).
XXXII. De menstruis feminarum (VIII. 7).
XXXIII. De aqua inter cutem (VIII. 8).
XXXIIII. De articulis et podagra (VIII. 10).
XXXV. De morbo comitiali (VIII. 11).
XXXVI. De torminibus (VIII. 13).
XXXVII. De levitate intestini (VIII. 14).
XXXVIII. De humeris (VIII. 14).
XXXVIIII. De singultibus (VIII. 15).
XL. De mulieribus (VIII. 16).
XLI. De quartanis aestatis (states *in marg. capitis*, stater *in indice* : VIII. 16).
XLII. De capitis dolore (VIII. 18).
XLIII. De febribus (VIII. 19).
XLIIII. De fistula urinae (fisae urinae *utroque loco*) (VIII. 20).
XLV. De pulmonum morbis (VIII. 22).
XLVI. De signis laterum (VIII. 22).
XLVII. De suppurationibus (VIII. 23).
XLVIII. De mulieribus (VIII. 25).
XLVIIII. De aqua inter cutem (VIII. 26).
L. De articulis (VIII. 28).

LIST OF CHAPTER HEADINGS

LI. De comitiali morbo (VIII. 29).
LII. De atra bili cum tormine (VIII. 31).
LIII. De liene (VIII. 34).
LIIII. De intestino tenuiore (VIII. 35).
LV. De mulieribus fetis (foetis *in indice*) (VIII. 35).
LVI. De coxis (VIII. 38).
LVII. De absolutione membrorum (de resolutione *in marg. capitis*) (VIII. 40).
LVIII. De feminis gravidis (VIII. 41).
LVIIII. De quartana autumnali (VIII. 42).
LX. De his qui suspendio liberantur (VIII. 43).
LXI. De curationibus (IX).
LXII. Quae sint communes curae (IX. 1).
LXIII. De sanguinis emissione per venam (X).
LXIIII. Quibus tolli debet sanguis (X. 1: *deest in marg. capitis*).
LXV. Quae indicia (inducia *cod*) sunt morbi quae sān emīs requirunt (X. 6: *in marg. capitis*: Quae sanguinis missionem requirunt).
LXVI. An aegris (aegro *in marg. capitis*) sanguis sit tollendus (tollendus sit *in marg. capitis*) (X. 7.)
LXVII. Quando sine ulla observatione emittendus (mittendus *in marg. capitis*) est sanguis (X. 8).
LXVIII. Cruđ sanḡ emittendus est sed non semper (Si cui post cibum vel quocumque casu sanguinem mittendus est *in marg. capitis*) (X. 9).
LXVIIII. Per biduum si necesse sit sanḡ subtrahi (subtrahendum per biduum si necesse est sanguinem *in marg. capitis*) (X. 12).
LXX. Contra eos qui dicunt longe ab eo loco qui laesus sit sanguinem emittendum (mittendum *in marg. capitis*) (X. 13).
LXXI. Capite fracto de brachio mittendus est sanguis (X. 14).

LIST OF CHAPTER HEADINGS

LXXII. Si in umero vitium est ex altero brachio mittendus est sanguis (X. 14: *deest in marg. capitis*).

LXXIII. Quae vena percuti debeat (X. 14).

LXXIIII. Quae periculosa (pericula *marg. capitis*) sint cum vena male percutitur (X. 15).

LXXV. Qui locus venae feriendus sit qualis color sanguinis (sanguis *in marg. capitis*) esse debeat (X. 16).

LXXVI. Tandiu est (ex *in indice*) tollendum quamdiu rubere coeperit vel perlucere (X. 19).

LXXVII. De cucurbitis medicinalibus (XI).

LXXVIII. De deiectione (XII).

LXXVIIII. Quae ad solvendum ventrem (venerem *in indice m* 1) veteres dabant (XII. 1 A).

LXXX. Qualiter debeat ad ventrem dari (Dari *ante* qualiter *in marg. capitis*) (XII. 1 B).

LXXXI. De subtrahendo ventre in quibus causis maxime necessarium est (*om in marg. capitis*) (XII. 1 B).

LXXXII. Quale sit quod venter (quae ventrem *in indice*) fecerit (XII. 2 A).

LXXXIII. Qualiter iniciendus observari debeat (XII. 2 B).

LXXXIIII. Quid ad solvendum ventrem per curam parandum sit (XII. 2 D).

LXXXV. De vomitu (XIII).

LXXXVI. De frictione (XIV).

LXXXVII. De gestatione (egestatione *in indice*) (XIV. 1).

LXXXVIII. Quid Hippocrates de frictione senserit (XIV. 2.

LXXXVIIII. Quid (quod *marg*) inter frictionem et unctionem intersit (XIV. 4).

LIST OF CHAPTER HEADINGS

XC. Contra eos qui unctionem numero implendum putarunt (potarunt *index*, poterant *marg. capitis*) (XIV. 9).
XCI. De utilitate gestandi (XV. 1).
XCII. Quot genera gestandi sint (XV. 3).
XCIII. De abstinentia (XVI).
XCIIII. De sudore eiciendo (eligendo *marg. capitis*) (XVII).
XCV. De balneis qualiter aegris competat (conueniat *marg. capitis*) (XVII. 2).
XCVI. De fomentis (XVII. 9).
XCVII. De cibis et potionibus (XVIII).
XCVIII. Quae annona fortior vel levior (XVIII. 4).
XCVIIII. De legumine (XVIII. 5).
C. De pomis (XVIII. 6).
CI. De avibus (XVIII. 6).
CII. Discretio levium et gravium locorum (XVIII. 9).
CIII. De potionibus (XVIII. 11).
CIIII. De aquis (XVIII. 12).
CV. Quae res pituitam facit (faciat *index*) uel qualem (qualis *index*) (XIX).
CVI. Quae res alvum movent (moveat *index*) (XXIX).
CVII. Quae res adstringat alvum (XXX).
CVIII. Quae res urinam movent (XXXI).
CVIIII. Quae somno apta sunt (sint *index*) (XXXII).
CX. Quae calefaciunt in cataplasmatis (cataplasmis *index*) (XXXIII. 5).

LIST OF CHAPTER HEADINGS

CAPITVLA LIBRI III IN CODICE *J* PRAESCRIPTA

I. De instantibus morbis (*deest in marg. capitis*).
II. De singulis morbis et (vel *marg. capitis*) curis (I. 3).
III. Qua (ut *index*) ratione morborum genus noscatur (II).
IIII. Haec (De *index*) signa in secundo libro require (II. 1).
V. De cura morbi levioris (II. 5).
VI. De cura veri morbi vel agnitione (*deest in marg.*) (II. 6).
VII. Genera febrium (III).
VIII. De quartana febre (III. 1).
VIIII. De tertiana febre (III. 2).
X. De quotidiana febre (III. 3).
XI. Quid sit frigus et quid sit horror (III. 3).
XII. De febrium cura (IV): (*add in marg. capitis :* quae supra exposuit nunc dicit).
XIII. Asclepiades quod officium medici dixerit (IV. 1): *deest in marg. capitis*.
XIIII. Observatio primae febris (IV. 4).
XV. Qualiter cibandus aeger sit (IV. 5).
XVI. Asclepiades quarta die cibum dabat (IV. 6).
XVII. Themison, quando desisset vel levata esset febris (IV. 6).
XVIII. Calidum caelum magis digeret (IV. 8): *add marg. capitis :* et ideo in Africa aeger abstinendus est.
XVIIII. Erasistrati de cibo dando (IV. 9): *deest in marg. capitis*.
XX. Ab uno medico multos non posse curari (medi-

LIST OF CHAPTER HEADINGS

cum multos simul non posse curare *marg. capitis*): (IV. 9).
XXI. De diebus aegrorum (IV. 11).
XXII. Supervacue imparem numerum dierum observari (observare *utroque loco*, impare numero *index*) (IV. 12).
XXIII. Pythagoras imparem numerum infaustum esse dixit parentem malum (IV. 15).
XXIIII. In qua febre quo tempore cibus dandus est (V).
XXV. Cum continua febris levior non fit et dari cibum necesse est (V. 4).
XXVI. Si aliqua gravioris nuntii aegro necesse est nuntiari, post cibum et somnum esse dicenda (et—dicenda *om marg. capitis*) (V. 11).
XXVII. Quo tempore potio febrienti dari debeat (VI).
XXVIII. Heraclidis Tarentini de potione (VI. 4).
XXVIIII. Feminam post partum febricitantem satis bibere non debere (VI. 4).
XXX. Quo genere ratio febris agnoscatur (VI. 6).
XXXI. Medicum contra sedere debere ut faciem aegri videat (VI. 8).
XXXII. Quando aeger cibandus est (VI. 8).
XXXIII. Varios cibos dandos aegro Asclepiades iussit (VI. 11).
XXXIIII. De fluido ventre (ventre fluido *in marg. capitis*) (VI. 15).
XXXV. De curandis febribus (VII).
XXXVI. Aliter pueros esse curandos (VII. 1 B).
XXXVII. De emitriteis (emitritis *in marg. capitis*) (VIII).
XXXVIII. De lentis febribus (IX).
XXXVIIII. Petro qualiter curabat (IX. 2).

LIST OF CHAPTER HEADINGS

XL. Si cum febre dolor capitis fuerit quid fieri oporteat (Cura capitis in febribus *marg. capitis* (X).
XLI. Hyris illiricis (*iris Illyrica*) ad capitis dolores (X. 2).
XLII. De notis inflammationum et cura (X. 3).
XLIII. De horroribus et curationibus eorum (eorum *om in marg. capitis*) (XII).
XLIIII. De (*om in marg. capitis*) * * Cleophantus ante Asclepiadem remedium tertianae (tertianis *in marg. capitis*) et quartanis (XIV): *scribendum* De ⟨eo quod adhibebat⟩.
XLV. De duabus quartanis (XVI).
XLVI. De quotidiana quae ex quartana accedit (accessit *in marg. capitis*) (XVII).
XLVII. De insania quam Graeci phrenesim vocant (XVIII).
XLVIII. Quid sit phrenesis (XVIII. 3).
XLVIIII. De cura insaniae (XVIII. 4).
L. Soranus nihil misceri aquae (atque *in marg. capitis*) sive oleo (oleum *l. s.*) iubet propter odorum noxam (XVIII. 8: *conf Cael. Aurel. acut. morb.* I. 9, 67. 15, 137).
LI. Soranus his repugnat ut (repugnatū *in marg. capitis*) supra notatum (est *add marg. capitis*) (XVIII. 12).
LII. De somno (XVIII. 12).
LIII. Asclepiades his repudiatis haec praecipit (XVIII. 14).
LIIII. Soranus haec matutinis horis fieri (maturioribus fieri oris *in marg. capitis*) iubet (XVIII. 14: *conf Cael. Aurel. acut. morb.* I. 15, 141).
LV. Gestatio omittenda (XVIII. 15).
LVI. De alio genere insaniae (quod ... *add in marg. capitis*) (XVIII. 17).

LIST OF CHAPTER HEADINGS

LVII. hoc in declinatione (XVIII. 17 *pars posterior*).
LVIII. hoc satis utile (XVIII. 18?).
LVIIII. De tertio genere insaniae (qui sunt melancolici *add in marg. capitis*) (XVIII. 19).
LX. hoc utile (XVIII. 20).
LXI. Remotis plagis omnia utilia (XVIII. 21).
LXII. De cardiacis (XIX).
LXIII. De alio morbo insaniae (et insania et lethargo *in marg. capitis*) (XX).
LXIIII. De hydrope (de hydropico *in marg. capitis*) (XXI).
LXV. Hydropis genera tria (hydropi tria genera tympana leucofleumatia tertium ascitem *in marg. capitis*) (XXI. 1).
LXVI. Metrodorus Epicur⟨e⟩us hoc morbo temptatus (temptantem *index*, sanatus est *in marg. capitis*) (XXI. 4).
LXVII. † Duriores siccioribus (XXI. 6): *durioris generis cibus?*
LXVIII. Exemplum Asclepiadis (XXI. 8: *deest in marg. capitis*).
LXVIIII. De hydrope leucoflecmantia (leocoflegmantia *index*) (XXI. 11).
LXX. Cura Tharriae (XXI. 14).
LXXI. De iecore et liene (XXI. 14).
LXXII. Erasistrato displicuit (XXI. 15).
LXXIII. De tabe (XXII).
LXXIIII. Atrophia (XXII. 1).
LXXV. Graeci cacexian (*ita marg. capitis :* Caecia *index*) (XXII. 2).
LXXVI. De ptysi (ΦΘ*ECIN in marg. capitis*) (XXII. 3).
LXXVII. Quemadmodum ptisici (ptysicus *index*) possunt (possit *index*) agnosci (XXII. 3).

476

LIST OF CHAPTER HEADINGS

LXXVIII. Densius caelum Alexandriae quam Italiae (XXII. 8).

LXXVIIII. De tussis remediis (XXII. 12).

LXXX. Omoplata (Umo planta *index*) imae partis (partes *index*) scapularum (XXII. 12).

LXXXI. De morbo comitiali: ipsi sunt lunatici (*deest in marg. capitis*) (XXIII).

LXXXII. De morbo arquato sive regio: ipse auriginosus (*ita index :* Morbus arquatus sive regius ipse est quem (que *cod*) auriginem de colore dicimus *in marg. capitis*) (XXIV).

LXXXIII. Exemplum Hippocratis (*priori titulo add marg. capitis*) (XXIV. 1).

LXXXIIII. Diocles (XXIV. 1).

LXXXV. Cura auruginis (*ita marg. capitis :* Auruginem *index*) (XXIV. 2).

LXXXVI. Asclepiadis (XXIV. 3).

LXXXVII. Unde regius morbus (morbus regius *index*) dictus est (XXIV. 5).

LXXXVIII. De elephantia (XXV).

LXXXVIIII. De adtonitis (XXVI).

XC. De resolutione nervorum (XXVII. 1).

XCI. De dolore nervorum (et remedio *add in marg. capitis*) (XXVII. 2).

XCII. De tremore nervorum vel internis suppurationibus (XXVII. 3).

LIST OF CHAPTER HEADINGS

CAPITVLA LIBRO IV IN COD *J* PRAESCRIPTA

De capite (I. 2).
sphragitides (phragitides *index*) (I. 2).
kariatides (Arteriae *index*) I. 2).
Glandulae (I. 2).
Arteria (*om in marg. capitis*) (I. 3).
Qualis sit arteria (I. 3 *med*) (*om marg.*).
Pulmo (I. 4).
Cor (I. 4).
Iecur (I. 5).
fel (I. 5).
lienis (Venis *cod*) id est splen (*ita index*, splen *marg. capitis*) (I. 5).
Renes (I. 5).
[Renes].
Stomachus (I. 6).
Ventriculus (I. 6).
Reticulus (I. 6).
Pylas intestini (intestinum *index*) (I. 7).
Intestinum ieiunum (I. 7).
Intestinum sinibus (senibus *index*) implicitum (I. 8).
Intestinum caecum (I. 8) (*deest in marg. capitis*).
Intestinum pervium (I. 9).
Rectum ideo dictum (*ita in marg. capitis m* 2, dicitur *m* 1 *et index*) quia perderigatur (perdigeratur *utroque loco cod*) (I. 9).
Omentum (I. 10).
Adeps (I. 10).
Adeps cerebrum medulla sensu carent (*deest in marg. capitis*) (I. 10).

LIST OF CHAPTER HEADINGS

Uretenae venae albae quae a renibus ad vesicam (a vesica *cod*) feruntur duae (*deest in marg. capitis*) (I. 10).

De vessica (*deest in marg. capitis*) (I. 11).

Aliter in viris aliter in feminis posita est vessica (posita vesica *in marg. capitis*) (I. 11).

Colem (I. 12).

Vulva (I. 12).

hunc canalem † non se sic auctori est (*ita index* : hunc canalem sensit auctoritas *in marg. capitis*) (I. 12).

Peritoneus (*deest in marg. capitis*) (I. 13).

de capite quibus doloribus subiaceat (*om index*) (II).

hydrocephalon cum in capite tumor est (*om index*) (II. 4).

De remediis doloris (doloris *om marg. capitis*) capitis (II. 4).

Si ex vino vel cruditate caput doleat (dolet quod fieri opus *marg. capitis*) (II. 5).

Si caput de calore dolet (II. 6).

Si de frigore caput dolet (II. 7).

sin humor cutem inflat in caput (*deest in indice*) (II. 9).

Cynas morbus in capite (*deest in marg. capitis*) (III).

De lingua (IV).

Destillatione (V) *periit in marg.*

Si caput in nares destillat (V. 1) *periit in marg.*

Si in fauces (V. 1) *periit in marg.*

Si in pulmones (V. 1).

De gravedine (V. 2).

Gravedinem Hippocrates corizam nominat (V. 2).

De distillatione, Katastagmon Graeci nominant (distillationem catastagmon Graeci vocant. hinc quaerenda origo catharri *marg. capitis*) (V. 2).

LIST OF CHAPTER HEADINGS

Qualiter curanda sunt quae ex (de *marg.*) capitis distillatione veniunt (venit *marg.*) (V. 3).

Si in nares vel in fauces caput distillat (V. 6).

Si in pulmonem caput distillat (V. 7).

De cervice (De vertice *index*, causam (causa *cod*) cervicis hinc quaerunt *in marg. capitis*) (VI).

Opistoton (*deest in marg. capitis*) (VI. 1).

Asclepiades dicit: mittendus sanguis: alii prohibent (*deest in marg. capitis*) (VI. 2).

hoc utilissimum est (VI. 4).

De faucibus. Latini ancenam, Graeci prout res est vocant (hinc sinangitis remedia utenda. — Latini anginam *in marg. capitis duo lemmata*) (VII).

Kynancen (Kynalicen *index*, cynanxin *in marg. capitis*) (VII. 1).

De remediis faucium (VII. 2).

Efficacissimus salis calidi sucus (VII. 2).

Vulgo audisse se dicit quod (quando *in marg. capitis*) inter remedia faucium ponit (VII. 5).

De alio dolore faucium (Remedia ad ista inquirenda *marg. capitis*) (VIII).

Si interiores partes faucium dolent hinc quaerenda ratio (*ita marg. capitis :* De interioris faucium partes asclepiades dicit *index*) (IX).

Asclepiades (*deest in indice*) (IX. 2).

De tussi (*om in marg. capitis*) (X).

De sanguine qui per os mitti solet (*ita index :* Si quis sanguinem expuit hinc quaerat remedia *marg. capitis*) (XI).

Solere tussem aut ex pulmonibus aut ex iocinere sanguinem excutere (XI. 2) (*periit in marg.*).

Si sanguinem (sanguis *marg. capitis*) pus sequatur (XI. 3).

Ibi spina dolet (XI. 4).

LIST OF CHAPTER HEADINGS

De gingivis (cruentis viri *add marg. capitis*) (XI. 5).
Si id mulieri (XI. 5).
Erasistrati (herasistratus *marg. capitis*) (XI. 6).
Asclepiadis contra (asclepiades *marg. capitis*) (XI. 6).
De stomacho (XII).
Remedia quae faucibus dedit prodesse stomacho exulcerato (vulnerato *index*) (XII. 5).
De lateribus (XIII).
Trixago herba (XIII. 3).
De pulmone (XIV).
Peripleumoniacen (XIV. 1).
De iocineribus (XV).
hypatican (XV. 1).   } *omissa in margine capitum*
De liene (XVI).
De *renibus* (XVII).
De intestinis (XVIII).
Colera (XVIII. 1).
Unde *colore* dicatur (XVIII. 1).
Herasistratus (XVIII. 4).
De longo intestino (XIX. 1).
Duo morbi inter ipsa (ipsa *om index*) intestina, unus in tenuiore, alius in pleniore (XX).
Diocles (XX. 1).
Herasistrato (erasistratus *marg. capitis*) placuit (XX. 2).
De intestino pleniore (XXI).
Ad extremas (extremae *index*) partes sunt crura et brachia (XXI. 1).
Cassius (Casius *index*) colice invenisse (iuenis se *marg. capitis*) gloriabatur (XXI. 2).
De disinteria (de disintericis *marg. capitis*) (XXII).
Themisonis (Themosonis *index*) (XXII. 4).
De levitate intestinorum post disinteriam (post disint. *om marg. capitis*) (XXIII).

LIST OF CHAPTER HEADINGS

Hippocratis (XXIII. 3).
De lumbricis (cā lumbricorum *marg. capitis*) (XXIV).
Si lati sint (XXIV. 1).
Si rotundi sunt (XXIV. 2).
Tenesmos (Tenasmos *index*) (XXV).
De fluxu ventris quomodo in initiis subveniatur (XXVI).
Medicamentum ex pomis (XXVI. 5).
Aliud medicamentum (XXVI. 6).
Tertium medicamentum (XXVI. 7).
Notandum hoc (XXVI. 9).
De volva (causa matricis *marg. capitis*) (XXVII).
Git (De git *marg. capitis*) (XXVII. 1 C).
Volva exulcerata est (*om in marg. capitis*) (XXVII. 1 D).
De vessica
De calculis in vessica } *Duo haec capita periere*
In omni dolore vessicae (XXVII. 1 E). } omissa
De naturalibus (XXVIII). } *in*
De extremis corporis partibus (XXIX). } *margine*
Valentissimum remedium (XXIX. 2). } *capitum*
De *genis* (XXX).
De podagricis (curae manuum vel pedum vel podagrae *marg. capitis*) (XXXI).
Erasistratus (XXXI. 9).

LIST OF ALIMENTA

BOOK II. 18–33.

Alimenta—Nutriment.[1]

Cibus—Food.   *Potio*—Drink.

I. Animal Food.   II. Plant Food.   III. Drink.

I. A. Food from Mammals, domestic and wild; Milk, Cheese, Butter-fat.
   B. Food from Birds and Eggs.
   C. Food from Fish, Fish-pickle and Fish Sauce; Shell-fish.

II. A. Food from Grain.
   B. Food from Pulse.
   C. Food from Vegetables, Roots, Bulbs, Cucumbers and Gourds, Asparagus and edible shoots, Salads.
   D. Food from Fruit and Nuts.
   E. Olive and other Oils.

III. A. Honey and honeyed drinks.
    B. Wine and Wine Products.
    C. Water and Mineral Waters.

---

[1] The materials were classsified after an original method, and many of them were in addition used as medicines. For the various uses of each item, consult the Index.

LIST OF ALIMENTA

Hippocrates and Diet.

The Hippocratic Treatises relating to the Diaitetic Regimen.

α. περὶ τροφῆς, *De Alimento*—Nutriment. Littré, IX. 94; L.C.L., I. 342.
β. περὶ διαίτης ὑγιεινῆς. *De Salubri Victu*—Regimen in Health. Littré, VI. 72; L.C.L., IV. 44.
γ. περὶ διαίτης. *De Victu*—Diet or Regimen. Littré, VI. 466; L.C.L., IV. 224.
δ. περὶ διαίτης ὀξέων. *De Victu in Acutis*—Regimen in Acute Disease. Littré, II. 224.
ε. περὶ ὑγρῶν χρήσιος. *De liquidarum usu*—Use of Mineral Waters, Wine and Vinegar. Littré, VI. 119.

Classification.

Celsus classifies food-stuffs as strong, medium and weak (II. **18**); also as having good or bad juices (II. **19, 20, 21**); as bland or acrid (II. **22**), and in various other ways.

I. Animal Food.

A. *Food from Mammals, wild and domesticated.*

Agnus, lamb. The ewe and she-goat were kept for milking; mutton, as in modern Italy, was hardly ever eaten.
Aper, wild boar. Caught and fatted in *vivaria* (Pliny, *N.H.* VIII. 52 (78)).
Bubula, beef,
Caprea and *Capreolus*, wild she-goat and kid. The meat was held to be more digestible than that of

LIST OF ALIMENTA

the domesticated variety, although the kid (*haedus*) of the domesticated goat was commonly eaten.

Cervus, deer (venison).

Lepus, hare. Still plentiful in Greece and Italy, and particularly good eating. Cf. Aristophanes, *Wasps*, 709.

Onager, wild ass. Found in the deserts of Syria and N. Africa. Xenophon (*Cyrop.* 1. 4. 7., *Anab.* 1. 5. 2.) says that the meat was more tender than venison; Celsus refers to it as supplying strong food.

Sus, Suilla. Pork and bacon were the chief animal food of the Greeks and Romans (see p. 194, note); sucking pig (*suilla*) was the first meat that a fever patient was to be given (p. 268). *Maialis*, the hog, was fattened and made into bacon (*lardum*).

"*Omnes Beluae Marinae*" (II. **18.** 2), *i.e.* whales (*cetus, balaena mysticetus*), dolphins (*delphinus communis* and *D. Tursio*) and seals (*Phoca vitulinus* and *Ph. monachus*) supplied a meat similar to beef, which was commonly used.

Vitulina, veal.

The milk used was generally from ewes and goats, less often from asses or cows. Cheese was made from ewes' and goats' milk, and generally eaten fresh. Butter is only referred to by Celsus as an emollient (IV. **22.** 3, **25.** 2). The animal fats to which he refers are *adeps*, lard, soft fat (from pig), and *sebum*, suet, hard fat (from ox). He also mentions goose fat, cats' fat, he-goats' fat, lion's fat, but only for external application.

LIST OF ALIMENTA

B. *Food from Birds and Eggs.*

Anser, goose (*anserinum iecur*, foie gras. Cf. Juvenal, V. 114).
Columba livia, domesticated blue-rock pigeon.
Columba palumbina, wood-pigeon.
Gallus domesticus, *gallina*, cock and hen. *Pullus gallinaceus*, chicken; specially tender when kept in a coop (cohortalis, II, **18.** 8).
Grus, crane. These were fattened in *vivaria* and esteemed a great delicacy. (Pliny, *N.H.* X. 23. (30).)
Pavo, peacock. These were brought from the east and fattened on curds and grasshoppers. (Pliny, *N.H.* X. 20 (22), 43.)
Phoenicopterus, flamingo. Only the fleshy tongue, which was esteemed a delicacy, was eaten, as the rest of the bird was too fishy in taste to be palatable.

Small birds and song birds were largely eaten, as they still are in Italy. They were caught by birding (*aucupium*) and then fattened.

Hirundo, swallow.
Morticella ficedula, fig-peckers.
Turdus coturnis, quail,
Turdus pilaris, field-fare.
Turdus merula, ouzel, blackbird.
Turdus musicus, mavis, song-thrush.
Turdus viscoperus, mistle-thrush.

Eggs. Celsus refers constantly to hens' eggs as an article of diet, *Ovum durum*, hard-boiled, *molle* or *sorbile*, soft-boiled or liquid, *crudum*, raw. He speaks of pigeons' eggs, but only as useful (when hard-boiled) for reducing inflammation.

LIST OF ALIMENTA

C. *Food from Fish, Shell-fish, etc.*[1]

(a) *Fish.*

Acipenser, ἀκκιπησίος, sturgeon.
Chrysophrys aurata, χρύσοφρυς, gilt-head.
Corvus, Corax, κορακῖνος, perhaps crowfish or gurnard.
Lacertus, σαῦρος, horse or rough-tailed mackerel.
Lupus, λάβραξ, bass.
Mullus, τρίγλα, red mullet.
Oculata, probably black-tail.
Plani, πλατεῖς, flat-fish.
Saxatiles, small rock fish.
Scarus, σκάρος, parrot wrasse.
Scomber, σκόμβρος, common mackerel.
Sparus, σπάρος, sea bream.
Sargus, σαργός, sargus.
Trygon pastinaca, τρυγών, sting-ray.

(b) *Molluscs* (conchylia).

Loligo, τεῦθις, calamary or squid.
Murex brandaris, πορφύρα, purple-shell; see p. 204, note.
Mytilus, μῦς, mussel.
Ostreum, ὄστρεον, oyster.
Polypus, octopus, πολύπους, poulpe or octopus.
Pecten, scallop.
Peloris, πελώρις, clam or the lite.
Purpura haemostoma, see Murex.
Sepia, σηπία, cuttle-fish. Cuttle-fish ink (*atramentum sepiae*) was used as an aperient.

[1] A list of fish and their probable modern equivalents will be found in the translation of Oppian, and of Pliny Natural History (Vol. VIII), published in the Loeb Classical Library.

LIST OF ALIMENTA

Cocleae, snails, were artificially fed and largely eaten. Large quantities of their shells mark Roman sites.

(c) *Crustacea.*

Locusta (Palinurus quadricornis), κάραβος, sea crayfish, spiny lobster.

(d) *Echinoderms.*

Echinus (Echinus esculentus), ἐχῖνος, sea-urchin.

(e) *Fish sauce and Pickles.*

Garum (γάρον), fish sauce made chiefly of decomposing fish fry (ἀφύη). *Garum scombri* was made from the blood and crushed roe of the mackerel. Tunny was also used; *garum* sometimes was strained and used as a liqueur.

Salsamentum (ταρίχος), was fish (or sometimes pork) pickled in strong brine (*muria*). The fish chiefly used were mackerel, tunny and saperda or *sciurus umbra* (*ombre*) from the Black Sea; it was a sort of " anchovy sauce."

II. FOOD FROM PLANTS.

A. *Frumentum*, grain, wheat, millet, barley were the principal grains.

(a) *WHEAT, triticum* (πυρός).
    Triticum hibernum, winter wheat.
    Triticum aestivum, spring wheat. Cf. p. 196, note.
    Spelta, spelt, an inferior wheat grown on a poor soil.

LIST OF ALIMENTA

Siligo and *simila*, white, well-bolted flours made from the best wheat.

Amylum (ἄμυλον), unmilled wheaten flour, especially from wheat grown in Chios. See p. 201, note.

Pollen, inferior flour left over when the best flour had been bolted.

Autopyrus (αὐτόπυρος), unbolted flour and the coarse bread made from it.

Cibarius panis, the grey bread eaten by the poor, made of spelt flour roughly ground and mixed with bran (*furfur*).

Far, ground or pounded spelt, only used if nothing else was available (III. 22. 11). Far was used in religious ceremonials, where its use was traditional, as spelt was the most primitive form of wheat grown before the finer varieties had been cultivated.

Alica elota, spelt groats from which spelt gruel was prepared.

Tragum (τράγος), frumenty or wheat porridge.

(b) *MILLET*, *milium* or *panicum* (κέγχρος, ἔλυμος, μελίνη). A small-sized grain, which was used instead of spelt or barley flour to make inferior bread.

(c) *BARLEY*, *hordeum* (κρίθη). This was the staple grain of Greece, but Celsus chiefly mentions its use as pearl (peeled) barley in making gruel. *Hordeacia farina* (barley meal), he only speaks of for making poultices (III. 27. 2, C).

Polenta, pearl barley. When ground up this was sprinkled into wine or over meat. In

LIST OF ALIMENTA

modern Italy the word is used of a porridge made from maize and chestnuts.

Tisana (πτισάνη), decoction of pearl barley.

Cremor (χύλος), the thick decoction.

Sorbitio, the decoction strained and diluted, barley water, (Sorbitio, though generally prepared from pearl barley, could also be made from spelt groats, rice or amylum.)

(d) *RICE* (*oryza*, ὄρυζα), husked or unhusked was also used; it was regarded as a weaker equivalent of barley.

B. *Legumina*, ὄσπρια, pulse.

Faba (*Vicia Faba*), κύαμος, bean. The *Faba Aegyptiaca*, used in several prescriptions (V. **23.** 2, **25.** 6, etc.), has not been indentified.

Pisum Sativum, πισός, pea. In general use for porridge.

Lenticula (*Ervum Lens*), φακός, Egyptian lentil cooked to make soup or porridge and eaten uncooked as a pickle.

Ervum Ervilia (bitter vetch, ὄροβος), and *Lupinus Alba* (Lupin, θέρμος) were only used as calefacients and cleaners.

C. *Olus*, λάχανα, pot herbs and roots; also bulbs, edible shoots, gourds and cucumbers, salads and condiments.

Allium (*Allium sativum*), garlic.
Amaranthus (*Amaranthus blitum*), βλίτον, blite.
Ambubeia, see *Intubus*.
Anisum (*pimpinella anisum*), ἄννησον, anise.

LIST OF ALIMENTA

Apium (*Petrosolinum sativum* and *Apium graveolens*), ὀρειοσέλινον or σέλινον, parsley and celery.
Anethum (*Anethum graveolens*), ἄνηθόν or ἄννητος, dill.
Asparagus (*Asparagus acutifolius*), ἀσφάραγος, asparagus. Asparagus was also a general name for edible shoots eaten green or pickled in vinegar.
Atriplex (*atriplex hortense*), ἀδράφαξυς, orache.
Beta (*Beta maritima*), τεῦτλον, beet.
Brassica (*Brassica Cretica*), ῥάφανος, cabbage.
Capparis (the buds of *Capparis spinosa*), κάππαρις, caper.
Cepa (*Allium cepa*), κρόμυον, onion.
Cinara (*Cynara scolymus*), κάκτος, artichoke.
Coriandrum (*Coriandrum sativum*), κορίαννον, coriander.
Cucumis (*Cucumis sativus*), σίκυος, cucumber, grown in glass houses (*specularia*); the wild variety (*cucumis agrestis*), squirting cucumber, was used to make a purge (*elaterium*).
Cucurbita (*Cucurbita maxima*), κολοκύντη, gourd or pumpkin.
Cuminum (*Cuminum cyminum*), κύμινον, cummin (caraway).
Eruca (*Eruca sativa*), εὔζωμον, rocket (yielding a mustard-like oil).
Foeniculum (*Trigonella foenum-graecum*), βουκέρας or τῆλις, fenugreek, fennel.
Hysopus, ὕσσωπος, hyssop; identification doubtful; perhaps Capparis spinosa or Origanum Aegypticum or Hyssopus officinalis.
Intubus (*Cichorium intybus*), κίχοριον, chicory, endive.

LIST OF ALIMENTA

Lactuca (*Lactuca sativa*), θρίδαξ, lettuce. The juice was sometimes used as a soporific (Martial III. 89.)
Lapathum (*Rumex acetosella*), sorrel.
Lapsanum, see *Sinapis*.
Laser (*Scorodosma foetida*), asafoetida.
Malva (*Malva silvestria*), μάλαχη, mallow.
Mentha (*Mentha viridis*), ἡδύσμον = μένθη, green mint, spearmint.
    Mentha aquatica, σισύμβριον, Bergamot-mint.
    Mentha piperita = peppermint.
    Mentha pulegium, βληχώ, fleabane, pennyroyal.
Napus (*Brassica napa*), navew.
Nepeta (*Satureia calamintha*), catmint.
Nasturcium (*Lepidum sativum*), κάρδαμον, cress (*quod nasum torquet;* Pliny).
Ocimum (*Ocimum Basilicum*), ὤκιμον, basil.
Pastinaca, parsnip (including carrots).
Piper (*Piper nigrum, album*), πέπερι, pepper, black and white. The former gathered ripe, the latter unripe and decorticated. *Piper longum*, used in pickles.
Porrum (*Allium Porrum*), πράσον, leek. *Porrum sectile*, the shredded leaves used as a medicine.
Portulaca (*Portulaca oloracea*), ἀνδράχνη, purslane.
Pyrethrum (*Pyrethrum parthenion*), παρθένιον, Bachelor's button = *Herba Muralis*, pellitory. The root is hot and when chewed relieves toothache.
Radix, Radicula (*Raphanus sativus*), ῥαφανίς, radish.
Radix dulcis, γλυκύρριζα, liquorice.
Radix pontica (*Rheum Ponticum*), ῥᾶ, Turkey rhubarb. The name Rha, used by Ammianus Marcellinus (22. 8. 28), was perhaps derived from the river Rha (Volga), near which it grew.

LIST OF ALIMENTA

Rapum (*Brassica rapa*), γογγυλίς, turnip or rape.
Cyma (κῦμα) = turnip-tops.
Ruta (*Ruta graveolens*), πήγανον, rue.
Satureia (*Satureia thymbra*), θύμβρα, savory.
Scilla (*Urginea maritima*), σκίλλα, squill or sea leek.
Sinapis (*Sinapis alba*), νᾶπυ, white mustard. Mustard oil was made from the mixed seeds of the black and white mustard. The leaves were used as a salad. *Lapsanum* (*Sinapis Arvensis*) = charlock.
Siser (*Sium sisarum*), σίσαρον, skirret or parsnip.
Urtica (*Urtica urens*), ἀκαλύφη, stinging nettle. Young nettle-tops were used as a vegetable and in salads.

The anodynes, poppy, mandragora, etc., and other herbs used solely as medicines will be found in the list of drugs (*medicamenta*) at the end of Volume II.

Verbena (vervains): Celsus used this word to describe the leaves and twigs of various aromatic plants which grew wild in the maquis or brushwood bordering on the Mediterranean. Among the most common were Myrtle, Mastick, Rosemary, Laurel and Cistus. Decoctions of these were employed as repressants; they were also in common use for strewing on floors, decorating temples, or carrying in ceremonial processions.

D. *Poma*, orchard fruit (including figs and dates) and nuts.

(a) *Fruit*.

Cherry, see *Prunus*.
Dactylus, or *Palmula*, δάκτυλος, date, the fruit of the *Phoenix Dactilifera*.

LIST OF ALIMENTA

Ficus, σῦκον, fig, the fruit of the *Ficus Carica*. Ficus Sycamorus, συκάμινος ἡ Ἀιγυπτία, sycaminus, was so called because its leaves resembled those of the white mulberry (III. **18.** 13).

Malum. This word was used for any fruit, fleshy outside and having a kernel within; the most common of these was the apple, μῆλον (Pyrus malus). Several varieties of this are mentioned by Celsus; *Malum Amerinum* from Ameria in Umbria; *Malum orbiculatum*, "round" apples, a choice species; *Malum Scandianum* (or *Scaudianum*), either called after the town Scandianum, near Rhegium, or after a man Scandius.

Other fruits to which the name *Malum* was applied were:

Malum Cotoneum, κυδώνιον, quince, the fruit of the *Cydonia vulgaris*.

Malum Persicum, μῆλον Περσίκον, peach, the fruit of *Prunus Persica*.

Malum Punicum, ῥόα, pomegranate, the fruit of the *Punica granatum*; there was a sweet or acid variety of this.

Pirum ἄπιος, pear, the fruit of the *Pyrus communis*. There were many varieties of this; Celsus mentions *Pirum Crustuminum*, from Crustumeria; *P. Tarentinum* or *Signinum*, from Tarentum or from Segni in Latium; *P. myrapium* (myrrh-scented), the French Poire Ombre.

Prunus, κοκκύμηλον, plum, the fruit of the *Prunus domestica*; *cerasus*, κέρασος, cherry (*Prunus avium*), called after Cerasus on the Black Sea, whence the fruit was introduced into Italy by

LIST OF ALIMENTA

Lucullus in 68 B.C. For *Prunus amygdalus*, see below, s.v. Nux.

Morum, συκάμινος, mulberry, the fruit of the *Morus nigra*.

Rubus (*Rubus ulmifolius*), βάτος, blackberry. As in modern Italy, the fruit was hardly used at all, but Celsus mentions the leaves as used for a decoction.

Sorbus, ὄη, the sorb or service apple, the fruit of the *Sorbus domestica*.

(b) *Nuts*.

Nuces Abellanae (*Corylus Avellana*), hazel nuts and filberts (καρύα).

Nuces Amarae and *Nuces Graecae* (*Prunus Amygdalus*), ἀμυγδάλαι, bitter and sweet almonds.

Nuces Juglandes (*Jovis glans, Juglans Regia*), καρύα Περσίκα, walnuts (from Persia).

Nuces Pineae, edible seeds of the pine tree, still commonly eaten in Italy (Pinocchi).

E. *Olive and other vegetable oils.*

Oleum olivum, ἐλαίον, olive oil. The best was made from the finest ripe olives which had been subjected to a minimum of pressure; an inferior oil was produced from the *amurca*, dregs or lees, which were treated with hot water. Besides its use as an article of diet, olive oil, especially old oil (*oleum vetus*), was used for cataplasms and for inunction; *oleum acerbum*, the oil from white olives, was used as a refrigerant and for hardening.

LIST OF ALIMENTA

Other oils were used along with olive oil, to add special perfumes to it, or as adulterants.

Oleum amygdalinum, from sweet almonds.

Oleum cyprinum, from the *Lawsonia inermis alba;* Egyptian privet oil, with the yellow pigment, henna, alkana.

Oleum irinum, from the rhizone of the *Iris Illyrica*, supplying orris perfume.

Oleum Syriacum (Susinum), from lilies grown near Susa.

Oleum myrteum, from the *myrtus communis*, with the myrtle perfume.

Oleum rosae, from *rosa gallica centiflora*.

Oleum sesamum, from the seeds of *sesamum*, a near equivalent of olive oil, used in Assyria instead of it.

Note.—A list of plants and their probable modern equivalents will be found in Theophrastus, *Enquiry into Plants*, Vol. II., translated by Sir F. Hort and published in the Loeb Classical Library.

III. DRINK (Potio).

A. *Honey and Honeyed Drinks.*

Mel, honey, was exceedingly important as the one source of sugar; raw honey acted as an aperient; old honey, especially when boiled in a leaden pot, as an astringent.

Mel despumatum, liquid honey, was used as a special nutriment in fever. A wax salve, *ceratum*, was made from the wax.

LIST OF ALIMENTA

Aqua mulsa, hydromel, consisted of honey and spring water boiled down to one-third to prevent decomposition.

Oxymel, ὀξύγλυκυ πότον, was freshest honeycomb, crushed and boiled to dissolve out the honey and then acidulated with vinegar.

Mulsum (vinum), honey-wine, metheglyn, generally consisted of one part of honey to two of wine.

B. *Vine products and wine.*

Vitis, the vine, included the *vitis sylvestris*, the wild vine, upon which was grafted the *vitis vinifera*. The fruit was eaten fresh—grapes, *uva* or *acinus*, or dried—raisins, *uva ex olla*.

Wine formed an important part of the dietetic regimen; the prescriptions of it are as numerous as those relating to food. Wine was the commonest pharmaceutical excipient in compounding prescriptions, and wine with oil and water was the commonest application for wounds.

Wines varied in taste and consistency (*austerum, dulce, asperum, leve, tenue*) and in colour (*nigrum, rubrum, album, fulvum*).

Celsus only refers to wine from five districts.

Vinum Albuele from the district round the Tiber (Celsus, *Agr. Frag.* IX. quoted by Pliny *N.H* XIV. 2 (4), 31).

Vinum Allobrogicum, from the country of the Allobroges, approximately the district where Burgundy is now produced.

Vinum Aminaeum from Aminaea in Apulia, a wine

LIST OF ALIMENTA

similar to that from the neighbouring *Ager Falernus* in Campania, whence came the Falernian wine constantly referred to by Horace (*C.* I. 27. 10, etc.).

Vinum Rhaeticum, from the Grisons or Rhaetian Alps near Verona.

Vinum Signinum, from Segni in Latium.

Celsus also constantly refers to Greek wines; these were generally doctored with salt, resin or myrrh (*vinum salsum, resinatum* or *myrrhinum*) in order that they should keep better and to enable them to stand export by sea.

Other varieties of drinks prepared from wine were:

Vinum absinthium, wine flavoured with wormwood.

Vinum acetum, posca, sour wine, almost vinegar, which when diluted with water formed the ordinary drink of the common people.

Vinum mulsum, see under Honeyed Drinks.

Vinum mustum, new or unfermented wine, must.
   Mustum defrutum was must boiled down to one-half; it was then used as an aperient, but if boiled in leaden vessels as an astringent.

Vinum passum, raisin wine, much used for invalids.

Vinum siliatum, wine flavoured with *seselis* (saxifrage), usually drunk in the middle of the day.

Water and mineral waters.

Celsus classified water as rain, spring, river, well or cistern, marsh. The differences which he recognised between them were due to variations in the inorganic contents or to organic contamination.

LIST OF ALIMENTA

Mineral waters were plentiful in Italy, and then, as now, the natural mineral waters were very largely used.

Celsus mentions the waters at Baiae and Cumae, where were the *Thermae Neronianae*, sulphur baths on the Bay of Campania; also the Simbruine springs, in the district of Simbruvium, near Tivoli (Tacitus, *Annals*, XI. 13, XIV. 22), and those at Cutilia (modern Lago di Contigliano), which were extremely cold and biting (owing to the presence of carbonic acid gas), they lay between the Via Nomentana and the Via Salaria (Pliny, *N.H.* XXXI. 2. 6).

*Printed in Great Britain by
Richard Clay (The Chaucer Press), Ltd.,
Bungay, Suffolk*

THE LOEB CLASSICAL LIBRARY

VOLUMES ALREADY PUBLISHED

Latin Authors

AMMIANUS MARCELLINUS. Translated by J. C. Rolfe. 3 Vols.

APULEIUS: THE GOLDEN ASS (METAMORPHOSES). W. Adlington (1566). Revised by S. Gaselee.

ST. AUGUSTINE: CITY OF GOD. 7 Vols. Vol. I. G. E. McCracken Vol. II. W. M. Green. Vol. III. D. Wiesen. Vol. IV. P. Levine. Vol. V. E. M. Sanford and W. M. Green. Vol. VI. W. C. Greene.

ST. AUGUSTINE, CONFESSIONS OF. W. Watts (1631). 2 Vols.

ST. AUGUSTINE, SELECT LETTERS. J. H. Baxter.

AUSONIUS. H. G. Evelyn White. 2 Vols.

BEDE. J. E. King. 2 Vols.

BOETHIUS: TRACTS and DE CONSOLATIONE PHILOSOPHIAE. Rev. H. F. Stewart and E. K. Rand.

CAESAR: ALEXANDRIAN, AFRICAN and SPANISH WARS. A. G. Way.

CAESAR: CIVIL WARS. A. G. Peskett.

CAESAR: GALLIC WAR. H. J. Edwards.

CATO: DE RE RUSTICA; VARRO: DE RE RUSTICA. H. B. Ash and W. D. Hooper.

CATULLUS. F. W. Cornish; TIBULLUS. J. B. Postgate; PERVIGILIUM VENERIS. J. W. Mackail.

CELSUS: DE MEDICINA. W. G. Spencer. 3 Vols.

CICERO: BRUTUS, and ORATOR. G. L. Hendrickson and H. M. Hubbell.

[CICERO]: AD HERENNIUM. H. Caplan.

CICERO: DE ORATORE, etc. 2 Vols. Vol. I. DE ORATORE, Books I. and II. E. W. Sutton and H. Rackham. Vol. II. DE ORATORE, Book III. De Fato; Paradoxa Stoicorum; De Partitione Oratoria. H. Rackham.

CICERO: DE FINIBUS. H. Rackham.

CICERO: DE INVENTIONE, etc. H. M. Hubbell.

CICERO: DE NATURA DEORUM and ACADEMICA. H. Rackham.

CICERO: DE OFFICIIS. Walter Miller.

CICERO: DE REPUBLICA and DE LEGIBUS; SOMNIUM SCIPIONIS. Clinton W. Keyes.

CICERO: DE SENECTUTE, DE AMICITIA, DE DIVINATIONE. W. A. Falconer.
CICERO: IN CATILINAM, PRO FLACCO, PRO MURENA, PRO SULLA. Louis E. Lord.
CICERO: LETTERS TO ATTICUS. E. O. Winstedt. 3 Vols.
CICERO: LETTERS TO HIS FRIENDS. W. Glynn Williams. 3 Vols.
CICERO: PHILIPPICS. W. C. A. Ker.
CICERO: PRO ARCHIA POST REDITUM, DE DOMO, DE HARUSPICUM RESPONSIS, PRO PLANCIO. N. H. Watts.
CICERO: PRO CAECINA, PRO LEGE MANILIA, PRO CLUENTIO, PRO RABIRIO. H. Grose Hodge.
CICERO: PRO CAELIO, DE PROVINCIIS CONSULARIBUS, PRO BALBO. R. Gardner.
CICERO: PRO MILONE, IN PISONEM, PRO SCAURO, PRO FONTEIO, PRO RABIRIO POSTUMO, PRO MARCELLO, PRO LIGARIO, PRO REGE DEIOTARO. N. H. Watts.
CICERO: PRO QUINCTIO, PRO ROSCIO AMERINO, PRO ROSCIO COMOEDO, CONTRA RULLUM. J. H. Freese.
CICERO: PRO SESTIO, IN VATINIUM. R. Gardner.
CICERO: TUSCULAN DISPUTATIONS. J. E. King.
CICERO: VERRINE ORATIONS. L. H. G. Greenwood. 2 Vols.
CLAUDIAN. M. Platnauer. 2 Vols.
COLUMELLA: DE RE RUSTICA. DE ARBORIBUS. H. B. Ash, E. S. Forster and E. Heffner. 3 Vols.
CURTIUS, Q.: HISTORY OF ALEXANDER. J. C. Rolfe. 2 Vols.
FLORUS. E. S. Forster; and CORNELIUS NEPOS. J. C. Rolfe.
FRONTINUS: STRATAGEMS and AQUEDUCTS. C. E. Bennett and M. B. McElwain.
FRONTO: CORRESPONDENCE. C. R. Haines. 2 Vols.
GELLIUS, J. C. Rolfe. 3 Vols.
HORACE: ODES AND EPODES. C. E. Bennett.
HORACE: SATIRES, EPISTLES, ARS POETICA. H. R. Fairclough.
JEROME: SELECTED LETTERS. F. A. Wright.
JUVENAL and PERSIUS. G. G. Ramsay.
LIVY. B. O. Foster, F. G. Moore, Evan T. Sage, and A. C. Schlesinger and R. M. Geer (General Index). 14 Vols.
LUCAN. J. D. Duff.
LUCRETIUS. W. H. D. Rouse.
MARTIAL. W. C. A. Ker. 2 Vols.
MINOR LATIN POETS: from PUBLILIUS SYRUS TO RUTILIUS NAMATIANUS, including GRATTIUS, CALPURNIUS SICULUS, NEMESIANUS, AVIANUS, and others with "Aetna" and the "Phoenix." J. Wight Duff and Arnold M. Duff.
OVID: THE ART OF LOVE and OTHER POEMS. J. H. Mozley.

OVID: FASTI. Sir James G. Frazer.
OVID: HEROIDES and AMORES. Grant Showerman.
OVID: METAMORPHOSES. F. J. Miller. 2 Vols.
OVID: TRISTIA and EX PONTO. A. L Wheeler.
PERSIUS. Cf. JUVENAL.
PETRONIUS. M. Heseltine; SENECA; APOCOLOCYNTOSIS. W. H. D. Rouse.
PHAEDRUS AND BABRIUS (Greek). B. E. Perry.
PLAUTUS. Paul Nixon. 5 Vols.
PLINY: LETTERS, PANEGYRICUS. Betty Radice. 2 Vols.
PLINY: NATURAL HISTORY.
  10 Vols. Vols. I.–V. and IX. H. Rackham. Vols. VI.–VIII. W. H. S. Jones. Vol. X. D. E. Eichholz.
PROPERTIUS. H. E. Butler.
PRUDENTIUS. H. J. Thomson. 2 Vols.
QUINTILIAN. H. E. Butler. 4 Vols.
REMAINS OF OLD LATIN. E. H. Warmington. 4 Vols. Vol. I. (ENNIUS AND CAECILIUS.) Vol. II. (LIVIUS, NAEVIUS, PACUVIUS, ACCIUS.) Vol. III. (LUCILIUS and LAWS OF XII TABLES.) Vol. IV. (ARCHAIC INSCRIPTIONS.)
SALLUST. J. C. Rolfe.
SCRIPTORES HISTORIAE AUGUSTAE. D. Magie. 3 Vols.
SENECA: APOCOLOCYNTOSIS. Cf. PETRONIUS.
SENECA: EPISTULAE MORALES. R. M. Gummere. 3 Vols.
SENECA: MORAL ESSAYS. J. W. Basore. 3 Vols.
SENECA: TRAGEDIES. F. J. Miller. 2 Vols.
SIDONIUS: POEMS and LETTERS. W. B. ANDERSON. 2 Vols.
SILIUS ITALICUS. J. D. Duff. 2 Vols.
STATIUS. J. H. Mozley. 2 Vols.
SUETONIUS. J. C. Rolfe. 2 Vols.
TACITUS: DIALOGUS. Sir Wm. Peterson. AGRICOLA and GERMANIA. Maurice Hutton.
TACITUS: HISTORIES AND ANNALS. C. H. Moore and J. Jackson. 4 Vols.
TERENCE. John Sargeaunt. 2 Vols.
TERTULLIAN: APOLOGIA and DE SPECTACULIS. T. R. Glover. MINUCIUS FELIX. G. H. Rendall.
VALERIUS FLACCUS. J. H. Mozley.
VARRO: DE LINGUA LATINA. R. G. Kent. 2 Vols.
VELLEIUS PATERCULUS and RES GESTAE DIVI AUGUSTI. F. W. Shipley.
VIRGIL. H. R. Fairclough. 2 Vols.
VITRUVIUS: DE ARCHITECTURA. F. Granger. 2 Vols.

Greek Authors

ACHILLES TATIUS. S. Gaselee.
AELIAN: ON THE NATURE OF ANIMALS. A. F. Scholfield. 3 Vols.
AENEAS TACTICUS, ASCLEPIODOTUS and ONASANDER. The Illinois Greek Club.
AESCHINES. C. D. Adams.
AESCHYLUS. H. Weir Smyth. 2 Vols.
ALCIPHRON, AELIAN, PHILOSTRATUS: LETTERS. A. R. Benner and F. H. Fobes.
ANDOCIDES, ANTIPHON, Cf. MINOR ATTIC ORATORS.
APOLLODORUS. Sir James G. Frazer. 2 Vols.
APOLLONIUS RHODIUS. R. C. Seaton.
THE APOSTOLIC FATHERS. Kirsopp Lake. 2 Vols.
APPIAN: ROMAN HISTORY. Horace White. 4 Vols.
ARATUS. Cf. CALLIMACHUS.
ARISTOPHANES. Benjamin Bickley Rogers. 3 Vols. Verse trans.
ARISTOTLE: ART OF RHETORIC. J. H. Freese.
ARISTOTLE: ATHENIAN CONSTITUTION, EUDEMIAN ETHICS, VICES AND VIRTUES. H. Rackham.
ARISTOTLE: GENERATION OF ANIMALS. A. L. Peck.
ARISTOTLE: HISTORIA ANIMALIUM. A. L. Peck. Vols. I.–II.
ARISTOTLE: METAPHYSICS. H. Tredennick. 2 Vols.
ARISTOTLE: METEOROLOGICA. H. D. P. Lee.
ARISTOTLE: MINOR WORKS. W. S. Hett. On Colours, On Things Heard, On Physiognomies, On Plants, On Marvellous Things Heard, Mechanical Problems, On Indivisible Lines, On Situations and Names of Winds, On Melissus, Xenophanes, and Gorgias.
ARISTOTLE: NICOMACHEAN ETHICS. H. Rackham.
ARISTOTLE: OECONOMICA and MAGNA MORALIA. G. C. Armstrong; (with Metaphysics, Vol. II.).
ARISTOTLE: ON THE HEAVENS. W. K. C. Guthrie.
ARISTOTLE: ON THE SOUL. PARVA NATURALIA. ON BREATH. W. S. Hett.
ARISTOTLE: CATEGORIES, ON INTERPRETATION, PRIOR ANALYTICS. H. P. Cooke and H. Tredennick.
ARISTOTLE: POSTERIOR ANALYTICS, TOPICS. H. Tredennick and E. S. Forster.
ARISTOTLE: ON SOPHISTICAL REFUTATIONS.
On Coming to be and Passing Away, On the Cosmos. E. S. Forster and D. J. Furley.
ARISTOTLE: PARTS OF ANIMALS. A. L. Peck; MOTION AND PROGRESSION OF ANIMALS. E. S. Forster.

ARISTOTLE: PHYSICS. Rev. P. Wicksteed and F. M. Cornford. 2 Vols.
ARISTOTLE: POETICS and LONGINUS. W. Hamilton Fyfe; DEMETRIUS ON STYLE. W. Rhys Roberts.
ARISTOTLE: POLITICS. H. Rackham.
ARISTOTLE: PROBLEMS. W. S. Hett. 2 Vols.
ARISTOTLE: RHETORICA AD ALEXANDRUM (with PROBLEMS. Vol. II). H. Rackham.
ARRIAN: HISTORY OF ALEXANDER and INDICA. Rev. E. Iliffe Robson. 2 Vols.
ATHENAEUS: DEIPNOSOPHISTAE. C. B. GULICK. 7 Vols.
BABRIUS AND PHAEDRUS (Latin). B. E. Perry.
ST. BASIL: LETTERS. R. J. Deferrari. 4 Vols.
CALLIMACHUS: FRAGMENTS. C. A. Trypanis.
CALLIMACHUS, Hymns and Epigrams, and LYCOPHRON. A. W. Mair; ARATUS. G. R. MAIR.
CLEMENT of ALEXANDRIA. Rev. G. W. Butterworth.
COLLUTHUS. Cf. OPPIAN.
DAPHNIS AND CHLOE. Thornley's Translation revised by J. M. Edmonds; and PARTHENIUS. S. Gaselee.
DEMOSTHENES I.: OLYNTHIACS, PHILIPPICS and MINOR ORATIONS. I.–XVII. AND XX. J. H. Vince.
DEMOSTHENES II.: DE CORONA and DE FALSA LEGATIONE. C. A. Vince and J. H. Vince.
DEMOSTHENES III.: MEIDIAS, ANDROTION, ARISTOCRATES, TIMOCRATES and ARISTOGEITON, I. AND II. J. H. Vince.
DEMOSTHENES IV.–VI.: PRIVATE ORATIONS and IN NEAERAM. A. T. Murray.
DEMOSTHENES VII.: FUNERAL SPEECH, EROTIC ESSAY, EXORDIA and LETTERS. N. W. and N. J. DeWitt.
DIO CASSIUS: ROMAN HISTORY. E. Cary. 9 Vols.
DIO CHRYSOSTOM. J. W. Cohoon and H. Lamar Crosby. 5 Vols.
DIODORUS SICULUS. 12 Vols. Vols. I.–VI. C. H. Oldfather. Vol. VII. C. L. Sherman. Vol. VIII. C. B. Welles. Vols. IX. and X. R. M. Geer. Vol. XI. F. Walton. Vol. XII. F. Walton. General Index. R. M. Geer.
DIOGENES LAERTIUS. R. D. Hicks. 2 Vols.
DIONYSIUS OF HALICARNASSUS: ROMAN ANTIQUITIES. Spelman's translation revised by E. Cary. 7 Vols.
EPICTETUS. W. A. Oldfather. 2 Vols.
EURIPIDES. A. S. Way. 4 Vols. Verse trans.
EUSEBIUS: ECCLESIASTICAL HISTORY. Kirsopp Lake and J. E. L. Oulton. 2 Vols.
GALEN: ON THE NATURAL FACULTIES. A. J. Brock.
THE GREEK ANTHOLOGY. W. R. Paton. 5 Vols.

GREEK ELEGY AND IAMBUS with the ANACREONTEA. J. M. Edmonds. 2 Vols.
THE GREEK BUCOLIC POETS (THEOCRITUS, BION, MOSCHUS). J. M. Edmonds.
GREEK MATHEMATICAL WORKS. Ivor Thomas. 2 Vols.
HERODES. Cf. THEOPHRASTUS: CHARACTERS.
HERODIAN. C. R. Whittaker. 2 Vols.
HERODOTUS. A. D. Godley. 4 Vols.
HESIOD AND THE HOMERIC HYMNS. H. G. Evelyn White.
HIPPOCRATES and the FRAGMENTS OF HERACLEITUS. W. H. S. Jones and E. T. Withington. 4 Vols.
HOMER: ILIAD. A. T. Murray. 2 Vols.
HOMER: ODYSSEY. A. T. Murray. 2 Vols.
ISAEUS. E. W. Forster.
ISOCRATES. George Norlin and LaRue Van Hook. 3 Vols.
[ST. JOHN DAMASCENE]: BARLAAM AND IOASAPH. Rev. G. R. Woodward, Harold Mattingly and D. M. Lang.
JOSEPHUS. 9 Vols. Vols. I.-IV.; H. Thackeray. Vol. V.; H. Thackeray and R. Marcus. Vols. VI.-VII.; R. Marcus. Vol. VIII.; R. Marcus and Allen Wikgren. Vol. IX. L. H. Feldman.
JULIAN. Wilmer Cave Wright. 3 Vols.
LIBANIUS. A. F. Norman. Vol. I.
LUCIAN. 8 Vols. Vols. I.-V. A. M. Harmon. Vol. VI. K. Kilburn. Vols. VII.-VIII. M. D. Macleod.
LYCOPHRON. Cf. CALLIMACHUS.
LYRA GRAECA. J. M. Edmonds. 3 Vols.
LYSIAS. W. R. M. Lamb.
MANETHO. W. G. Waddell: PTOLEMY: TETRABIBLOS. F. E. Robbins.
MARCUS AURELIUS. C. R. Haines.
MENANDER. F. G. Allinson.
MINOR ATTIC ORATORS (ANTIPHON, ANDOCIDES, LYCURGUS, DEMADES, DINARCHUS, HYPERIDES). K. J. Maidment and J. O. Burtt. 2 Vols.
NONNOS: DIONYSIACA. W. H. D. Rouse. 3 Vols.
OPPIAN, COLLUTHUS, TRYPHIODORUS. A. W. Mair.
PAPYRI. NON-LITERARY SELECTIONS. A. S. Hunt and C. C. Edgar. 2 Vols. LITERARY SELECTIONS (Poetry). D. L. Page.
PARTHENIUS. Cf. DAPHNIS and CHLOE.
PAUSANIAS: DESCRIPTION OF GREECE. W. H. S. Jones. 4 Vols. and Companion Vol. arranged by R. E. Wycherley.
PHILO. 10 Vols. Vols. I.-V.; F. H. Colson and Rev. G. H. Whitaker. Vols. VI.-IX.; F. H. Colson. Vol. X. F. H. Colson and the Rev. J. W. Earp.

PHILO: two supplementary Vols. (*Translation only.*) Ralph Marcus.

PHILOSTRATUS: THE LIFE OF APOLLONIUS OF TYANA. F. C. Conybeare. 2 Vols.

PHILOSTRATUS: IMAGINES; CALLISTRATUS: DESCRIPTIONS. A. Fairbanks.

PHILOSTRATUS and EUNAPIUS: LIVES OF THE SOPHISTS. Wilmer Cave Wright.

PINDAR. Sir J. E. Sandys.

PLATO: CHARMIDES, ALCIBIADES, HIPPARCHUS, THE LOVERS, THEAGES, MINOS and EPINOMIS. W. R. M. Lamb.

PLATO: CRATYLUS, PARMENIDES, GREATER HIPPIAS, LESSER HIPPIAS. H. N. Fowler.

PLATO: EUTHYPHRO, APOLOGY, CRITO, PHAEDO, PHAEDRUS. H. N. Fowler.

PLATO: LACHES, PROTAGORAS, MENO, EUTHYDEMUS. W. R. M. Lamb.

PLATO: LAWS. Rev. R. G. Bury. 2 Vols.

PLATO: LYSIS, SYMPOSIUM, GORGIAS. W. R. M. Lamb.

PLATO: REPUBLIC. Paul Shorey. 2 Vols.

PLATO: STATESMAN, PHILEBUS. H. N. Fowler; ION. W. R. M. Lamb.

PLATO: THEAETETUS and SOPHIST. H. N. Fowler.

PLATO: TIMAEUS, CRITIAS, CLITOPHO, MENEXENUS, EPISTULAE. Rev. R. G. Bury.

PLOTINUS: A. H. Armstrong. Vols. I.–III.

PLUTARCH: MORALIA. 16 Vols. Vols. I.–V. F. C. Babbitt. Vol. VI. W. C. Helmbold. Vols. VII. and XIV. P. H. De Lacy and B. Einarson. Vol. VIII. P. A. Clement and H. B. Hoffleit. Vol. IX. E. L. Minar, Jr., F. H. Sandbach, W. C. Helmbold. Vol. X. H. N. Fowler. Vol. XI. L. Pearson and F. H. Sandbach. Vol. XII. H. Cherniss and W. C. Helmbold. Vol. XV. F. H. Sandbach.

PLUTARCH: THE PARALLEL LIVES. B. Perrin. 11 Vols.

POLYBIUS. W. R. Paton. 6 Vols.

PROCOPIUS: HISTORY OF THE WARS. H. B. Dewing. 7 Vols.

PTOLEMY: TETRABIBLOS. Cf. MANETHO.

QUINTUS SMYRNAEUS. A. S. Way. Verse trans.

SEXTUS EMPIRICUS. Rev. R. G. Bury. 4 Vols.

SOPHOCLES. F. Storr. 2 Vols. Verse trans.

STRABO: GEOGRAPHY. Horace L. Jones. 8 Vols.

THEOPHRASTUS: CHARACTERS. J. M. Edmonds. HERODES, etc. A. D. Knox.

THEOPHRASTUS: ENQUIRY INTO PLANTS. Sir Arthur Hort, Bart. 2 Vols.

THUCYDIDES. C. F. Smith. 4 Vols.

TRYPHIODORUS. Cf. OPPIAN.
XENOPHON: CYROPAEDIA. Walter Miller. 2 Vols.
XENOPHON: HELLENICA. C. L. Brownson. 2 Vols.
XENOPHON: ANABASIS. C. L. Brownson.
XENOPHON: MEMORABILIA AND OECONOMICUS. E. C. Marchant. SYMPOSIUM AND APOLOGY. O. J. Todd.
XENOPHON: SCRIPTA MINORA. E. C. Marchant and G. W. Bowersock.

IN PREPARATION

Greek Authors

ARISTIDES: ORATIONS. C. A. Behr.
MUSAEUS: HERO AND LEANDER. T. Gelzer and C. H. Whitman.
THEOPHRASTUS: DE CAUSIS PLANTARUM. G. K. K. Link and B. Einarson.

Latin Authors

ASCONIUS: COMMENTARIES ON CICERO'S ORATIONS. G. W. Bowersock.
BENEDICT: THE RULE. P. Meyvaert.
JUSTIN–TROGUS. R. Moss.
MANILIUS. G. P. Goold.

DESCRIPTIVE PROSPECTUS ON APPLICATION

London   WILLIAM HEINEMANN LTD
Cambridge, Mass.   HARVARD UNIVERSITY PRESS